ERGEBNISSE DER PHYSIOLOGIE

BIOLOGISCHEN CHEMIE UND
EXPERIMENTELLEN PHARMAKOLOGIE

REVIEWS OF PHYSIOLOGY

BIOCHEMISTRY AND
EXPERIMENTAL PHARMACOLOGY

BAND 57

SPRINGER-VERLAG BERLIN HEIDELBERG GMBH

Ergebnisse der Physiologie

biologischen Chemie
und experimentellen Pharmakologie

Reviews of Physiology

Biochemistry
and Experimental Pharmacology

Herausgegeben von K. KRAMER, O. KRAYER,
E. LEHNARTZ, A. v. MURALT und H. H. WEBER

54. Band

Die nervöse Steuerung der Atmung

Von OSCAR A. M. WYSS

Mit 92 Abbildungen. IV, 479 Seiten
Gr.-8°. 1964. Ganzleinen DM 98,—

55. Band

Mit Beiträgen von M. KLINGENBERG,
H. H. LOESCHCKE, H. J. SCHATZMANN
und H. H. WEBER

Mit 52 Abbildungen und 1 Porträt.
IV, 208 Seiten Gr.-8°. 1964.
Ganzleinen DM 58,—

Inhaltsübersicht: In Memoriam Hans Winterstein †. Erregung und Kontraktion glatter Vertebratenmuskeln. Muskelmitochondrien. Namen- und Sachverzeichnis.

56. Band

Mit Beiträgen von zahlreichen Fachleuten

Mit 43 Abbildungen. IV, 380 Seiten
Gr.-8°. 1965. Ganzleinen DM 98,—

Inhaltsübersicht: I. Ziegler, Pterine als Wirkstoffe und Pigmente. H. Isliker, H. Jacot-Guillarmod u. J. C. Jaton, The Structure and Biological Activity of Immunoglobulins and their Subunits. R. Gaunt, J. J. Chart u. A. A. Renzi, Inhibitors of Adrenal Cortical Function. J. A. Olson, The Biosynthesis of Cholesterol. A. Leaf, Transepithelial Transport and its Hormonal Control in Toad Bladder. W. H. McShan u. M. W. Hartley, Production, Storage and Release of Anterior Pituitary Hormones. D. Rudman, The Adipokinetic Property of Hypophyseal Peptides. Namen- und Sachverzeichnis.

Springer-Verlag Berlin
Heidelberg GmbH

ERGEBNISSE DER PHYSIOLOGIE
BIOLOGISCHEN CHEMIE UND
EXPERIMENTELLEN PHARMAKOLOGIE

REVIEWS OF PHYSIOLOGY
BIOCHEMISTRY AND
EXPERIMENTAL PHARMACOLOGY

HERAUSGEGEBEN VON

K. KRAMER   O. KRAYER   E. LEHNARTZ
MÜNCHEN   BOSTON   MÜNSTER/WESTF.

F. LYNEN   A. v. MURALT
MÜNCHEN   BERN

U. G. TRENDELENBURG   H. H. WEBER   O. WESTPHAL
BOSTON   HEIDELBERG   FREIBURG/BR.

BAND 57

MIT BEITRÄGEN VON

ST. W. KUFFLER · J. G. NICHOLLS · R. STEELE · H. WIEGANDT

MIT 41 ABBILDUNGEN

SPRINGER-VERLAG BERLIN HEIDELBERG GMBH 1966

ISBN 978-3-662-31067-0   ISBN 978-3-540-37147-2 (eBook)
DOI 10.1007/978-3-540-37147-2

Originally published by Springer-Verlag, Berlin · Heidelberg in 1966
Softcover reprint of the hardcover 1st edition 1966

Library of Congress Catalog Card Number 62-37142

Titel-Nr. 4777

Inhaltsverzeichnis

The Physiology of Neuroglial Cells*

By

STEPHEN W. KUFFLER and JOHN G. NICHOLLS**

With 33 Figures

Table of Contents

---

* From the Neurophysiology Laboratory, Department of Pharmacology, Harvard Medical School, Boston, Massachusetts, U.S.A.

This research was supported by a grant (NB 02253—06) from the National Institutes of Health, Bethesda, Maryland, U.S.A.

** Present address: Department of Physiology, Yale University Medical School, New Haven, Connecticut, U.S.A.; Fellow of the National Multiple Sclerosis Society.

I. Introduction

Glial cells have been a subject of considerable interest to neuroanatomists for more than 100 years. They constitute a special class of cells in the brain, more abundant than the neurons and differing from them in many characteristic ways. In the mammalian cortex, for example, the glial cells greatly outnumber the neurons (NURNBERGER and GORDON 1957) and may make up about half the total volume. Our knowledge of the morphology and fine structure of glial cells has made considerable progress over the years; in fact, most of the current ideas about their function have been derived from purely anatomical considerations. In the last decade, however, a beginning has been made in determining the physiological properties of glial cells in the central nervous system.

The purpose of this review is to summarize recent physiological studies and to re-examine some of the current hypotheses. Sections III--VII deal mainly with the experiments made by the authors and their colleagues, Drs. ORKAND, POTTER and WOLFE. The remaining sections are concerned with more general considerations of the physiology of glial cells.

At present there is much discussion and speculation about the role of neuroglia. Each proposal necessarily attributes certain properties to glial cells. For instance, it is commonly stated that glial cells are the channels through which various materials have to pass on their way from the blood to the nerve cells, or in the reverse direction; this statement implies that the glial cell membranes are relatively permeable to those substances or that there is some system of transport through the cells. Since much new information about glial cells has recently been obtained, several hypotheses such as this can now be examined rigorously. In some instances it will be shown that functions which have been attributed to glial cells are no longer compatible with their known anatomical and physiological properties.

In this article particular emphasis will be placed on experiments concerned with the membrane properties of glial cells, and with the interaction between neurons and glia. Other related problems concern the relative importance of glial cells and extracellular spaces for the movement of substances through the nervous system. It will be evident that in the selection of material we have deliberately decided to concentrate on recent work carried out on the nervous system of the leech *(Hirudo medicinalis)* and on the optic nerves of the mud puppy *(Necturus maculosus)*, and the frog *(Rana pipiens)*. Many interesting areas of study have been completely omitted, for instance those dealing with the histochemistry and the metabolism of glial cells, with edema of the brain and with the cytology of glial cells.

There are several recent reviews that cover extensively much of the old and the new literature (GLEES 1955; WINDLE 1958; DE ROBERTIES and GERSCHENFELD 1961; HORSTMANN 1962; NAKAI 1963; MUGNAINI and WALBERG 1964. For a recent bibliography on neuroglia see LITTLE and MORRIS 1965).

II. Early hypotheses of glial function

It is appropriate to start with several quotations from a wide-ranging lecture series on pathology delivered by RUDOLF VIRCHOW in 1858. He had originally coined the term neuro-glia in 1846 and was aware of the fundamental implications of his discovery.

"... hitherto, considering the nervous system, I have only spoken of the really nervous parts of it. But ... it is important to have a knowledge of that substance also which lies *between the proper nervous parts*, holds them together and gives the whole its form ..." (p. 310, English translation of VIRCHOW, 1859). Speaking of the ependyma, covering the surface of the ventricles he continues: "This peculiarity of the membrane, namely, that it becomes continuous with the interstitial matter, the real cement, which binds the nervous elements together, and that in all its properties it constitutes a tissue different from the other forms of connective tissue, has induced me to give it a new name, that of *neuro-glia* (nerve cement)" (p. 315).

Later on he states "Now it is certainly of considerable importance to know that in all nervous parts, in addition to the real nervous elements, a second tissue exists, which is allied to the large group of formations, which pervade the whole body, and with which we have become acquainted under the name of connective tissues. In considering the pathological or physiological conditions of the brain or spinal marrow, the first point is always to determine how far the tissue which is affected, attacked or irritated, is nervous in its nature, or merely an interstitial substance ... Experience shows us that this very interstitial tissue of the brain and spinal marrow is one of the most frequent seats of morbid change, as for example, of fatty degeneration ... Within the neuroglia run the vessels, which are therefore nearly everywhere *separated from the nervous substance* by a slender intervening layer, and are not in immediate contact with it" (p. 317, our italics).

At the turn of the century numerous theories about glial function had been formulated on the basis of histological observations and several have retained their usefulness. We will briefly summarize some of these suggestions concerning function.

1. The role of glia as passive structural support or 'nerve-glue' was formulated by WEIGERT in 1895 (referred to by CAJAL 1952 edition), and this aspect still seems essential today. He pointed out that the spaces not taken up by neurons are occupied by neuroglia. Further, whenever neurons disappear as a result of injury or disease they are replaced by glial cells (cf. also CAJAL 1928). This in turn suggests that neuroglia has a function in processes of repair or regeneration (see later).

2. Glial cells might isolate neurons from each other in order to prevent undesirable interaction due to current flow during impulse activity. While the necessity of eliminating such 'cross-talk' as proposed by CAJAL (1952) seems less acute, there has recently been discussion of glia as a barrier to the spread of chemical transmitters at synapses (PETERS and PALAY 1965).

3. A nutritive-supporting function which was formulated in GOLGI's writings about 1883 (p. 460, 1903 b) seems worth quoting.

"I find it convenient to mention that I have used the term connective (tissue) with regard to neuroglia. I would say that 'neuroglia' is a better term, serving to indicate a tissue which, although connective because it connects different elements and for its own part serves to distribute nutrient substances, is nevertheless different from ordinary connective tissue by virtue of its morphological and chemical characteristics and its different embryological origin" (our translation).

GOLGI (1903 a, p. 40) in particular emphasized that glial cells are interposed between blood vessels and neurons and his histological demonstrations of this have had a profound effect on subsequent ideas about glial function.

The notion became firmly established that glial cells are the channels by which substances pass from the blood to neurons (see below and Fig. 2).

4. Secretory function. After lucidly discussing other theories, including LUGARO's view of an excretory function of neuroglia, NAGEOTTE (1910) stated that the results suggest "... que la névroglie est une glande interstitielle annexée au système nerveux". While this view has been largely forgotten it seems worth considering once again in the light of electron microscopy.

5. Dynamic neuron-glia interaction. This concept was stated clearly by CAJAL in 1913 (CAJAL 1928, p. 459), who spoke of "neuro-neuroglial symbiosis", and considered the possible mechanism by which neurons might influence glial cells. For instance, he stated:

"In a normal state, that is, when the reciprocal actions are in equilibrium, the satellites are few. They abstain from proliferating and they respect the neuron morphology. This quiescence is perhaps due to the paralyzing action of some principle which is liberated, under normal conditions, by the young and robust neurons. When these become fatigued, however, or when they weaken or die, the antimitogenic check is moderated or suspended and the satellite cells therefore multiply and press upon the periphery of the neuronal soma, forming in it pits and even holes, handles, fenestration, etc. ..."

The strength of some of the ideas of GOLGI and CAJAL is shown by the following quotation from a book on neuroglia that appeared in 1963 (HOSO-KAWA and MANNEN 1963).

"Functions of astrocytes: connective and supporting functions. Isolation of nerve elements proper. Repair of injuries (scar formation). Transport of metabolic substances between blood vessels and neurons. Participation in the mechanism of blood-brain barrier."

III. Structural aspects of neuroglia

1. Types of neuroglial cells. The identification of glial cells in the mammalian brain has depended since their discovery by VIRCHOW on their staining properties, their shapes and configurations and on their characteristic position between and around neurons. Unlike neurons, glial cells do not have axons and they retain the ability to divide throughout life. The identification is usually made without difficulty, and in addition a clear subdivision into two main classes can be made in the mammalian brain, on the basis of light microscopy and electron microscopy. These are the astrocytes and the oligodendroglia (PENFIELD 1932; FARQUHAR and HARTMAN 1957; PALAY 1958a; MUGNAINI and WALBERG 1964; SCHULTZ 1964). No survey will be attempted of the numerous detailed findings that electron microscopy has added in recent years, except to mention that they have mitochondria, an endoplasmic reticulum, ribosomes, fat, glycogen, and in general they present

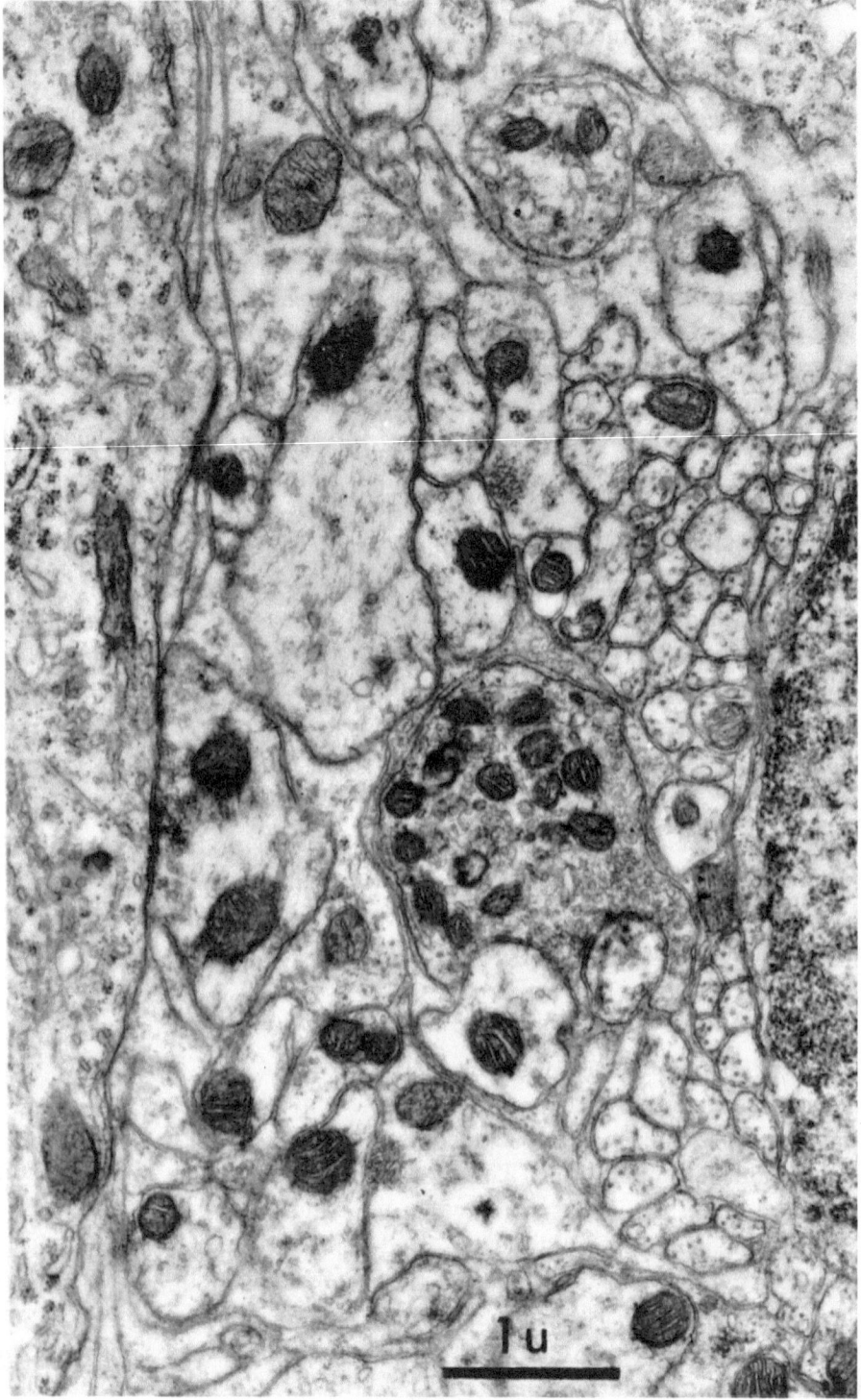

Fig. 1 A

Fig. 1A and B. Relationship of neurons and glia in the cerebellum of a rat. Two identical prints were useol. In B some of the structures are marhed for eosier identification. Left: portion of a Purkinje cell (*P.C.*). Right: a small granule cell (*Gr.C.*) with a narrow band of cytoplasm around its nucleus. The mid portion of picture is tightly packed by axons (*A*) and dendrites (*D*); only a few of these have been marked. The

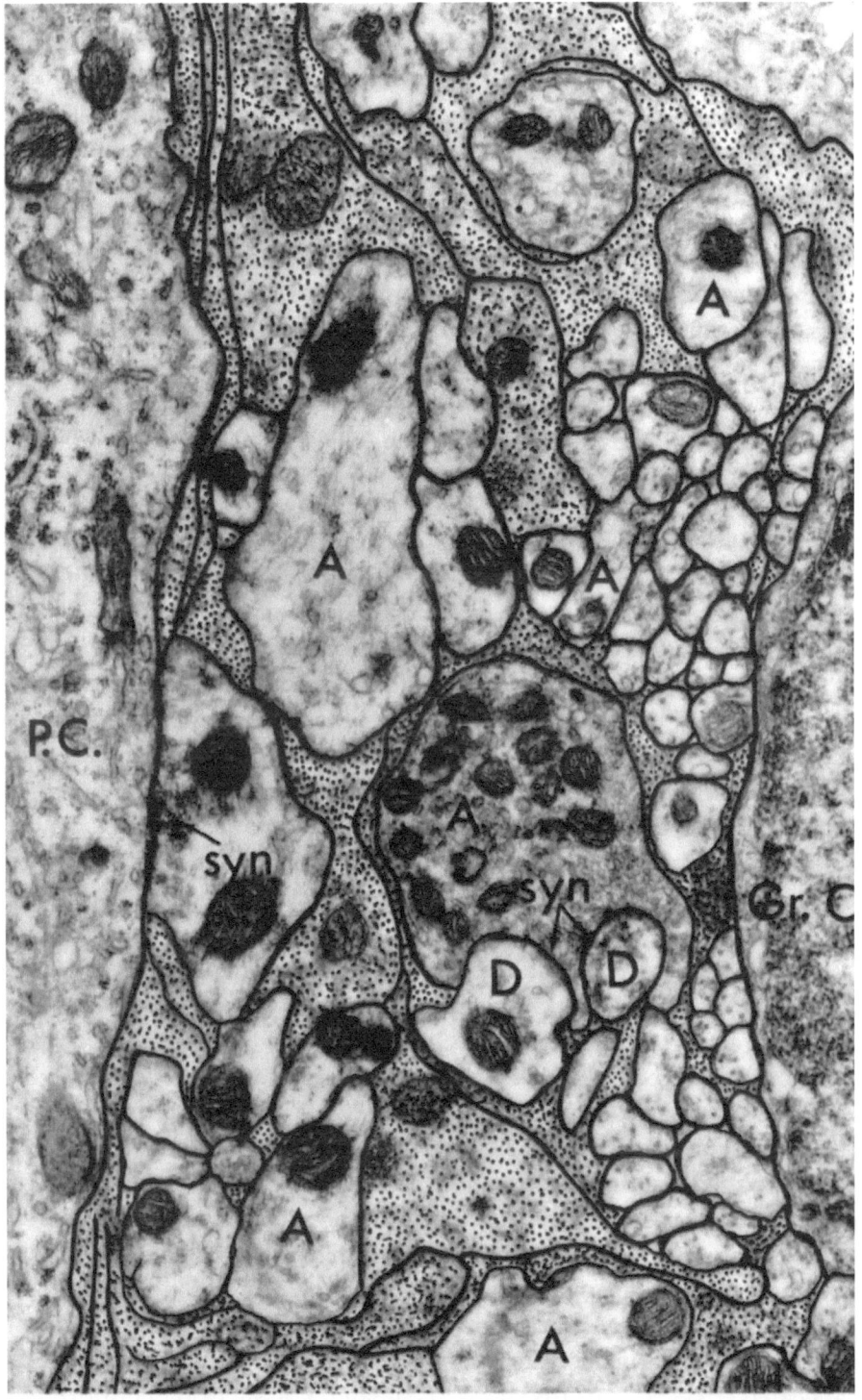

Fig. 1 B

cell surfaces have been outlined by tracing the various intercellular clefts. Arrows indicate synapses (*syn.*). Astrocytic glial processes have been stippled and are seen around neurons. Many of the axons face each other's surface without glia interposed between them. All the cell surfaces are separated by clefts of about 150 Å (kindly provided by Dr. S. L. PALAY, Harvard Medical School)

the picture of metabolically active cells. A few electron micrographs of verte-brate and invertebrate glial cells are shown below.

Astrocytes exhibit a wide variety of shapes and fine structure. They are generally divided into two groups: 1. fibrous astrocytes, which contain large bundles of filaments in their cytoplasm and are most prevalent in the white matter; 2. protoplasmic astrocytes, which contain less fibrous material and are more frequently found in the gray matter. Both types of astrocytes have numerous desmosomes and make extensive 'endfoot' contacts with blood vessels and neurons, as we shall discuss in detail below (GOLGI 1903a; PENFIELD 1932; GLEES 1955; NAKAI 1963).

Oligodendroglial cells have smaller nuclei than astrocytes and their peri-karya are occasionally associated with capillaries, but unlike astrocytes they do not form endfeet as seen by light microscopy (PENFIELD 1932; CAM-MERMEYER 1960). Oligodendrocytes are predominantly located in the white matter and are responsible for myelin formation (MATURANA 1960; PETERS 1960; BUNGE, BUNGE and PAPPAS 1962). Their processes usually wrap around more than one axon as noted originally by RIO HORTEGA (1928).

In the peripheral nervous system neurons are associated with Schwann cells. Like the oligodendrocytes Schwann cells are concerned with the formation of myelin around the axons (GEREN 1954). They form a thin envelope or covering around the non-myelinated axons or around nerve endings. They also cover the cell bodies in peripheral ganglia. Whereas the Schwann cells arise embryo-logically from the neural crest both the astrocytes and the oligodendroglia are derived directly from the ependymal cells which line the inner surfaces of the brain, namely the ventricles and canals facing the cerebrospinal fluid. All these cells are of ectodermal origin (RUGH 1964).

In lower vertebrates and invertebrates there also is an abundance of glial tissue which in many respects resembles that of mammals. However, there is usually less agreement concerning a classification into distinct subgroups.

The microglial cells, which were established as a separate group by RIO HOR-TEGA (1932), should not really be classed as glial cells. They are of mesodermal origin and are macrophages which have entered the nervous system from the blood (KONIGSMARK and SIDMAN 1963). Their function is related to patho-logical processes and phagocytosis. They will not be considered in this article.

2. Electron microscopy of neuron-glia relationship. The advent of electron microscopy changed many of the morphological concepts about the inter-relationship between cells in the nervous system. The relatively large spaces between cells and around blood vessels seen in light microscopy practically disappeared (PALAY 1956, 1958b; WYKOFF and YOUNG 1956; SCHULTZ, MAY-NARD and PEASE 1957). The cells were so closely packed that the only remaining space consisted of very narrow clefts that were usually 100—200 Å wide and divided the various elements from each other (HORSTMANN and MEVES 1959).

The electron micrograph of Fig. 1 from the cerebellum of the rat will serve as an introduction to several typical features of the neuron-glial relationship. Wherever neurons border on glia (stippled areas in Fig. 1 B), they are directly apposed. Extensions of astrocytic glial cells, at times in very thin and tenuous sheets, tend to be interposed between dendrites and around synapses (but not in the synaptic clefts). The illustration also gives an impression of the variety in the appearance of the cytoplasm of several neurons with their various organelles.

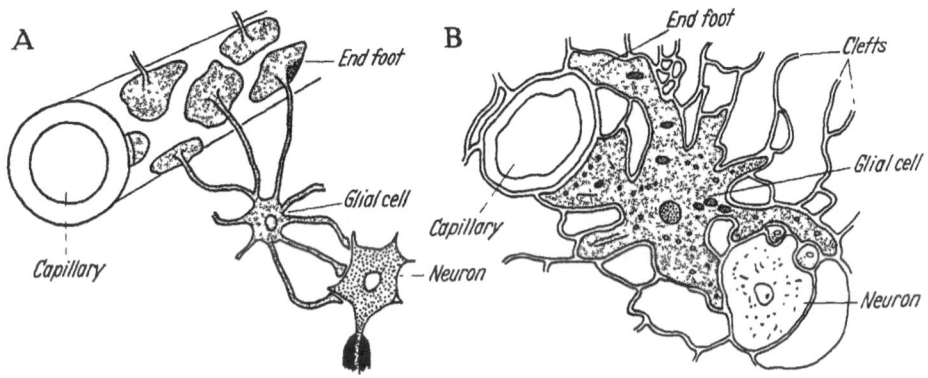

Fig. 2 A. Schematic diagram of neuron-glia-capillary relationship in the vertebrate brain as seen in the light microscope. Capillary walls are closely invested by 'endfeet' of astrocytes which also make contact with neurons. Only one astrocyte is shown bridging the space between a neuron and a capillary. This type of intimate relationship served as the basis for GOLGI's concept of glia being a channel for the distribution of nutritive materials between the blood and neurons

Fig. 2 B. Sketch of neuron-glia-capillary relationship as seen in the electron microscope. Again only one astrocyte (shaded) is shown, interposed between the endothelium of the capillary and the neuron. All the cells, axons, dendrites and astrocytes are tightly packed with narrow clefts (their dimensions greatly exaggerated in this scheme) of about 150 Å between them (see also Fig. 1 and other electron microscopic illustrations)

Before proceeding, it seems useful to present a diagram of the scheme that came to be accepted after GOLGI's original observations. Fig. 2 A shows a blood vessel covered with 'endfeet' or 'sucker processes'. Just one astrocyte is shown bridging the space between a capillary and a neuron. One should note that in this scheme, based on light microscopy, there are ample spaces between the cells and around the vessel by way of which nutrients could traverse the nervous system without the necessity of postulating transport through glial cells. A similar scheme that incorporates the findings of recent electron microscopy indicates why GOLGI's original idea became even more attractive during the past few years. In Fig. 2 B a capillary is tightly invested by a number of astrocytes, separated by clefts 100—200 Å wide. Again, only one astrocyte (shaded) is fully shown; wherever it borders another cell a similar narrow cleft, whose dimensions are exaggerated in this diagram, intervenes between the two membranes. In this scheme very little space is available for movement of fluid between cells and until recently it was not clear whether such spaces were open for diffusion. It was even proposed that intercellular clefts were not functional pathways whereby substances reached neurons from the blood (Gerschenfeld, Wald, Zadunaisky and De Robertis 1959; De Robertis and Gerschenfeld 1961; Sjöstrand 1961; Cummins and Hydén 1962; De Robertis 1962; Luse 1962; Davson 1963).

On closer examination of Fig. 2B the necessity for postulating transport through the glial cell loses some of its force. A particle in the lumen of a capillary on its way to a neuron would have two alternative pathways: 1. After it had escaped from the lumen into the extracellular space, occupied by the basement membrane, it could traverse the glial cell membrane. It could then move through the glial cytoplasm by diffusion or by active transport. Next it would have to cross the glial membrane a second time in order to enter the extracellular space where the glia is apposed to the neuron. From there it could enter the neuron by active transport, or by diffusion. This is the complex pathway to and from the neuron that follows from the hypothesis of glia as a channel of distribution. 2. After crossing the capillary endothelium the particle could take the longer tortuous route through the intercellular clefts and thus reach the neuronal membrane by diffusion. Although the route is longer, this actually is a simpler scheme because it does not require uptake, transport and secretion by glial cells. Similar considerations, of course, apply for the movement of substances in the opposite direction.

One of the major tasks in the study of neuroglia that emerges from electron microscopic considerations is to determine the relative importance of the two possible pathways for the movement of substances. First there is the question whether the intercellular clefts are open for diffusion. They could contain materials that impede diffusion either by virtue of fixed charges (Treherne 1962a, b) or by occluding the space (Hess 1962). Alternatively, the apposed membranes might be fused at critical locations, barring diffusion (Gray 1961, 1964; Peters 1962). If this were the case and the intercellular spaces did not communicate with the blood or cerebrospinal fluid, one would have to postulate that the glial cells themselves act to provide the immediate environment of neurons, and probably contain a high Na^+ concentration. Since this ion is needed for impulse propagation, it was widely assumed that the Na^+ which was present in the brain was contained in glial cells, which in effect were the extracellular spaces of neurons (Katzman 1961; De Robertis 1962; Koch, Ranck and Newman 1962). Incidentally, this could not be the case in many places in the nervous system where whole groups or bundles of axons or dendritic processes adjoin each other, with 100—200 Å spaces in between, *without* interposed glial cells (see Fig. 1, 6, 9, 10). If narrow clefts were not an effective extracellular space in such situations, the neurons would have to provide each others' extracellular space.

A related question concerns the properties of the glial membrane, whether it serves as an effective barrier to diffusion, what type of transport system operates across it and what type of intracellular milieu does it maintain. If glial cells have membrane potentials, are these due to K^+ accumulation in the cytoplasm as in most nerve and muscle cells? From the very fact that such elementary questions had to be asked a few years ago, it was apparent that there was little information about the function of glial cells.

3. Recent preparations used for the study of neuroglia. In order to study the physiology of neuroglial cells, preparations with certain well-defined properties are needed. In this context some thoughts by CAJAL formulated around 1900 are of interest. Inquiring into the role of neuroglia, he wrote: "What is the function of the neuroglia in the nervous centers? Nobody knows at present, and what is more serious, is that the problem seems to be insoluble for a long time to come, because physiologists lack a direct method to attack it" (CAJAL 1952 ed., p. 246). CAJAL's prediction proved correct because reasonably direct physiological techniques did not become available until about ten years ago. Until now it has been too difficult to investigate glial cells in the mammalian brain *in situ*, and only one preliminary study has so far appeared (TASAKI and CHANG 1958). But perhaps the greatest hindrance has been the lack of identification of cells by physiological criteria. Recently, several suitable preparations have been discovered.

a) Tissue culture. Glial cells in tissue cultures from mammalian brains have been successfully studied with intracellular electrodes by HILD and TASAKI (1962) and WARDELL (1964). In cultures, individual neurons can also be identified under direct visual observation and their potentials can be recorded with intracellular and extracellular electrodes (CRAIN 1956; HILD and TASAKI 1962). Tissue cultures are also promising because immunological methods can be used to investigate the formation of myelin (see Section XII). POMERAT (1958a, b), LUMSDEN and POMERAT (1951), LUMSDEN (1958), HILD (1957), and BORNSTEIN and MURRAY (1958) have developed this field by their vivid presentations of nerve and glial cells (cf. GEIGER 1963, for a review).

b) Retina. While studying the visual system in the fish retina, SVAETICHIN and his collaborators observed slow potential changes, which apparently arose in the horizontal cells in the outer plexiform layer and also deeper inside the retina (see SVAETICHIN, NEGISHI, FATECHAND, DRUJAN, and SELVIN DE TESTA 1965, for references). These cells have generally been classed as glial cells (they will be discussed in Section VII, 3).

Subsequent to the successful use of the retina and tissue cultures, other suitable nervous systems for the study of neuroglia have been found. Two preparations which have been extensively used for the study of glial cells are the central nervous system of the leech and the optic nerves of amphibia (KUFFLER and POTTER 1964; KUFFLER, NICHOLLS, and ORKAND 1966). These tissues have the following advantages: 1. The preparations survive well in isolation. They are, however, also suitable for physiological work within the animal when the circulation is intact. 2. They have large glial cells that can be used for experiments with microelectrodes and for chemical studies. 3. In addition the leech provides a transparent nervous system in which all the component cells are visible. Moreover, in both preparations the anatomical relationship between neurons and glia is intimate, as in the mammalian brain, and therefore physiological results promise to be of general interest. In order to give an adequate background for the physiological discussion a description of the structure of the central nervous system of the leech and of the amphibian optic nerve will be given.

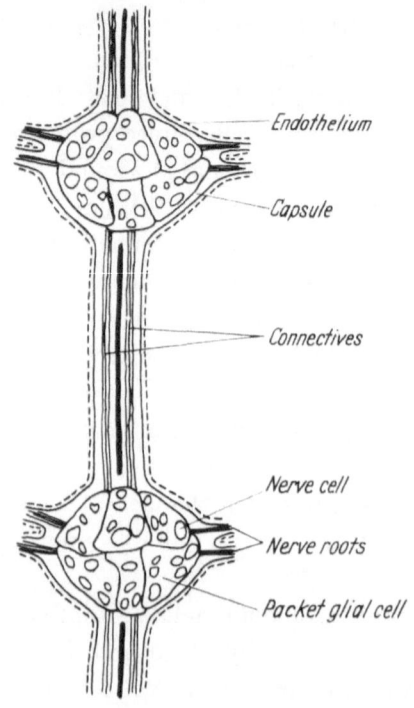

c) *The nervous system of the leech.* A schematic presentation of the central nervous system of the leech *(Hirudo medicinalis)* is given in Fig. 3. For a full description see GRAY and GUILLERY (1963); COGGESHALL and FAWCETT (1964). The ventral nerve cord consists of a chain of 23 ganglia (two shown in Fig. 3) joined by connectives. It lies in a blood sinus, without itself being penetrated by blood vessels. Nutrients from the blood, therefore, have to reach neurons and glial cells by diffusion. Each ganglion contains about 350 unipolar nerve cells which are grouped in six separate clusters or 'packets'. A remarkable feature is that

Fig. 3. Diagram of a portion of the leech central nervous system. Only two ganglia, viewed from the ventral side, are shown. A ganglion is made up of six separate clusters (packets) of nerve cells. Each 'packet' contains one large glial cell and about 60 neurons. Ganglia are joined by two connectives, each made up of many unmyelinated axons and one glial cell. A fine tenuous medial connective (FAIVRE's nerve) has been omitted. The nervous system is surrounded by a connective tissue capsule which is covered by an endothelial cell layer (dashed line) facing the blood. The entire structure lies in a blood sinus

all the nerve cells and their initial axonal processes within one packet are embedded in a single large glial cell. In the connectives, many hundreds of unmyelinated axons are surrounded by another single large glial cell whose length extends over many millimeters. Because the glial cells in the leech are so large and contain fibrillar material that can be readily stained they were extensively studied by a roster of eminent histologists: APÁTHY, BIEDERMANN, HERMANN, RIO HORTEGA, GASKELL, ODURIH, RETZIUS (for references see ITO 1936; COGGESHALL and FAWCETT 1964). In fact, many of the characteristic features of vertebrate glial cells were initially most thoroughly studied in the

leech. In the electron micrographs the cell boundaries between neurons and glia
are clear, and clefts of about 150 Å always separate their membranes in the
ganglia and in the connectives. Fig. 4 illustrates a ganglion as seen under the
dissecting microscope, with the individual nerve cells clearly visible within the
clusters. In this dorsal view of the ganglion the lighter area in the center is

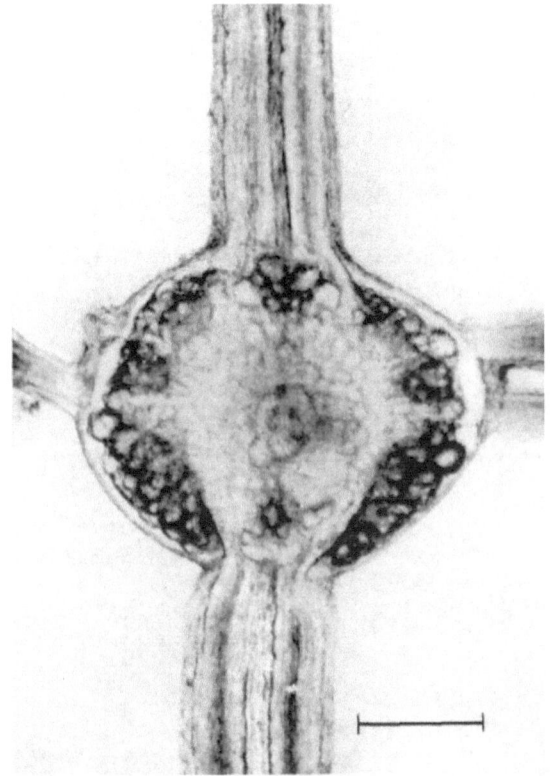

Fig. 4. Photograph of a living leech ganglion seen in isolation with transmitted illumination during experi-
mentation. Dorsal view, showing the centrally located neuropil (pale area) which receives processes from
the neurons in the 'packets' and from other regions of the body. Individual neurons are easily distinguished.
The numerous nerve fibers in the connectives and roots appear as dark strands. Glial cells are transparent
and surround the neurons. Scale is 200 μ (from KUFFLER and POTTER 1964)

the neuropil where synaptic contacts are made by the various axonal processes.
Portions of the connectives are seen above and below the ganglion while the
roots emerge on both sides. Glial cells cannot be seen as such; they are transparent and fill the clear spaces between the nerve cells. The electron micrograph of Fig. 5 presents a portion of the neuropil in a ganglion showing neuronal
processes which are intermingled with the many processes of one glial cell.
Neurons and glial tissue are always closely apposed and the appearance is
strikingly similar to electron micrographs in the vertebrate brain (see above,
Fig. 1). In this illustration a great deal of fibrillar material is present in the
glial cytoplasm. A transverse section of part of a connective shows the glia-
neuron relationship with greater clarity. In Fig. 6 the axons are relatively light

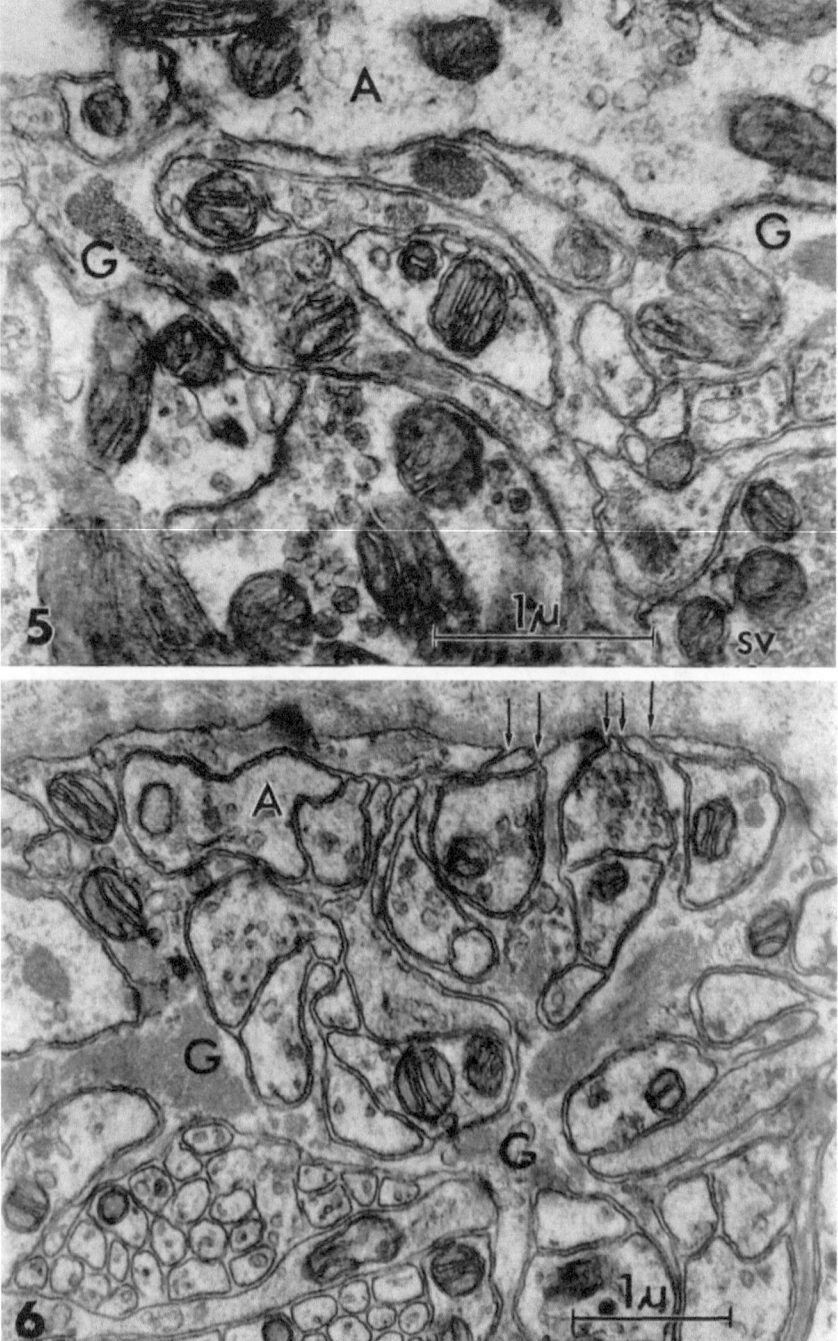

Fig. 5. Electron micrograph of a portion of the neuropil in a leech ganglion. On top is part of one of the larger axons (A), the lower portion is made up of small nerve processes, some filled with synaptic vesicles (SV in lower right corner). Slender processes, all part of one glial cell (G), are between the axons. Glia and neurons are separated from each other by intercellular clefts of about 150 Å width. Scale 1 μ

Fig. 6. Electron micrograph of a portion of a connective in the leech. On top surface is the collagenous material of the connective tissue capsule. Lighter structures, irregularly shaped, are axons (A). They are surrounded everywhere by the gray cytoplasmic processes of the connective glial cell (G), containing fine fibrillar material. Axons and glia are separated by 150 Å clefts which open at numerous places onto the large extracellular space of the capsule (arrows). In the lower left corner is a bundle of small axons which face each other, without interposed glial extensions (see also Fig. 1). The endothelium and the blood sinus lie beyond the capsule. Scale 1 μ (Figs. 5 and 6 kindly provided by Drs. R. E. COGGESHALL and D. W. FAW-CETT, Harvard Medical School)

and contain mitochondria, whereas the cytoplasm of the glial cells is more uniform, consisting mainly of fine fibrillar material. The intercellular clefts open onto the surface at numerous distinct places; some of the mesaxon openings are indicated by arrows. It is here that direct communication is made between the tortuous cleft system within the connective and the wider extracellular space of the connective tissue capsule. The latter is covered by an endothelium which faces the blood. Nowhere are there spaces of more than about 200 Å width within the connective. In the ganglia on the other hand there are in addition larger extracellular spaces consisting of spicules of basement lamina and lacunae or collections of

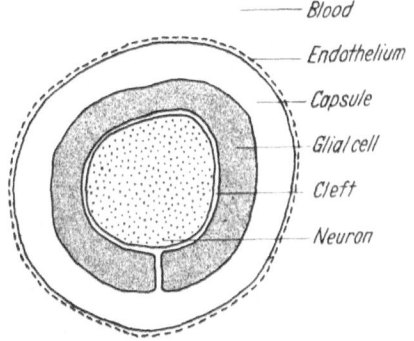

Fig. 7. Diagramatic presentation of the constituents of the leech nervous system. One neuron is depicted, surrounded by one large glial cell (shaded). The narrow cleft separating the two cells leads into the large capsular space which is covered by an endothelium facing the blood. The pathway between blood and neurons is essentially similar in the leech and in vertebrates

what may be extracellular fluid (GRAY and GUILLERY 1963; COGGESHALL and FAWCETT 1964).

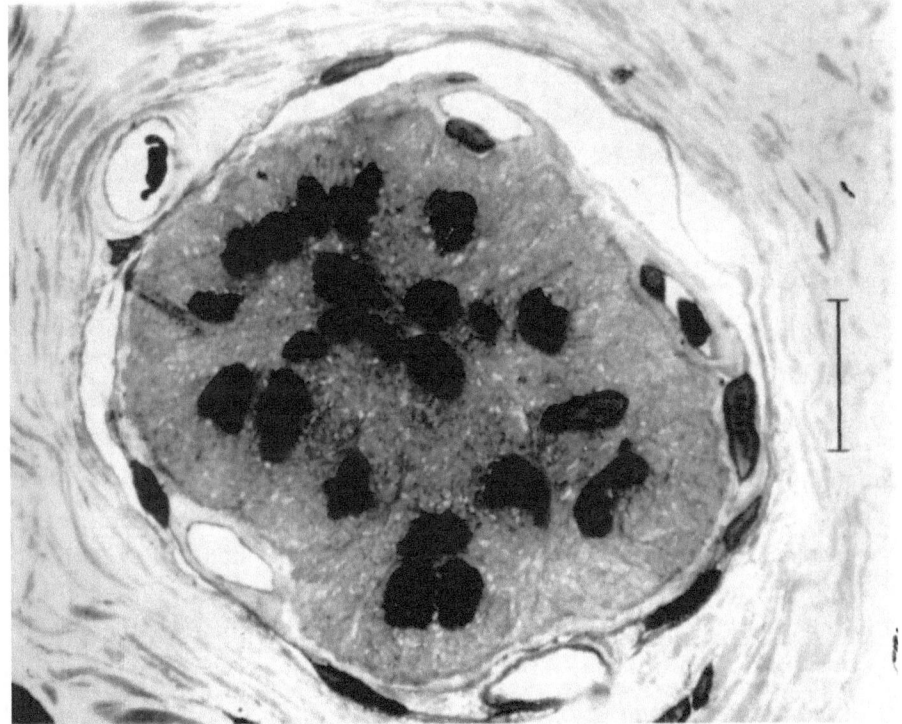

Fig. 8. Cross section of the *Necturus* optic nerve, stained with toluidine blue. The nuclei of glial cells are darkly stained; the outlines of the glial cytoplasm and of the bundles of non-medullated axons cannot be distinguished. Blood vessels run close to the surface in the pia-arachnoid covering of the optic nerve. Scale 50 μ (kindly provided by Dr. D. E. WOLFE, Harvard Medical School)

Another diagramatic presentation in Fig. 7 simplifies the constituents of the leech nervous system, the neurons, glia and extracellular spaces. This scheme, with minor modifications, is also helpful in discussing the various pathways from the blood to neurons in the vertebrate nervous systems. It will be seen that the problems concerning cell constituents and 'barriers' are essentially

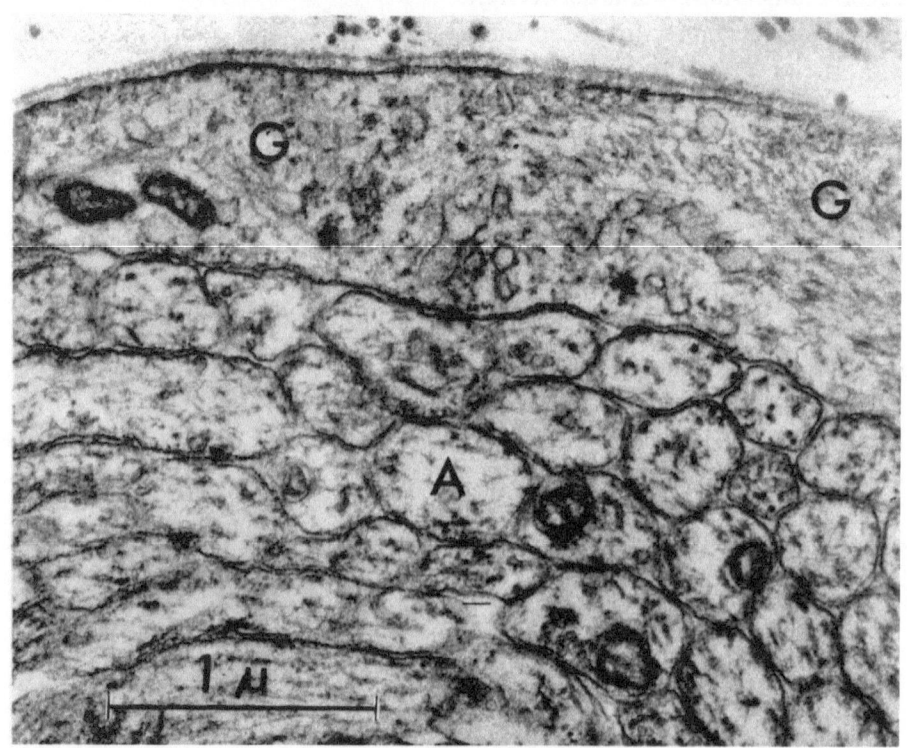

Fig. 9. Electron micrograph of the optic nerve of the *Necturus*. The pia-arachnoid covering has been removed, leaving only the basement lamina on the surface of the nerve. The glial cytoplasm (G) surrounds a group of axons (A) some of which are cut obliquely. As in the mammalian (Fig. 1) and in the leech central nervous systems (Fig. 5 and 6) the neurons and glia are closely apposed, separated by clefts approximately 150 Å wide. The same spacing is seen between the membranes of the axons within the bundle. (Kindly provided by Dr. D. E. WOLFE, Harvard Medical School)

similar (compare with Fig. 2B). Note that the sketch of Fig. 7 shows only one of the many possible intercellular clefts.

d) The optic nerves of amphibia. The optic nerves of the mud puppy *Necturus maculosus* and of the frog have proved to be convenient for the study of neuroglial cells in the vertebrate. The glial cells in optic nerves are, of course, part of the central nervous system. They are large and intimately surround the nerve fibers (MATURANA 1960; WOLFE 1966). The optic nerve is about 0.15 mm in diameter in *Necturus* and in medium sized *Rana pipiens* it is about 0.5 mm. Both nerves are covered by a pia-arachnoid layer containing blood vessels which run parallel to the surface. In *Necturus* no blood vessels are seen within the neural tissue (WOLFE, in preparation). Only occasional capillaries within

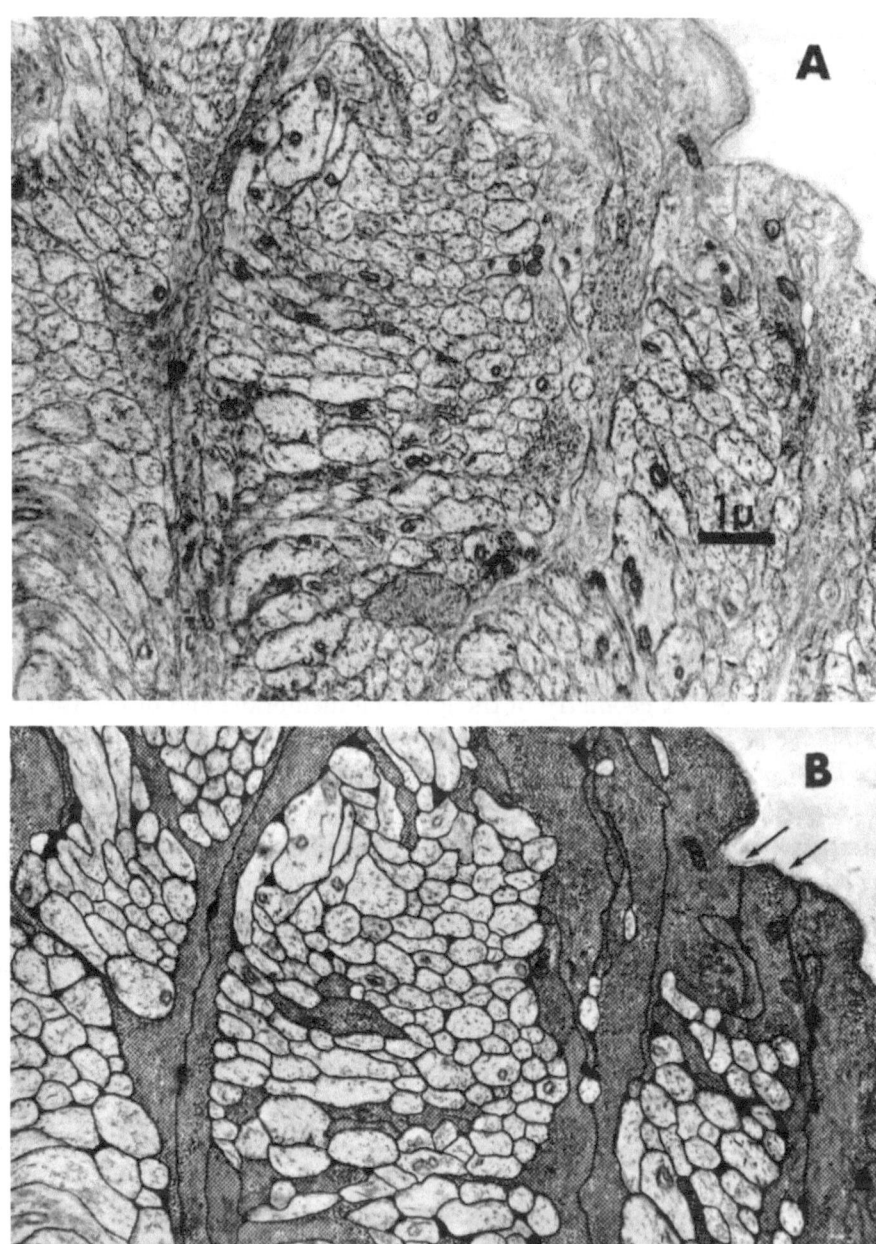

Fig. 10 A and B. Glial cytoplasm, axons and extracellular space in a cross section of a portion of the optic nerve of the *Necturus*. As in Fig. 9 the pia-arachnoid covering has been removed. A and B are identical prints. In B all the intercellular spaces have been traced in black, emphasizing the widespread tortuous cleft system. Two cleft openings reaching the surface of the optic nerve are marked by arrows. The glial processes in print B are stippled, thereby showing more clearly the neuron-glia relationship and the extent of the glial cytoplasm. The cross-sectional areas of axons and glia are approximately equal (WOLFE, in preparation)

the optic nerve of the frog have been reported by MATURANA (1960). Thus, as in the leech, the neurons and glia receive their nutrients principally from the surface. Fig. 8 presents a cross section of the *Necturus* optic nerve stained with toluidine blue. The glial nuclei appear dark and prominent, while the glial cytoplasm remains unstained. The outlines of the bundles of non-medullated axons (fiber diameter 0.1 to 1.0 µ), are too small to be seen in this light micrograph. Small blood vessels run close to the surface of the nerve which is covered by loose connective tissue. The neuron-glia relationship is illustrated in Fig. 9, showing spaces 100—200 Å wide between the cells, as in the mammalian system (Fig. 1) and in the leech (Fig. 5 and 6). One also sees part of a cluster of tightly packed axons, where glia surrounds the bundle, but not individual axons.

In both the central nervous system of the leech and the optic nerve of *Necturus* the glial cells occupy 35 to 55 % of the total volume (KUFFLER and POTTER 1964; WOLFE, in preparation). This can be readily appreciated by examining Fig. 10. To emphasize graphically the glial components, they have been stippled in Fig. 10B (compare with Fig. 1). Glial cytoplasm permeates the whole nerve, usually surrounding axons that are grouped in tightly packed bundles. The complex geometry of the glial cell membrane and of the tortuous intercellular clefts are also demonstrated in Fig. 10B where the intercellular space has been traced in black ink (cf. also WOLFE 1965). Note that there are two places where the clefts ('mesaxons') open to the outside (arrows). The connective of the leech appears strikingly similar (KUFFLER and POTTER 1964).

IV. Physiological properties of neuroglia

1. **The membrane potential and K^+ content of glial cells.** The first resting potentials from glial cells were obtained in tissue cultures (HILD and TASAKI 1962). The potentials were 50—70 mV and the membrane resistance was low, about 3—10 ohms cm² (see, however, later in this section). The resting potential was abolished by isotonic K^+ solutions added to the bathing fluid. Studies on the retina in which resting potentials have been measured will be discussed in Section VII, 3.

In the leech, the glial resting potentials were relatively high, about 75 mV, compared with the neuronal potentials which were no more than 50 mV (KUFFLER and POTTER 1964). The low value found in neurons is not due to damage, because the cell bodies (see Fig. 3 and 4) are large and repeated penetrations with microelectrodes frequently cause little deterioration (see also HAGIWARA and MORITA 1962; ECKERT 1963). Similar potentials are recorded in the circulated nerve cord *in situ* (KUFFLER and POTTER 1964). The glial cells, however, are difficult to impale because in most places their cytoplasm consists of folds and fine processes. The evidence that the higher resting potentials were recorded from the glial cells was that (i) the microelectrode tip could be seen under the dissecting microscope to be next to but not in a neuron (ii) when substances were electrophoretically injected from the microelectrode they were seen histologically to be in a glial cell. In the *Necturus* optic nerve, penetration of glial cells was easier and in a large series of measurements the glial membrane potential was close to 90 mV in both isolated and normally circulated nerves (KUFFLER, NICHOLLS, and ORKAND 1966). No measurements of neuronal membrane potentials in the optic nerve could be made because the unmyelinated axons are too thin for penetration (less than 2 μ diameter). The high, stable resting potentials must therefore have been recorded when the microelectrode tip was in a glial cell; marking experiments confirmed that this was so. High resting potentials of glial cells were also obtained in the optic nerve of the frog. Measurements on spinal neurons (ARAKI and OTANI 1955; KATZ and MILEDI 1963) have shown that 70 mV is probably close to the normal resting potential of neurons in the amphibian CNS. Consequently, the difference between neuronal and glial membrane potentials is likely to apply to amphibia as well as leeches (see also later, Fig. 25). The membrane potentials measured in Schwann cells that surround squid axons have been of the order of 33—46 mV, i.e. lower than those of the axons (VILLEGAS, VILLEGAS, GIMENEZ and VILLEGAS 1963). It is possible that the Schwann cell potentials were low because the cytoplasm is reduced to very fine processes that are difficult to penetrate with microelectrodes (see below).

A consistent finding was that the resting potential of the glial cells in *Necturus* was at E_K, the K^+ equilibrium potential, and changed as predicted by the Nernst equation over a range of 1.5 mEq/liter to 75 mEq/liter (KUFFLER, NICHOLLS and ORKAND 1966). According to the Nernst equation

$$E = RT/F \ln K_i/K_o \tag{1}$$

where E is the membrane potential in mV, R is the gas constant, F is the faraday, T is the absolute temperature, and K_i and K_o are the intracellular and external concentrations of K^+. Thus in optic nerve glial cells the ratio between the internal (K_i) and the external potassium (K_o) determines the membrane potential. Other ions make only a negligible contribution. This is in contrast to most muscle or nerve cells that have been studied, where a considerable deviation from the Nernst equation is seen in the physiological range of K^+ concentration (for references see ADRIAN 1956).

Fig. 11 presents a series of resting potentials plotted against K_o concentration measured in glial cells of the optic nerve of *Necturus*. Each point is the mean of 6 or more measurements, ± the S.D. The solid line has a slope of 59 mV for a ten-fold change in K_o (temp. 24°C). One should note that there was little or no deviation from the expected relationship even when K_o was halved from the normal value of 3.0 mEq/liter to 1.5 mEq/liter.

Knowing the relationship between the membrane potential of glial cells and K_o one can estimate the intracellular concentration of K^+, because whenever K_o and K_i are equal the resting potential should be zero. K_i found in this manner was about 100 mEq/liter. Similar measurements of K_i in leech glial cells gave a value of about 110 mEq/liter (NICHOLLS and KUFFLER 1964). In the leech the results could be confirmed directly by flame photometric measurements of K and also of Na (NICHOLLS and KUFFLER 1965). The amounts of Na

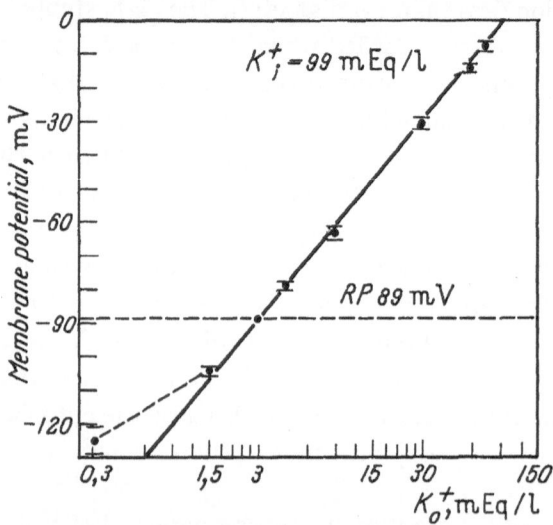

Fig. 11. Relation between the glial membrane potential and the K^+ concentration in the bathing fluid (K_o) in optic nerves of *Necturus*. Forty-two measurements were made with K^+ concentrations below or above normal. The mean of the resting potentials in Ringer's solution (3 mEq/liter K^+) was 89 mV, indicated by the horizontal dotted line. Horizontal bars indicate $\pm$ SD. of mean. The solid line has a slope of 59 mV for a ten-fold change in K_o according to the Nernst equation (see text). It fits the observed points accurately between 1.5 and 75 mEq/liter. Only at 0.3 mEq/liter K^+ is there a marked deviation (dotted curve) from the curve predicted by the Nernst equation. The membrane potential is zero when the internal K^+ concentration (K_i) equals K_o. K_i therefore is 99 mEq/liter. The membrane potential of glial cells can be used as an accurate K^+ electrode (see text, from KUFFLER, NICHOLLS and ORKAND 1966)

and K in the nerve cord are approximately equal. A large part of the Na is extracellular, located in the extensive connective tissue capsule. After this extracellular Na has been replaced by sucrose the K:Na ratio is more than 8:1. Knowing that the glial and neuronal volumes are approximately equal one can estimate that there must be at least four times more K than Na in the neuroglial cells. Thus the intracellular Na was not more than about 25 mEq/liter.

The glial cells, therefore, have a predominance of K^+ in their cytoplasm. The direct measurements in the leech of K and Na also precluded the possibility that Na may be sequestered in large quantities in cell compartments, e.g. in vesicles or in the endoplasmic reticulum.

In the leech it was also interesting that the equilibrium potential for K^+ was similar in neurons and glial cells (about 85 mV), in spite of the different resting potentials (NICHOLLS and KUFFLER 1964). The difference between neuronal and glial resting potentials would arise if the permeability to ions other than K^+ were relatively higher in neurons than in glia, or if the metabolic ion pumps in the two cells were different.

2. Na and K in mammalian glial cells. There is as yet little direct evidence of the Na^+ and K^+ concentrations in mammalian glial cells. Considerable quantities of Na (up to 35 %) (DAVSON 1955) have been found in pieces of mammalian brain but the distribution of this ion between neurons, glia and extracellular space is uncertain (KATZMAN 1961; see, however, discussion of section V, 5). In the study by KOCH, RANCK and NEWMAN (1962) an attempt was made to obtain a predominantly 'glial' population by cutting the afferent and efferent connections of the lateral geniculate body in cats. Na was found to constitute about 50 % of the total cation content, but this result does not on its own necessarily indicate that the glial cells were high-Na cells because 1. the extracellular space in this tissue was not known; 2. many neurons were still present but were abnormal because of the operative procedures; 3. there was presumably invasion by microglial cells.

An attempt has also been made to compare the K^+ content of clumps of neuroglia dissected from the mammalian brain with that of single neurons (HAMBERGER and RÖCKERT 1964). The disadvantages of this technique in terms of the integrity of the cells will be fully discussed in section XI. The results have been interpreted to show that the K^+ content of neuroglial cells and neurons is similar. One may mention here that the values are expressed as K^+ per nerve cell or per equivalent volume of 'glia' and not per unit weight. Thus, the concentration of K^+ in the neurons cannot be determined and might be reduced by the dissection procedure. Further, without Na^+ measurements one does not know either the $K^+ : Na^+$ ratio or the K^+ concentration in the cells. We shall see later that the glial sample is probably contaminated by neurons. It should be noted that if a cell has a high Na^+ content, its resting potential could still depend on K^+, for example if the Na^+ were present in a bound form.

3. Lack of signalling mechanism in neuroglia. The absence of long processes, such as axons, makes it unlikely that glial cells are used for long-range signalling. They might, however, generate signals which in turn could initiate or modify neuronal activity. Experiments have been made to determine whether a regenerative process, similar to that in neurons, occurs in glial cells of the leech (KUFFLER and POTTER 1964), *Necturus*, and frog (KUFFLER, NICHOLLS and ORKAND 1966). The action potential generating mechanism in neurons is initiated by a change in membrane potential (HODGKIN 1961). Accordingly, the glial membrane potential was displaced by current pulses delivered through an intracellular electrode while the changes in membrane potential were recorded with a second microelectrode. The membrane potential could be displaced by 100 mV or more in each direction. Over the entire range the cell membrane behaved passively, like an ohmic resistance, without giving an 'active' response.

Such results show that the neuroglial cells in the leech, *Necturus*, and frog are not "sensitive" to currents and lack a mechanism for generating signals which resemble those in neurons.

It should be recalled that HILD and TASAKI (1962) found that glial cells in tissue cultures gave a slow 'response', accompanied by a conductance change when large currents were passed with extracellular electrodes. These experiments were repeated and confirmed by WARDELL (1964) with doublebarrelled intracellular recording and current-passing electrodes. The 'responses' were obtained with both anodal and cathodal currents or with mechanical stimuli. They were seen not only in glial cells but also in fibroblasts (WARDELL, personal communication). Membrane potential displacements of the order of 200 mV or greater were, however, needed to obtain the slow changes which WARDELL regards to be result of a dielectric breakdown. There seems to be no conflict between the tissue culture results and those reported in the leech and in amphibian optic nerve. In neither case were 'responses' evoked by changes of membrane potential within the physiological range. Another difference concerns the estimated specific membrane resistances (R_m). In the leech glia a minimum value of $1000 \, \Omega \, cm^2$ was obtained for R_m, similar to the value found in neurons (KUFFLER and POTTER 1964). HILD and TASAKI'S (1962) low estimate of R_m ($3-10 \, \Omega \, cm^2$) may be due to referring the measured imput resistance to single cells only, without considering the possibility of electrical coupling to other glial cells (see below section IV, 5). This difficulty was encountered in the *Necturus* optic nerve where glial cells are extensively linked to one another by low resistance connections. No estimate of R_m could be made because of uncertainties of the membrane area. Input resistances ranged from 0.5—5.0 MΩ.

Criticism concerning the lack of responses in the experiments on leech glia has appeared in a recent report by SVAETICHIN, NEGISHI, FATEHCHAND, DRUJAN and SELVIN DE TESTA (1965) who stated that the experiments were made on preparations in 7—8 % alcohol solutions and at low temperatures; they asserted that these conditions were not only unphysiological but would prevent the appearance of responses. In the original paper (KUFFLER and POTTER 1964) it was stated that contractions of the smooth muscle fibers in the capsule surrounding the nervous system could best be prevented by 7—8 % ethyl alcohol. This, however, was done only in some of the experiments; most of the results on the leech nervous system and afterward on the amphibian optic nerves were made without the addition of drugs and at a wide range of temperatures (4—24° C). Many were also done in normally circulated preparations.

4. Neurons deprived of glia. The finding that glial cells do not signal does not rule out the possibility that their presence is essential for neuronal impulse generation. There are instances where the nerve membrane normally lacks a glial covering (Section X, 4). At the node of Ranvier the action potential mechanism operates across the nerve membrane without participation of the Schwann

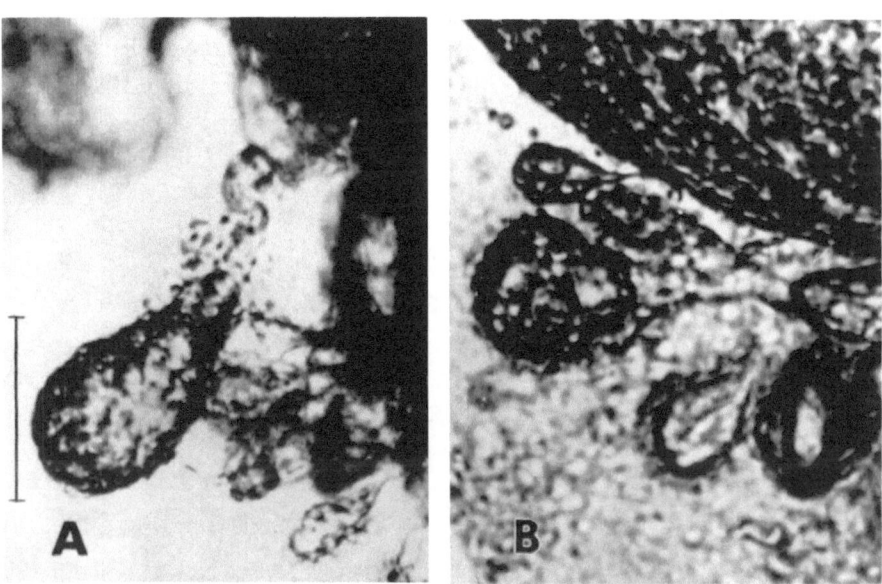

Fig. 12 A and B. Neurons deprived of glia in a leech ganglion. A cut was made in the capsule of a 'packet' in a ganglion containing neuronal cell bodies. Glial cytoplasm around the neurons escapes. A. One of the larger nerve cells and part of its axonal process are almost entirely freed of their surrounding glia. B. Same preparation just after opening the capsule before the glial cytoplasm has been washed away. Cytoplasm appears as a gray granular material around the nerve cells. Dark mass to the top right is the neuropil into which axonal processes run. Scale: 50 μ (from KUFFLER and POTTER 1964)

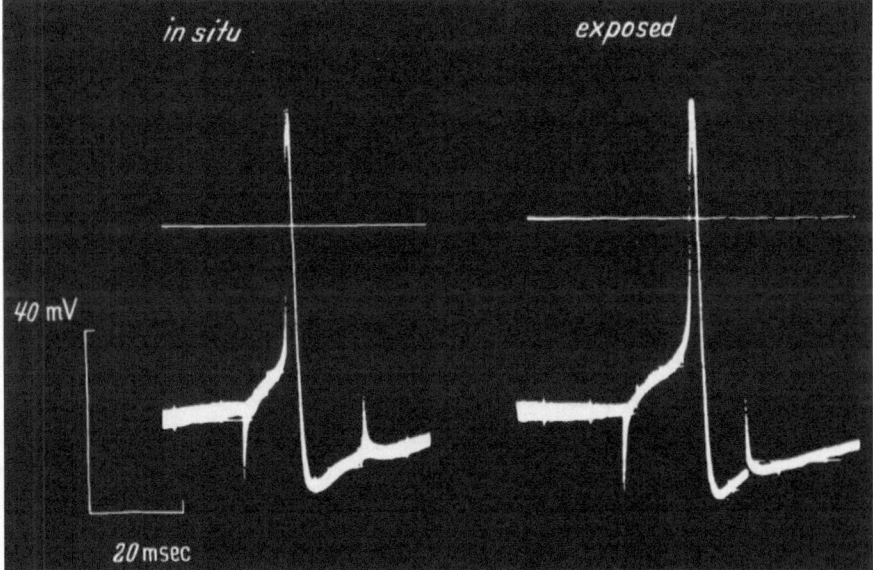

Fig. 13. Action potential from the same neuron *in situ* and half an hour later after removal of the surrounding glial cytoplasm in a leech ganglion (as shown in Fig. 12 A). Exposure to the bathing fluid did not appreciably alter the resting or action potentials. Stimuli were applied by the intracellular recording electrode (from NICHOLLS and KUFFLER 1964)

cell. It was of interest to determine whether neurons that are normally surrounded by a glial cell can signal after the glial cell has been removed. A direct test of the importance of glia can be made in the leech ganglion where the glial cell in a packet can be destroyed, leaving neurons exposed to the bathing fluid (KUFFLER and POTTER 1964). Fig. 12 illustrates nerve cells in a leech

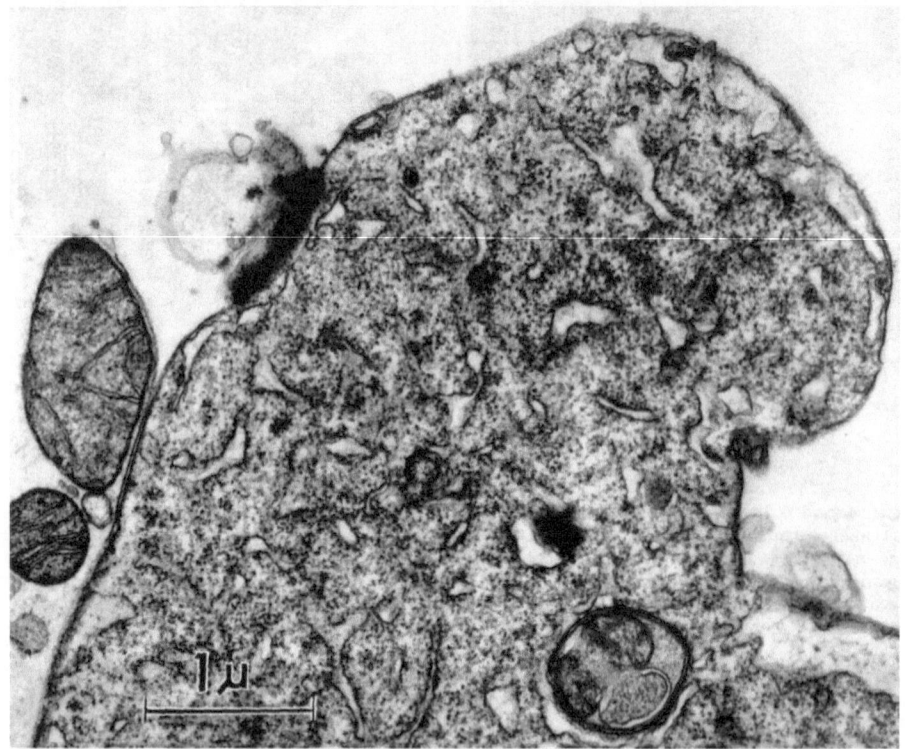

Fig. 14. Electron micrograph of a portion of a neuron from a leech ganglion that had been deprived of its surrounding glial cytoplasm. Over large areas the neuronal surface membrane is directly exposed to the bathing fluid. At other areas there are patches where a band of glial cytoplasm is left and the usual 150 Å cleft between the neuron and glial surface membranes is maintained. Note on the left the glial mitochondrion partially covered by glial membrane which had peeled off from the neuronal surface and a desmosome attached to a fragment of basement lamina. At other places (not shown), the glial membrane alone, practically without glial cytoplasm, still remained apposed to the neuron. Scale: 1.0 μ (kindly provided by Dr. D. E. WOLFE, Harvard Medical School)

ganglion after the capsule had been opened by a cut, letting glial cytoplasm escape. In Fig. 12 B the granular material around neurons is the remainder of the cytoplasm in the process of being washed away. After the preparation is rinsed for several minutes the cell bodies appear to be devoid of their normal glial surrounding (Fig. 12 A). Such neurons can still conduct impulses for many hours; an example of overshooting action potentials is given in Fig. 13 from the same neuron before and after removal of its glial cell.

The possibility remained that the neurons retained their signalling ability because the glial membrane which is normally apposed had not been removed together with the cytoplasm. Electron micrographs made by Dr. D. E. WOLFE from naked, exposed neurons, which were known to give normal signals, have shown that only small fragments of glia remained attached to the cell. Fig. 14 presents a portion of such a neuron. The essential feature here is that large areas of the neuronal surface are directly exposed to the bathing medium. Significantly, there are also patches where the usual intercellular cleft of about 150 Å is still maintained, although the glial cytoplasm is practically absent. At other sites the glial membrane is partially peeled off. We shall see later (Section IX, 2) that this persistence of the narrow clefts in preparations fixed after the removal of the glial cell is relevant to the question of the size of the extracellular spaces in the central nervous system. From these experiments it can be concluded that glial cells are not essential for neuronal signalling over periods of a few hours.

5. **Special connections between glial cells.** Although glial cells do not give propagated responses it is of interest to know whether they are linked electrically either to each other or to neurons. A convenient test for electrical interaction is to pass current through one cell while recording the potential changes in its neighbors. In the leech nervous system and in *Necturus* and frog optic nerves, membrane potential changes spread between glial cells (KUFFLER and POTTER 1964; KUFFLER, NICHOLLS and ORKAND 1966). There is no rectification but, as one might expect, the potential is attenuated with distance; it can be detected over a stretch of more than 1 mm. Similar low resistance connections have now been seen in many tissues, for example between heart muscle fibers, smooth muscle fibers, electrically coupled pre- and postsynaptic neurons (for references see FURSHPAN 1964); and more recently between gland cells and epithelial cells (LOEWENSTEIN and KANNO 1964; LOEWENSTEIN, SOCOLAR, HIGASHINO, KANNO and DAVIDSON 1965). In all these cases there are regions where the intercellular clefts of 100—200 Å are either reduced or even obliterated. These are the zonulae and maculae occludentes, commonly known as tight junctions (FARQUHAR and PALADE 1963). Similar intimate contact areas have been found between glial cells in the optic nerves and elsewhere in the nervous system (GRAY 1961, 1964; PETERS 1962). An example from the *Necturus* is presented in Fig. 15. It seems likely that the low resistance connections between glial cells occur at these specialized regions. As yet such contacts have not been observed between glial cells in the leech.

The functional significance of the link between glial cells is not known, but if currents and therefore ions can pass between the cells, the same is probably true for a variety of small molecules. This opens the possibility that metabolic interaction occurs between glial cells. Connections between the cells are also essential for the generation of current flow by glia (Section VII,1).

6. **The absence of electrical interaction between glial cells and neurons.** Low resistance bridges do not occur between neurons and glia. In the leech a direct test has been made by passing current in the manner described in the previous

section (KUFFLER and POTTER 1964). If, for instance, the membrane potential of a neuron is displaced by as much as 80—100 mV in the hyperpolarizing or depolarizing direction, no significant changes of more than 1—2 mV are recorded in the neighboring glial cell. Similarly, alterations of glial membrane potential over a large range do not change the neuronal resting potential. In

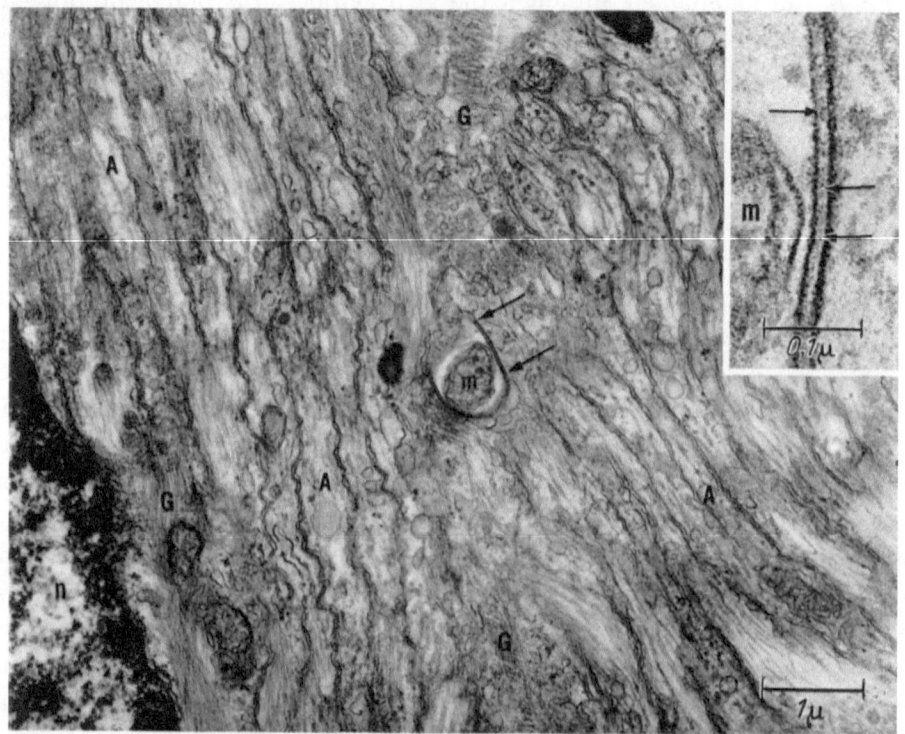

Fig. 15. Tight junction between glial membranes in the optic nerve of *Necturus*. In this oblique section axons (A) are separated from each other and from glial cytoplasm (G) by clefts of about 150 Å width (see Fig. 9). To the left is part of a glial nucleus (n). In the middle of the print is a boundary between glial processes where the intercellular cleft is obliterated (two arrows). Close by is a mitochondrion (m). Part of this area is seen at a much higher magnification in the inset (upper right). The outer lamellae of the unit membranes are fused as indicated by the thin intermediate line, marked by the tips of the three arrows, thus forming a tight junction or zonula occludens. Scale: 1 μ and 0.1 μ (kindly provided by Dr. DAVID WOLFE, Harvard Medical School)

Necturus optic nerves the currents associated with a maximal nerve action potential lead to no detectable change in the membrane potential of the surrounding glial cells (see Section VI, Fig. 22). There is also evidence that current does not spread between the giant axon of the squid and its neighboring Schwann cell (VILLEGAS, VILLEGAS, GIMENEZ, and VILLEGAS 1963).

A simple explanation for the failure of current to pass between glial cells and neurons is that glial cells and neurons are separated by intercellular spaces of 100—200 Å that are in continuity with the bathing fluid. Apparently such clefts are an effective pathway for current flow and they isolate the two types of cells from each other electrically. Other experiments, fully discussed in section V, support this assumption.

V. Pathways for the movement of substances through the nervous system

The evidence presented so far has shown that the glial and neuronal cell membranes are distinct structures with special properties of their own. Both have a high electrical resistance but are separated from each other by low resistance pathways that communicate with the outside fluid. This, therefore, raises the question of the relative importance of the intercellular clefts and glial cells as pathways for the movements of small molecules through the nervous system. Do substances move through the glial cells or around them by way of the narrow intercellular clefts? Numerous electron microscopic studies leave little doubt that large molecules can enter the spaces between cells. As a start we shall consider the following points: 1. Would spaces, if open, be large enough to allow diffusion of substances such as Na^+ or sucrose? 2. What is the rate of movement of substances through the nervous system of the leech and of the optic nerve of *Necturus*? 3. What is the evidence that the movement of certain substances does not occur through the glial cell? Subsequently we shall deal with the effect of nerve impulses on the K^+ concentration in the clefts and the role of glia in regulating the ionic composition of the clefts. In section VIII we shall discuss related problems of the blood-brain barrier.

1. Theoretical rates of diffusion in narrow clefts. A frequent source of confusion is the notion that a space 150—200 Å in width is too narrow to allow the rapid diffusion of molecules the size of, say K^+ and Na^+. In fact the hindrance to diffusion under these conditions is only slight (PAPPENHEIMER 1953). The relative coefficient of diffusion D'/D is given by

$$D'/D = \frac{(1 - a/r)^2}{1 + 2.4\, a/r} \tag{2}$$

where D is the coefficient of diffusion, r is the radius of the tube and a is the size of the particle. Taking a value of 4.4 Å for the radius of sucrose and 150 Å as the width of the tube, the relative coefficient of diffusion turns out to be only 18 % less than that in free solution. The size becomes limiting only for molecules having a radius of more than about 15 % of the cleft width. Hence the actual dimensions of the clefts would not necessarily prevent the rapid movement of ions and small molecules. We shall, in Section IX, present evidence that the dimensions of the clefts are probably of the same order as seen in electron micrographs. The above considerations would not apply if the intercellular clefts were 1. filled with material that prevented or slowed diffusion (HESS 1962; TREHERNE 1962a), or 2. were closed off at certain points (FARQUHAR and PALADE 1963).

2. Electron dense markers in the intercellular spaces. Electron microscopic studies indicate that relatively large molecules can pass through the inter-

cellular clefts. Lasansky and Wald (1962) soaked isolated frog retina preparations in ferrocyanide and showed that this substance was present in the
narrow intercellular spaces, while relatively little was seen within the cells.
Similarly Rosenbluth and Wissig (1964) found that, in the toad, ferritin
molecules, which are approximately 100 Å in diameter, could enter the spaces
between dorsal root ganglion cells and their surrounding Schwann cells. The
same type of experiment was done in the squid by Villegas and Villegas
(1964) who used thorium dioxide, and by Baker (1965) who, in crab nerve
detected iodide which had rapidly diffused into the mesaxons. In the rat Brightman (1965) injected ferritin into the ventricles and followed its distribution
through the ependyma. Numerous analogous studies on other non-nervous
tissues are also available to support the idea that spaces of 100—200 Å width
seen in the electron microscope are open to molecules of relatively large size
(Miller 1960; Kaye and Pappas 1962; Farquhar and Palade 1963). At the
same time generalizations about access to all spaces cannot be made because
some may be closed off at certain critical areas (Section VIII, 1).

3. Rates of movement through nervous tissue. The observations discussed so
far suggest that the intercellular clefts are open and that their small diameter
will not drastically retard the diffusion of molecules such as sucrose. The
problem remains whether the intercellular clefts are adequate in number and
distribution to account for the rates of movement of substances observed to
occur through the nervous system. In both the central nervous system of the
leech and the optic nerve of *Necturus* the half time for the exchange of Na^+
with sucrose or K^+ is less than 12 seconds (Nicholls and Kuffler 1964;
Kuffler, Nicholls and Orkand 1966).

The technique for measuring diffusion times depends on the use of neurons
as indicators of the Na^+ and K^+ in the environment. If the Na^+ in the environment of neuron is replaced by sucrose or choline, the action potential becomes
smaller. Increased K^+ concentrations on the other hand reduce the membrane
potential. The experimental situation, sketched in Fig. 16A, has been used in
the connectives of the leech and in the optic nerve of *Necturus*. The preparation
lies in a chamber consisting of three compartments, sealed off from each other
by vaseline. The side compartments contain Ringer's fluid and are used for
stimulating the nerve and for recording its action potential. The narrow central
chamber can be perfused with various test solutions. Fig. 16 B shows that
when the solution perfusing the central chamber is changed to Na^+-free
Ringer's solution that contains sucrose, conduction rapidly fails in the *Necturus*
optic nerve. Within 12 seconds all the impulses are blocked when they reach
the central compartment. If Na^+ is introduced again into the perfusion fluid,
the conducted potential is fully restored to its control size after 10 sec (Fig. 16 B
at right). The results in the leech connectives are very similar, when either
sucrose or choline is used to replace Na^+.

The above experiments have been interpreted as follows: When Na+ is replaced by sucrose, the ion moves out of the nerve and sucrose takes its place. Block of conduction results because Na+, which is needed for the impulse mechanism, is lost in the immediate environment of the axons. By testing different concentrations of Na+ it was found that complete block occurred when 60—75 % of the Na+ in the central perfusion compartment had been

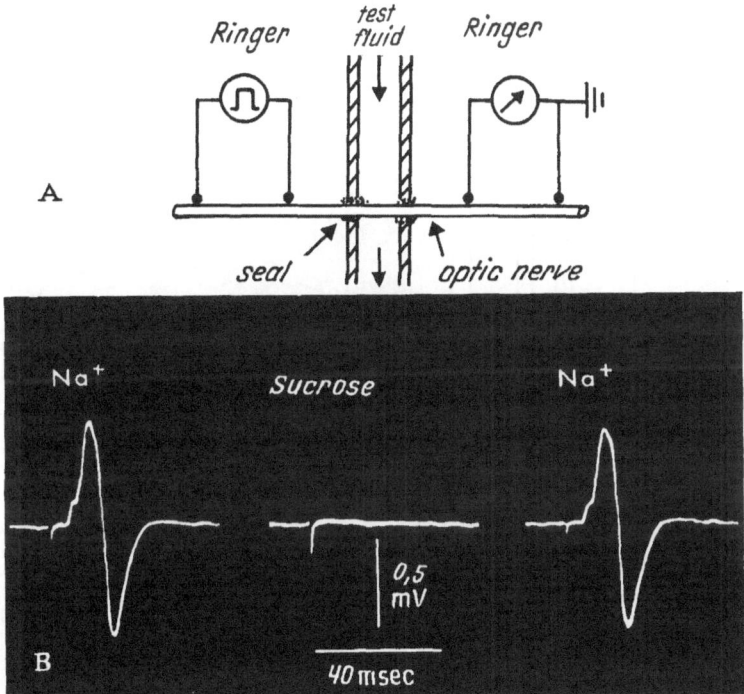

Fig. 16 A and B. Movement of Na+ and sucrose through the optic nerve of *Necturus*. Reversible conduction block produced by replacing Na+ with sucrose. A. Scheme of the 3-compartment chamber in which the optic nerve lies. The two side chambers contain normal Ringer's solution and are sealed off from the narrow central compartment by vaseline. The test solution flows through the middle part of the chamber. B. Left record: A compound nerve action potential set up by maximal stimulation in the left chamber was recorded in the right chamber. Middle record: 12 sec after perfusion with Na+-free sucrose-Ringer, conduction through the middle portion of the nerve was blocked. Right: Recovery was complete 10 sec after returning to the normal Ringer's solution (from KUFFLER, NICHOLLS and ORKAND 1966)

replaced by sucrose. One can therefore conclude that within 12 seconds in Na+-free solution at least 60 % of the Na+ in the optic nerve had been exchanged for an equivalent amount of sucrose.

A more accurate measure of the rate of movement through the nervous system can be obtained by recording intracellularly from neurons in leech ganglia. To measure the rate of penetration of K+, for example, the relation between K+ concentration and the membrane potential of a neuron is first established (Fig. 17A). From this one can translate any membrane potential into an equivalent K+ concentration. Next a large concentration of K+ is applied to the ganglion and the cell becomes depolarized in a few seconds

.(Fig. 17 B). One can now estimate the equivalent K⁺ concentration around the neuron at any instant after the K⁺ has been applied to the bathing fluid. In this way, plots for the influx and efflux of K⁺ have been made. By the same principle the rates of movement of Na⁺, sucrose and choline have been measured, using the size of the action potential as an indicator of concentration. These results have the advantage that a complete diffusion curve can be constructed. The points fall on a reasonably straight line when plotted on a logarithmic scale against time as one would expect for a diffusion process. The half time for the exchange of Na⁺ with K⁺ through the leech ganglion is about 4 sec, while for Na⁺ with sucrose it is 10 sec.

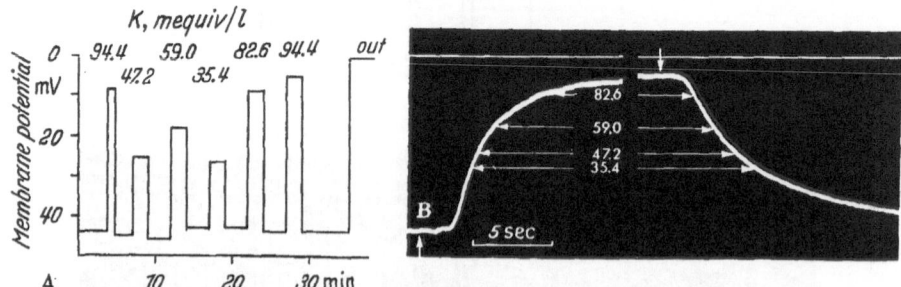

Fig. 17 A and B. Determination of the rate of movement of K⁺ through the leech ganglion by correlation of the membrane potential in a ganglion cell with the K⁺ concentration in the bathing solution. A. Steady-state membrane potentials in five solutions with increased K⁺ content, plotted against time. These data permit a conversion of membrane potential into K⁺ concentration as shown in B, illustrating the time course of depolarization when 94.4 mEq/liter K⁺ was introduced (first arrow) into the perfusion fluid, and repolarization when normal K⁺ was re-introduced at the second arrow. This is the record of the last trial shown in A and it is aligned accordingly. One hundred sec, indicated by a gap, were cut out from the record. The K⁺ concentrations corresponding to the varying membrane potentials, established in A, are given in mEq/liter K⁺. Half-times for K⁺ influx and efflux are 3.7 sec and 5.0 sec (from NICHOLLS and KUFFLER 1964)

It is of interest to compare these times with the values predicted on the assumption that simple diffusion is occurring through narrow channels. Knowing the diffusion coefficients for NaCl and sucrose, the half time for diffusion within an individual cleft in the nervous system can be calculated from the formula for linear diffusion (HITCHCOCK 1945):

$$\frac{C}{C_o} = 1 - \frac{2}{\sqrt{\pi}} \int_0^y e^{-y^2 dy}.$$ (3)

In this equation C_o is the initial concentration and $y = x/2\sqrt{Dt}$, x being the distance (cm), and D the coefficient of diffusion (cm²/sec). The distance x, estimated by measuring the length of the mesaxons, is probably not more than 50 μ while in the connectives it is probably not more than 30 μ. Using these values for maximum distance, the half time for the exchange of Na⁺ with sucrose would be about 3.6 sec in a leech ganglion packet and about 1.3 sec in a connective. In each case the half times measured experimentally were considerably larger (by a factor of over 2) than those predicted from equation 3. This is to be expected since the calculations at best yield only a rough approximation (see NICHOLLS and KUFFLER 1964). But they do indicate that simple diffusion would probably be rapid enough to account for the rates of movement of Na^+, K^+ and sucrose observed in the nervous system.

It is quite possible that fixed charges either within the clefts or on the membrane might affect diffusion rates. The results presented above are probably not sufficiently precise to disclose such effects. Tracer experiments might be expected to be more useful, but the system of intercellular clefts is too small a compartment to be studied in this way (NICHOLLS and WOLFE, in preparation).

4. Exclusion of glia as a pathway for rapid diffusion. The question remains whether Na^+ and sucrose (or choline) exchange by taking a pathway through the glial cells or around them, through the cleft system. The initial expectation is that movement through intercellular channels would leave the glial membrane potential unchanged whereas passage through the cells would alter it. Glial membrane potentials were therefore measured while isotonic Na^+ and sucrose moved through the nervous system (NICHOLLS and KUFFLER 1964; KUFFLER, NICHOLLS and ORKAND 1966). It was found that the glial potentials remained practically unaffected by the exchange of Na^+ and sucrose-Ringer. One can therefore conclude that neither Na^+ nor sucrose takes the pathway through the glial cells. It is worthwhile to examine the reasons for such a conclusion more closely; it has already been shown that the glial membrane potential depends on the distribution of K^+ on the two sides of the membrane (Fig. 11). If K_i for any reason were reduced, an immediate fall of the resting potential would occur. If, therefore, isotonic Na^+ or sucrose moved through the cytoplasm by free diffusion, the K^+ in the cytoplasm would be diluted, leading to a fall in the resting potential. Any other scheme, such as sucrose and Na^+ being kept at a low concentration in the cell during the exchange, while keeping the cleft concentrations high, needs a complicated system of Na^+ and sucrose and choline pumps. Similar arguments apply to the movement of K^+. K_i would have to change if K^+ moved through the glial cell in concentrations ranging from that in Ringer's fluid to isotonic K_o. If K_i did change, the relation between membrane potential and K_o would not obey the Nernst equation which depends on K_i remaining constant. The only remaining possibility is that substances are carried through the glial cytoplasm in 'sealed' packets that prevent their mixing with the cytoplasm. A process such as this would require energy and should be sensitive to changes in temperature (CHAPMAN-ANDRESEN 1962; RYSER 1963). In the leech the evidence suggests that a metabolically linked pinocytotic process for ions or small molecules is not essential for rapid transport through glial cells (NICHOLLS and KUFFLER 1964).

The conclusion, therefore, seems compelling that during a rapid exchange, as illustrated in Fig. 16 and 17, Na^+, K^+, sucrose and choline do not pass through the glial cells but around them. In the case of the *Necturus* optic nerve and the leech connectives the intercellular spaces all appear remarkably uniform with an approximate width of 150—200 Å. In the leech ganglia there are in addition occasional expansions of the extracellular space. Hence the results obtained on optic nerves and the leech connectives, rather than on ganglia, are the decisive ones in showing the importance of the narrow intercellular clefts for rapid diffusion. So far we have discussed only a few substances. It is possible that metabolites such as glucose and amino acids are actively transported across the glial cells, i.e. in through one surface and out of the other. If this were so, the relative importance of the two pathways would have to be determined (see Section X).

5. Glial cells and the ionic composition of intercellular spaces. The finding that substances can move rapidly through intercellular spaces focuses attention upon the ionic composition of the intercellular fluid. Is the composition of intercellular fluid in the central nervous system different from that of the blood and/or the C.S.F., due to an activity of the glial cells (Tschirgi 1960, 1962; Cummins and Hydén 1962; De Robertis 1962)? In this section evidence is presented that the glial cells in the leech do not modify the K^+ or Na^+ content of intercellular fluid. The larger and more difficult question of the nature of the blood-brain barrier is discussed in Section VIII.

The composition of the 'interstitial' fluid of the brain has long been a matter for conjecture; a quantitative evaluation of the chemical composition of the fluid in minute spaces which might be no more than several hundred Å wide is well beyond the range of current analytical techniques. Although the intercellular spaces are pathways for the rapid diffusion of various small molecules or ions, the possibility remains that the clefts do not have an ionic composition similar to that of the bathing fluid (Treherne 1962a). In the leech one can test directly the possibility that neuroglial cells or fixed charges can modify the environment under resting conditions and maintain an intercellular composition that differs significantly from that of the bathing medium (Nicholls and Kuffler 1964). Once again the membranes of the neurons themselves can be used as sensitive and accurate detectors of Na^+ and K^+ in the immediate environment. As we have seen, the neuronal membrane potential depends on the distribution of K^+ between the inside (K_i) and the outside of the cell (K_o) and its action potential depends on the distribution of Na^+. One therefore can make use of the observation already shown in Figs. 12 and 13 that nerve cells can still behave normally for many hours after much of the glia that usually surrounds them has been removed. If such 'exposed' cells behave like those *in situ* in respect to changes in K_o, it follows that the presence of glia around the cell does not significantly modify the K^+ concentrations around these neurons. When K⁺ was increased in the bath, the cells without glia, in direct contact with the bathing fluid, showed the same membrane potential changes as those with their glia intact. A difference of 15 % in K⁺ concentration would have been detected (Fig. 12 in Nicholls and Kuffler 1964). An example is illustrated in Fig. 18. The membrane potentials of two neurons were recorded simultaneously with intracellular electrodes (Fig. 18 A).

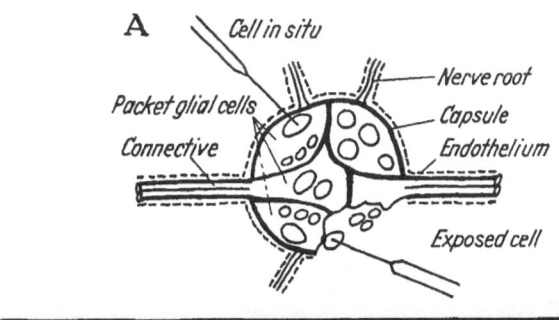

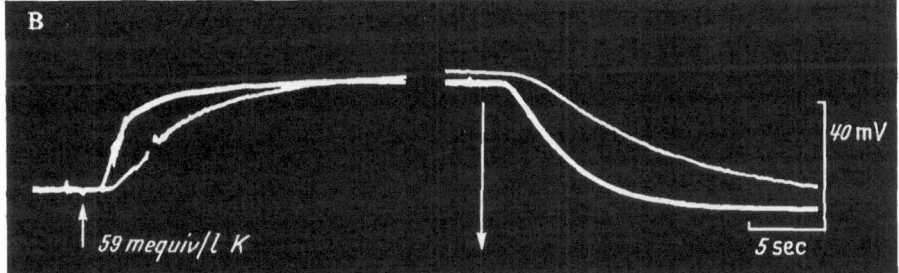

Fig. 18 A and B. A. Diagram of arrangement for simultaneous recording from a cell which is *in situ* and from one which has been exposed to the bathing fluid, in the same ganglion from a leech. One packet, shown on the lower right, has been opened by slitting the capsule. The endothelium is indicated by the dotted line. B. Time course of depolarization and repolarization of *in situ* and exposed nerve cells in the same ganglion. Simultaneous records are superimposed. At first arrow 59 mEq/liter K⁺ was added to the perfusing solution and at second arrow normal Ringer's solution was reintroduced 2 min. later. Depolarization in the exposed cell starts earlier and rises more rapidly. Similarly, the latency for repolarization is shorter and the recovery is more prompt. Note that both cells depolarize to a similar steady state level. Irregularities on the rising phases were due to impulses. The gap corresponds to 1 min. The difference between the rise times of the potentials is due to the tissue barriers which surround the neuron *in situ* (from Nicholls and Kuffler 1964)

One of the neurons was exposed directly to the bathing fluid, the other was left intact in its 'packet', surrounded by its normal glial and connective capsule covering. Fig. 18 B illustrates the effect of increasing K⁺ from 4 to 59 mEq/liter in the fluid which flows past the preparation. The important point is that both cells depolarized to about the same steady level (within 3 mV). The more rapid changes of potential in the 'exposed' cell are due to the absence of tissue 'barriers' such as capsule, glia and clefts, which delay the attainment of the final K⁺ concentration (see above). Similar results were obtained in a large number of cells in K⁺ concentrations varying from 4—120 mEq/liter.

The same principle has been used to determine the Na⁺ concentrations around neurons, by means of the known relation between Na⁺ and action potential size. In Fig. 19 the relative Na^+ concentration in the bathing fluid is plotted against the decrease in impulse size for *in situ* and exposed cells, No significant difference in the relation between spike size and external Na^+ was seen, indicating that the Na^+ concentrations around both are probably similar ($\pm 15\%$).

From these experiments one can conclude that the K^+ and Na^+ composition of the intercellular fluid in the ventral nerve cord of the leech *in vitro* is similar to that in the bathing medium. In the animal, however, the nerve cord is

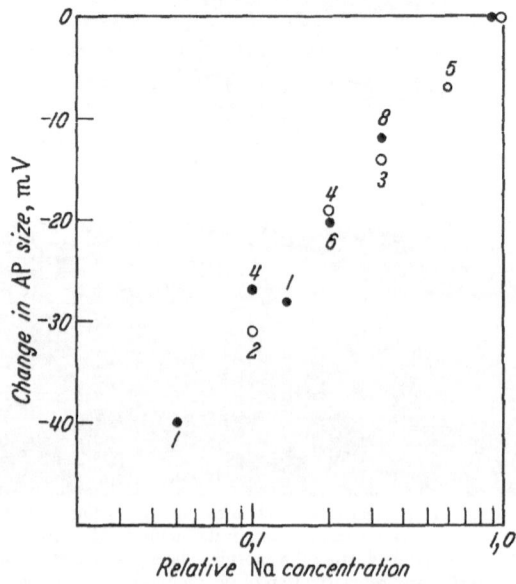

Fig. 19. Relation between Na^+ concentration and impulse size in neurons exposed to the bathing fluid (open circles) and neurons *in situ* (closed circles). Each point is the mean of several tests, indicated by numbers. The mean SD for experiments was about ± 6 mV for both *in situ* and exposed cell. Low Na^+ solution was made by replacement with isotonic sucrose (from Nicholls and Kuffler 1964)

surrounded by a continuous layer of endothelial cells thath separates it from the blood. Removal of the preparation from the leech inevitably damages the endothelium over the roots where they are cut. It could therefore be that the intercellular fluid in the nerve cord is normally maintained at some different value by the endothelial cells. What the above experiments show is that the glial cells do not regulate the composition of the intercellular clefts with regard to K^+ and Na^+ under resting conditions. While fixed charges within the clefts might influence rates of diffusion, they do not seem to determine the steady state ionic composition of the intercellular fluid.

The situation in the vertebrate is less clear since it has not been possible to compare nerve cells with and without glia. During neuronal activity, however, the cleft composition with respect to K^+ is markedly altered (see Section VI).

We do not yet know whether the intercellular fluid in *Necturus* optic nerves resembles the plasma or the C.S.F. with respect to ions. This problem is of course of general importance since it concerns the chemical environment of the neurons in the vertebrate brain. One aspect has recently been investigated by PAPPENHEIMER, FENCL, HEISEY and HELD (1965) and FENCL, MILLER and PAPPENHEIMER (1966), who have obtained evidence that the intercellular fluid surrounding a group of respiratory neurons in the mammalian brain resembles C.S.F. rather than blood with respect to its H^+, HCO_3^- and Cl^- content. They have used the alveolar ventilation rate of a goat as an index of the H^+ concentration surrounding its respiratory neurons during chronic acidosis or alkalosis. Their results clearly showed that small changes in the C.S.F. concentration of these ions led to marked changes in respiration whereas changes in the plasma had far less effect; i.e. the ventilation corresponded to the H^+ concentration in the ventricles, rather than in the blood plasma. The simplest interpretation of these results was that the concentration of H^+, HCO_3^- and Cl^- in the intercellular fluid around the respiratory neurons was in equilibrium with the C.S.F. These experiments again raise the question of the nature of the bloodbrain barrier (Section VIII).

VI. A potassium mediated effect of nerve activity on glia

In the optic nerve of *Necturus* and the frog, nerve impulses consistently led to a decrease in the membrane potential of the glial cells (ORKAND, NICHOLLS and KUFFLER 1966). The effect will first be described and evidence presented to show that it is not mediated electrically, but is caused by K^+ leaking out of the unmyelinated fibers during activity. In leech ganglia and connectives, tests for interaction were made by stimulating neurons repetitively while recording from glial cells. The results were inconclusive because contractions of the smooth muscles in the connective tissue capsule of the nervous system tended to dislodge the microelectrode and made the recording situation unfavorable.

1. **Characteristics of the glial depolarization.** Fig. 20 presents the type of phenomenon that is seen when a glial cell in a *Necturus* optic nerve is impaled and the axons are maximally stimulated; these results are readily obtained either in isolated optic nerves or in normally circulated preparations in anesthetized animals. Fig. 20 A (upper record) shows the transient depolarization of a glial cell by a single nerve volley; the membrane potential was 86 mV. The potential rose to a peak of 2.4 mV in about 0.15 sec and declined to half in about 2 sec. Repeating the axonal stimulation three times at 1/sec (Fig. 20A, lower record) caused a summation of glial depolarization. Fig. 20 B shows the depolarization of another preparation in which the stimulation was maintained for a longer period, each nerve volley adding to the potential built up by the preceding ones.

The magnitude of the glial depolarization was influenced by the following three parameters: the resting potential of the glial cell, the number of axons excited, and the frequency and duration of stimulation. The glial cell depolarization decreased in size if the glial resting potential fell for any reason, such as damage by the microelectrode. It could never be observed in the absence of a resting potential. In the *Necturus* optic nerve, submaximal stimuli, exciting a smaller number of axons, gave rise to smaller depolarizations. The results in the frog optic nerve, which contains a number of myelinated fibers (MATU-

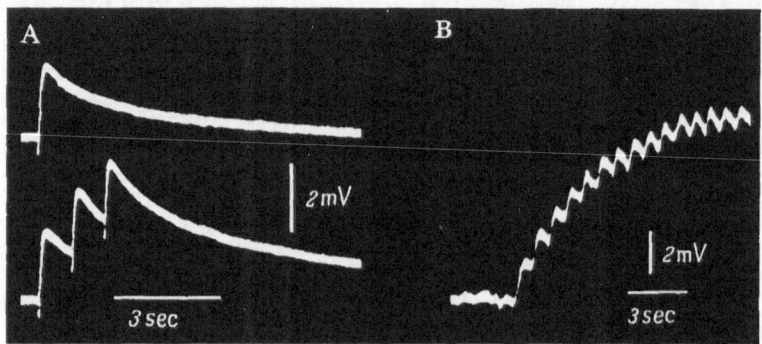

Fig. 20 A and B. Depolarization of glial cell by nerve impulses. Intracellular recordings from glial cells in an isolated optic nerve of the *Necturus*. A. Resting potential 86 mV; upper record: depolarization following a single maximal nerve stimulus. Note the slow rise time (0.15 sec) and half-time of decline (2 sec). Lower record: three stimuli at 1/sec; potentials sum. B. Another glial cell, but stimulation at 1/sec is prolonged
(ORKAND, NICHOLLS and KUFFLER 1966)

RANA 1960), are interesting in this connection. No measurable glial depolarization could be recorded with submaximal stimulation that activated only the myelinated fibers. The glial depolarization only appeared with stimuli strong enough to activate the unmyelinated fibers. As one might expect from the slow time course of the glial depolarization, the frequency of nerve stimulation had a pronounced effect, as shown in Fig. 21. At a frequency of 5/sec or higher, the effects of individual nerve volleys became fused. The summed depolarizations that resulted could reach a plateau of up to 48 mV and decay with half-times as long as 14 sec (Fig. 7, ORKAND, NICHOLLS and KUFFLER 1966).

By what mechanism do the nerve impulses affect the glial membrane potential? Is it an electrical effect caused by current flow in the axonal membranes, or is it mediated by the liberation of a substance(s) during impulse activity? The first alternative seems *a priori* less likely as the interposition of fluid-filled spaces between neurons and glial cell membranes greatly attenuates current spread (KUFFLER and POTTER 1964; ORKAND, NICHOLLS and KUFFLER 1966). The two cells could, however, interact by a chemical mechanism because the dimensions of the intercellular spaces are probably narrower than those of some synaptic clefts, for example at neuromuscular junctions. One of the most striking features of the change in glial membrane potential is its slow time course. It is measured in seconds, whereas the peak of the current flow during the neuronal action potential lasts less than 25 msec. Fig. 22 A illustrates recordings made from the *Necturus* optic nerve with a microelectrode in two successive positions: just outside a glial cell, recording no membrane potential ($V_m = 0$), and inside the glial cell, registering a resting potential of —84 mV. With the electrode outside the cell, the extracellular currents associated with the nerve impulses set up a brief triphasic potential. With the

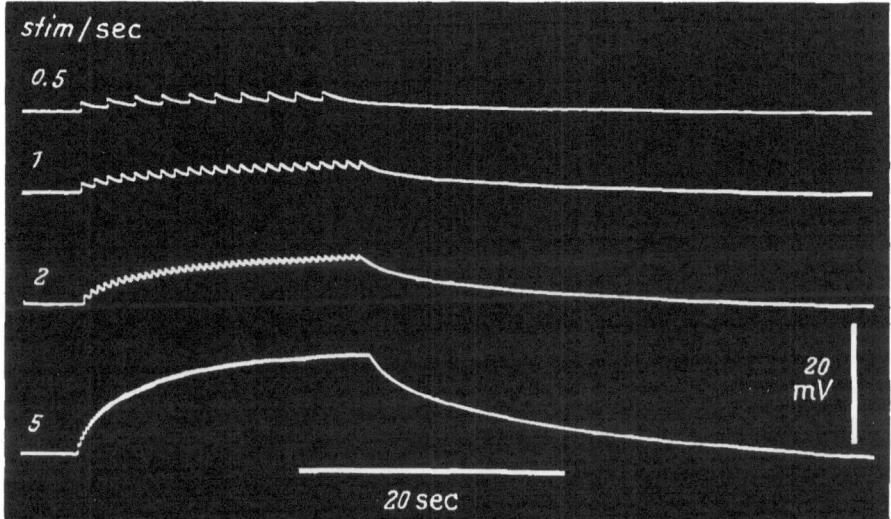

Fig. 21. Effect of frequency of nerve impulses on the depolarization of glial cells. Optic nerve stimulation at 0.5 to 5.0/sec while recording intracellularly from a glial cell in the *Necturus* optic nerve. With sustained stimulation a plateau of depolarization is built up (from ORKAND, NICHOLLS and KUFFLER 1966)

electrode inside the cell, the nerve volley once again sets up a triphasic potential that was almost identical to that in the extracellular position. It was followed, however, by a slowly rising depolarization. This is analogous to the rising phase of the slow potentials already seen in Fig. 20 on a slower time scale. A simple consideration of the recording situation shows that the triphasic rapid potential was not a voltage drop across the glial membrane since records taken from the inside and outside of the glial membrane were practically identical. Additional evidence that axonal current flow is not the cause of the glial depolarizations was the observation that the extracellular potentials were frequently quite small or not detectable. For example in Fig. 22 B, registered at a greater amplification than 22 A in a different preparation, the current flow was not recorded at all either outside the glial cell or inside the cell when a large slow depolarization was set up by the nerve impulses. Another consideration comes from the direct experimental test of applying current pulses across the glial membranes. Current flow did not produce depolarizations which were comparable to those following nerve impulses (Section IV, 3).

In conclusion, the effect of axonal stimulation on glial cells is not triggered or maintained directly by current flow generated by nerve impulses (ORKAND, NICHOLLS and KUFFLER 1966).

2. K+ liberation: the mechanism of the effect of neuronal activity on glia. Once the direct action of electric currents is excluded, a different mechanism has to be postulated. There are two possibilities: 1. a change in the extracellular concentration of an ion such as K+ that contributes to the glial membrane

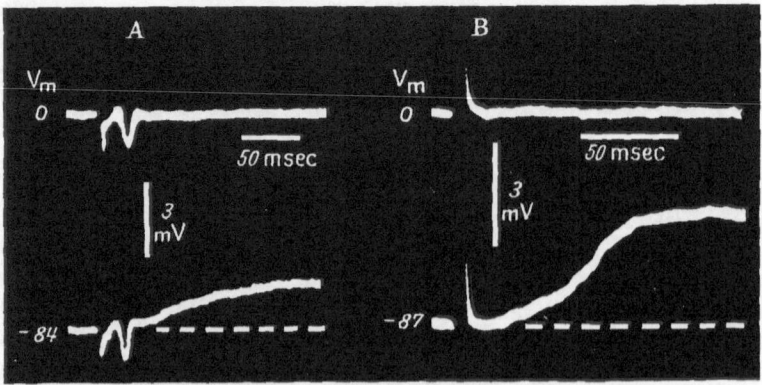

Fig. 22 A and B. Lack of effect of current flow generated by axons on the membrane potential of glia cells. Potentials recorded with a microelectrode in the optic nerve of *Necturus*, stimulated maximally with external electrodes. A. The position of the microelectrode tip is extracellular (Vm = 0), outside a glial cell in an isolated optic nerve. The triphasic potential indicates the current flow generated by a single volley of nerve impulses. In the lower record the electrode tip has been advanced into the glial cell and records a membrane potential of 84 mV. Maximal nerve stimulation sets up the same triphasic potential which, however, is followed by a slowly rising depolarization. B. Same experiment as in A performed in the optic nerve of an anesthetized circulated animal. Although the amplification is higher, no potentials are recorded during nerve impulse activity. The glial depolarization in the lower record rises to a plateau in about 75 msec. These experiments show that the glial depolarization does not result directly from current flow generated in the neurons (see text) (from ORKAND, NICHOLLS and KUFFLER 1966)

potential; 2. the release of a transmitter that changes the permeability of the glial membrane. The experiments of FRANKENHAEUSER and HODGKIN (1956) have shown that the K+ that leaks out of a squid giant axon during the action potential accumulates in the clefts between the axon and Schwann cells. The K+ leakage accounts quantitatively for the negative after-potentials recorded across the axonal membrane. Other observations have been made in unmyelinated peripheral nerve, in sympathetic ganglia and in skeletal muscle (GREENGARD and STRAUB 1958; BLACKMAN, GINSBORG and RAY 1963; FREYGANG, GOLDSTEIN and HELLAM 1964). An attractive hypothesis for the effect of nerve on glia in the optic nerve is that during each nerve impulse the concentration of K+ in the intercellular clefts increases, and that this in turn leads to a depolarization of the adjoining glial cells.

This K+ hypothesis can be checked experimentally in the optic nerve of *Necturus*. The test depends on the fact that the relation between the glial membrane potential and external K^+ is known to be logarithmic over a wide range (Fig. 11). This means, of course, that the addition of a given concentration of K^+ to the fluid around a glial cell produces a smaller depolarization when the background level of K^+ (K_o) is high and a larger depolarization when K_o is low. If, therefore, this effect of nerve impulses is mediated by K^+ accumulation, one should be able to predict how the glial depolarization changes in Ringer's fluid containing high or low K^+. The glial depolarization should behave as though the K^+ liberated by the axons were adding logarithmically to the K_o already present in the bathing fluid. For this prediction to be correct, it has to be assumed that the level of external K^+ in the bathing fluid has little effect on the amount of K^+ liberated per nerve impulse.

These predictions agreed quantitatively with observed results in a number of glial cells. Tests were made in Ringer's fluid containing the normal K_o concentration (3 mEq/liter), half the normal concentration of K_o (1.5 mEq/liter), and one and a half times the normal K_o concentration (4.5 mEq/liter). In each case, the effect of a constant train of nerve volleys on the glial membrane potential was reduced in high K_o and increased in low K_o as expected.

An example is illustrated in Fig. 23. In this figure, the solid circles indicate the glial resting potentials in three different K_o concentrations: 1.5, 3.0 (normal), and 4.5 mEq/liter. The size of the glial depolarization after a constant brief train of maximal nerve volleys at 10/sec in each medium is shown as an open circle joined to the solid circle by a vertical line. The glial cell in this experiment had a resting potential of 89 mV in the normal Ringer's fluid containing 3 mEq/liter K_o. Stimulating the nerve maximally at a rate of 10/sec for 1 sec in this solution set up a summed glial depolarization that reached a peak of 12.1 mV (middle open circle, Fig. 23). Reducing K_o in the bathing fluid to 1.5 mEq/liter increased the resting potential to 105 mV and the same volley of impulses caused a larger depolarization of 18.5 mV. On the other hand, in a background concentration of 4.5 mEq/liter K_o, the membrane potential fell to 78 mV and the effect of nerve stimulation was smaller, only 9.3 mV. In the normal Ringer's solution (containing 3 mEq/liter K^+), the peak depolarization of 12.1 mV can be matched by adding 1.8 mEq/liter K^+ to the bathing fluid. Next, one can calculate how large a depolarization an increment of 1.8 mEq/liter K^+ would produce if it were added to bathing solutions that contained 1.5 or 4.5 mEq/liter K^+. This calculation can be read off directly from the curve in Fig. 23 (see also Fig. 11) and is indicated by the horizontal dotted lines. The calculated values of depolarization (dotted lines) and those observed after nerve stimulation (open circles) are in excellent agreement (for details, see ORKAND, NICHOLLS and KUFFLER 1966).

Additional evidence for the K^+ hypothesis was obtained by establishing that successive glial potentials set up by a train of nerve volleys also summed as one would predict. They should sum as if successive identical quantities of K^+

were added to each other. Once again the expected and the observed results were in good agreement. For example, in Fig. 20 B, successive volleys to the nerve set up progressively smaller depolarizations. When each depolarization is converted to an equivalent K+ concentration, the values turn out to be practically identical. Thus the depolarizations have become smaller by exactly the same amount as one would expect if each nerve volley liberated a constant

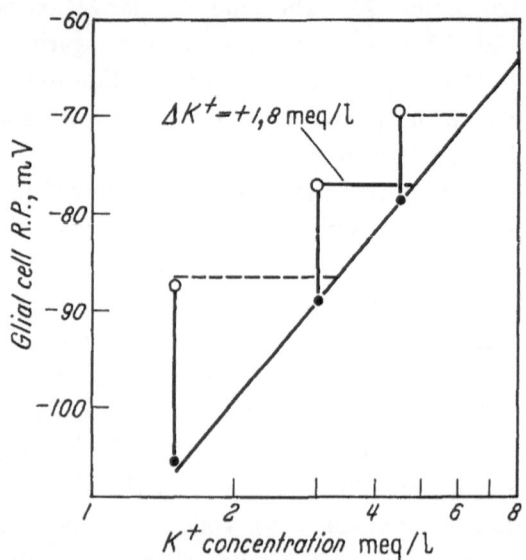

Fig. 23. Test for the liberation of K+ by nerve impulses as the cause of glial depolarization. All measurements were made on one glial cell in the optic nerve of *Necturus*. The K+ concentration — glial membrane potential curve is plotted at three concentrations (solid circles, as in Fig. 11). The solid line is the theoretical curve predicted by the Nernst equation with a slope of 59 mV for a ten-fold change in K_o. At the normal resting potential (89 mV) in Ringer's solution (3.0 mEq/liter K+) the optic nerve is stimulated by a standard train of impulses which cause a glial depolarization of 12.1 mV (middle open circle). This depolarization can be matched by the addition of 1.8 mEq/liter K+ added to the bathing fluid. If this is close to the amount of K+ liberated by the train of nerve impulses in Ringer's solution one can predict the size of glial depolarizations which will be set up by the same train of impulses when the K+ content in the bath is lower or higher. With 1.5 mEq/liter of K+ in the bath, the standard train of nerve impulses depolarized the glial cell by 18.5 mV (lower open circle), while the calculated effect of 1.8 mEq/liter K+ was almost the same (lower dotted line). The same good agreement between calculated and observed values is seen when the bath contained 4.5 mEq/liter K+, i.e., the potential changes produced by the addition of known fixed amounts of K+ to the bathing solution can be matched by the amounts of K+ liberated by identical trains of nerve impulses (from ORKAND, NICHOLLS and KUFFLER 1966).

amount of K+ into the intercellular clefts (Fig. 6, in ORKAND, NICHOLLS and KUFFLER 1966).

Although the experiments reported above are in agreement with the hypothesis that K+ is the substance accumulating in the clefts, they do not rule out the possibility that some other substance is released by the nerve. Theoretically such a substance might act like a "transmitter", driving the glial membrane potential toward an equilibrium level. In this case the effect of nerve stimulation on glial membrane potential would become smaller at low resting potentials and larger at high potentials in a manner that could also be non-linear. Since, however, it is known that K+ is liberated by small unmyelinated axons (cf. KEYNES and RITCHIE 1965), and in view of the good agreement between the K^+ hypothesis and the above experiments, an alternate assumption for a different substance seems at present unnecessary.

3. Conduction block and the accumulation of K^+ in the intercellular spaces. An interesting corollary to the results of nerve stimulation was the observation that prolonged maximal stimulation at 10/sec or more could lead to block of conduction in optic nerve fibers in *Necturus*. At the same time, the glial cell could be depolarized by as much as 48 mV. It is therefore necessary to establish whether these observations can be explained by the K^+ hypothesis. For example, what concentration of K^+ blocks axonal transmission? Is this concentration compatible with the extent of the glial depolarization observed? Experiments have indicated that the block of conduction and glial depolarization can be attributed to K^+ accumulation in the intercellular clefts.

The K^+ concentration estimated from the glial depolarization during nerve stimulation is an average concentration gradient in contact with the whole glial membrane. Only part of each glial cell is apposed to neuronal membranes; the remainder borders on other glial membranes or is exposed to the bathing fluid at the surface of the optic nerve. Whenever K^+ is liberated it will be most concentrated in the intercellular spaces around the active axons which are grouped in bundles (Figs. 9 and 10). There will be a concentration gradient towards the surface of the bundle where it is surrounded by glia. Therefore the actual concentration of K^+ around some of the active axons must have been greater by an unknown amount than our calculations indicate. A reduction of the glial membrane potential by 48 mV corresponds to a surprisingly high value of 20 mEq/liter K^+ overall concentration. The concentration which has to be added to the Ringer's fluid to block conduction should be even higher. Therefore, tests were made to block conduction by the application of known amounts of K^+. 30 mEq/liter K^+ in the bathing solution always blocked, while 23 mEq/liter K^+ did not lead to block of conduction in the whole nerve. If 20 mEq/liter or less had blocked, the K^+ hypothesis would have been untenable. These experiments are consistent with the conclusion that K^+ is the agent that depolarizes glial cells following nerve impulses (for details see ORKAND, NICHOLLS and KUFFLER 1966).

4. Removal of K^+ from the extracellular spaces. If the glial cell depolarization is due to K^+ accumulation, its declining phase should result from a combination of diffusion out of the clefts and from active uptake of K^+ by neurons and/or glia. We know from experiments on the leech and on the *Necturus* that K^+ applied to the bathing fluid diffuses rapidly in and out of the nervous system by way of the intercellular channels (NICHOLLS and KUFFLER 1964; KUFFLER, NICHOLLS and ORKAND 1966). We also know that neurons have to regain the K^+ that they have lost during impulse activity. Another factor may also have to be considered. If currents flow through the

glial membrane as a result of neuronal activity (Section VII), K$^+$ will enter the cell in one region and leave it in another. That this occurs under certain circumstances (see below) can be predicted but the quantitative significance of such a distribution of K$^+$ cannot be evaluated (cf. also HERTZ 1965). At present, then, we are not in a position to discuss quantitatively the different mechanisms which are responsible for the removal of K$^+$ which has accumulated in the intercellular spaces during neuronal activity. Nevertheless the over-all rate of K$^+$ disappearance can be calculated by converting glial membrane potentials to equivalent K$^+$ concentrations. The half-time for K$^+$ disappearance

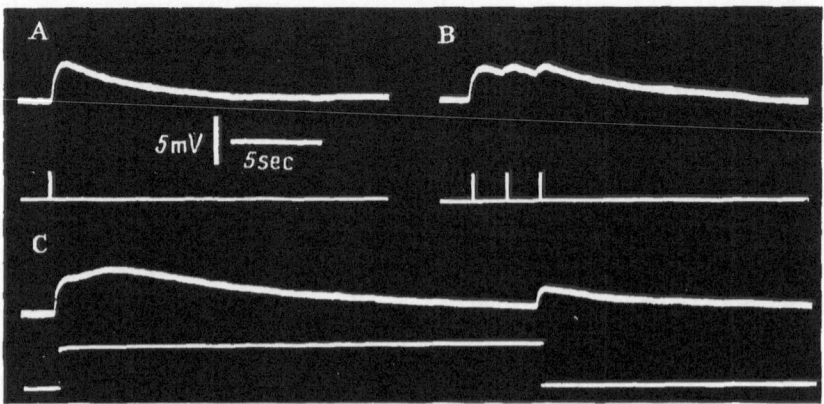

Fig. 24 A - C. 'Natural' nerve stimulation. Light flashes were shone into the eye while recording intracellularly from a glial cell in a circulated *Necturus* optic nerve. Lower beams monitor current flow during flashes. A. Single light flash of 100 msec duration sets up transient depolarization. B. Three flashes. C. Light stimulus is maintained for 27 sec. The initial glial depolarization declines due to adaptation of neuronal discharges. An 'off' burst of impulses, when the light is turned off, sets up another glial depolarization
(from ORKAND, NICHOLLS and KUFFLER 1966)

was about 1.7 to 6.3 sec after nerve stimulation and about 8 sec after the K$^+$ concentration had been decreased in the bath (ORKAND, NICHOLLS and KUFF-LER 1966).

5. **Glial depolarization after natural stimulation in vivo.** The glial depolarization that results from neuronal activity can also be elicited by physiological stimulation of the eye by light. In all the experiments described so far, the stimulation consisted of massive synchronous excitation of a large number of axons in isolated nerves, or in the animal with intact circulation. Recordings with microelectrodes have also been made in glial cells in lightly anesthetized *Necturus*, with their eyes intact and their circulated optic nerves exposed. Fig. 24A shows a record in which one flash of white light of 100 msec duration was shone into the eye. A single flash was followed by a glial depolarization which rose to a peak of 4 mV in about 0.5 sec and declined in a further 4 sec. In Fig. 24B three flashes were given, the second and third stimulus adding less than the first. This is probably due to adaptation of the neuronal discharges, because if the light stimulus is maintained for longer, e.g., 27 sec as in Fig. 24C, the glial depolarization almost ceases within 20 sec. When the light is turned

off, however, another burst of impulses once more depolarizes the glial cells. As one would expect, the effects are graded with light intensity, and are strongly influenced by the background illumination. The significance of these results is that "natural" asynchronous activity of neurons can be effective in depolarizing glial cells. Ganglion cell discharges in the *Necturus* retina were recently studied by BORTOFF (1965).

6. Physiological consequences of K⁺ accumulation in intercellular spaces. It has been shown above that a flash of light leads to discharges in nerve fibers which liberate enough K⁺ to depolarize glial cells. This indicates that under physiological conditions the K⁺ concentration in clefts will rise and fall over a considerable range, depending on the intensity of neuronal activity.

Although the K⁺ concentration is unlikely to reach a blocking concentration for axons (23—30 mEq/liter), one might suppose that smaller quantities would have a considerable effect. A depolarization of even a few mV will have important consequences on nerve terminals and synapses. It is, therefore, of interest that the neuronal membranes seem designed to minimize the effects of K⁺ fluctuations. A clear example is taken from measurements of the effect of various K⁺ concentrations on cell bodies and their surrounding glia in the leech, shown in Fig. 25. When the K⁺

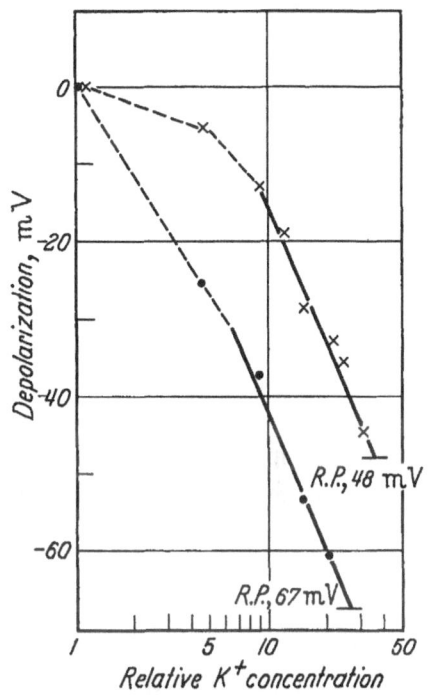

Fig. 25. Difference in the effect of increasing the K⁺ concentration in the bathing fluid on the membrane potential of neurons and of glial cells in the leech nervous system. The neurons (crosses) have a mean resting potential (RP) of 48 mV and glial cells (full circles) have a mean RP of 67 mV. The change of RP (depolarization) is plotted against the logarithm of the relative K⁺ concentration in the bath. The solid lines have a slope of 58 mV for a ten-fold change in K_o. Neurons are less sensitive to increments of K⁺ concentrations. Thus, increasing K⁺ five times (from 4 to 20 mEq/liter) depolarizes the neurons by 5 mV and the glial cells by 25 mV (graph adapted from NICHOLLS and KUFFLER 1964)

concentration is increased by a factor of five, from the normal value of 4 mEq/liter to 20 mEq/liter, the neurons become depolarized by only 5 mV, while the glial potentials change by 25 mV. It is also shown in Fig. 25 that the membrane potential in neurons is lower than in glial cells, a difference which is not due to injury or deterioration (KUFFLER and POTTER 1964); at the same time, both types of cell have similar K⁺ equilibrium potentials. Whereas the glial cells obey the prediction made by the Nernst equation for K_o over a wide range, neurons do so only in high K_o concentrations, i.e., they are relatively insensitive to concentration changes near the physiological range (NICHOLLS and KUFFLER 1964). Similar observations, demonstrating a low K+ sensitivity near the physiological range are well known in isolated axons and muscle fibers. It seems likely that the differences between neurons and glia shown in Fig. 25 also apply to amphibia (KUFFLER, NICHOLLS and ORKAND 1966).

It is probably significant that neurons are protected from the effects of fluctuations of K+ concentrations in their environment oder a considerable range. This might enable them to continue to function better in a K+-rich medium while they reabsorb the K+ they have lost during activity.

7. **Significance of K+ as a signal.** The question arises whether K+ accumulation acts as a physiological signal between neurons and glial cells. The glial depolarization may be used as an index for registering neuronal discharges, because each impulse leaves behind an increment of K+ in the clefts and that in turn will cause a depolarization of the glial membrane potential. For example, if we place an electrode within a glial cell in the optic nerve, the level of the membrane potential will indicate the amount of impulse traffic in the vicinity, as has been demonstrated in Fig. 24 with light flashes. The glial cell cannot discriminate between excitatory and inhibitory axons. If we recorded from an astrocyte in the cortex the situation would be similar, but in addition nerve terminals and synaptic elements would also be included. Here again one would register the amount of general, non-specific activity in the neighborhood of a glial cell in the brain. The depolarization by K+ could be the type of signal one might wish to postulate as a stimulus for a trophic response from glial cells (see Section X).

One may also speculate along different lines: the glial cells, while registering over-all neuronal activity also create current flow (Section VII) which in turn might feed back onto the neurons and influence their discharges. At best one can expect a small effect since neurons are well isolated from glial cells by intercellular clefts (KUFFLER and POTTER 1964; ORKAND, NICHOLLS and KUFFLER 1966).

VII. Contribution of neuroglia to potentials recorded with surface electrodes

It has been shown that glial cells can be depolarized as a result of nerve impulses and that slow potentials can be recorded from them with intracellular electrodes. The question now arises whether glial cells also contribute to slow electrical changes which are recorded with surface electrodes from various regions of the nervous system. The following considerations show that in certain situations they will contribute to slow potentials to an unknown extent.

1. **Optic nerve.** The compound action potential in the optic nerve of *Necturus* or frog is followed by a negative after-potential that can be registered with conventional extracellular electrodes if the recording is 'monophasic' (ORKAND,

NICHOLLS and KUFFLER 1966). There is reason to assume that a part of the after-potentials recorded in this way is contributed by neuroglial cells which become depolarized by K+ leakage from the axons. An example where the time course of an extracellularly recorded after-potential can be compared with glial depolarization is seen in Fig. 26. The record of Fig. 26A shows the compound action potential, contributed by the fast conducting medullated axons followed by the slower non-medullated fibers, and last the negative after-potential. To obtain large effects the negative after-potentials were summed by stimulating the nerve repetitively (Fig. 26B), on a slower sweep speed

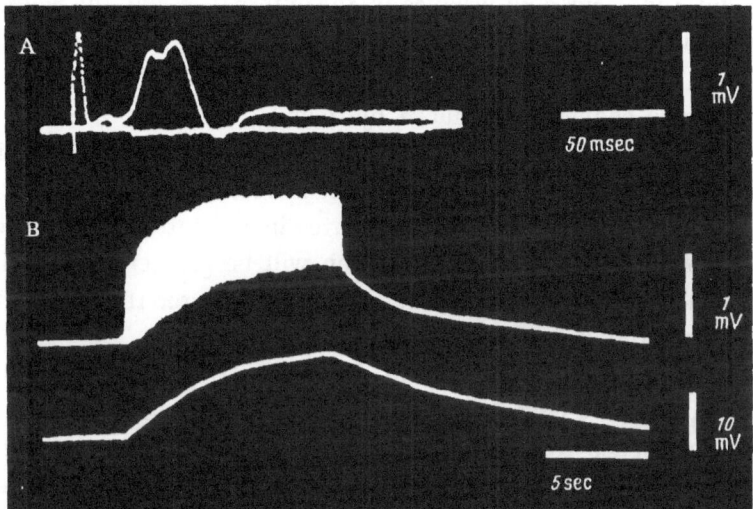

Fig. 26 A and B. Intracellular and extracellular recordings from frog optic nerve. A. Extracellular 'monophasic' record with a suction electrode near the cut end of the nerve. Note the compound action potential set up by a single maximal nerve stimulus; the short-latency deflexion is due to medullated axons and is followed by the more slowly conducting non-medullated fibers. Last is the small negative after-potential. B. (Upper record) Same recording conditions as in A, but stimulation at 10/sec displayed at a slower sweep speed. Negative after-potentials sum. Lower record, simultaneous recording registered with an intracellular electrode in a glial cell. Note the similar time course of the summed negative after-potential with surface leads and the glial depolarization. However, a discrepancy is seen in the falling phase of the two potentials. Calibrations: A and B (upper trace) same amplification, different time bases. B same time base, different amplifications.

so that the fast components of Fig. 26A appear fused. The rising phase of these summed negative after-potentials, but not the falling phase, recorded monophasically with external electrodes, has a similar time course to the glial membrane potential change recorded simultaneously with an intracellular electrode (Fig. 26B, lower record).

The records of Fig. 26 are interpreted as follows: the glial cells near the 'active' electrode become depolarized by K+ that has leaked out of the axons; since the recording is monophasic, action potentials do not reach the second electrode where the glial resting potentials remain unaffected. The depolarized glial cells, however, are linked by low resistance pathways (see Section IV, 5) to the normally polarized cells in the regions of the nerve where conduction has failed and they will draw current, in a similar way as axons, creating a potential difference in the recording circuit. In other words, glial cells have "cable" properties causing potentials set up in one group of glial cells to be distributed decrementally by electrotonic spread to neighboring cells to which they are electrically coupled (Kuffler and Potter 1964; Kuffler, Nicholls and Orkand 1966). This would not be the case if glial cells were small and electrically independent of each other. In the latter situation the depolarization of cells would not cause appreciable current flow. From examples such as Fig. 26 one can see that the current flow generated by the glial depolarization has the right time course to produce in the external circuit the voltage drop that is recorded by the surface electrodes. The remaining question is how much current flow do the glial cells contribute and what proportion of the resulting voltage change is produced by them?

It has already been shown in Fig. 25 in the leech that for a given increase in the external concentration of K$^+$, glial cells become more depolarized than neurons. Consequently when K$^+$ accumulates in the intercellular spaces as a result of nerve impulses the depolarization will be greater in the glial cells. A similar result is likely in the optic nerve because here too the neuronal resting potential is probably not at E_k.

If we assume that glial cells contribute 40—50% of the cross-sectional area of the optic nerve (Fig. 10) and if the total glial and neuronal resistances as well as the 'space constants' of the tissues are similar, glial cells could contribute more than neurons to potentials recorded with surface electrodes. This is essentially a quantitative argument and it must be emphasized that there is as yet no evidence for the extent to which glial depolarization contributes to 'after-potentials'. Changes in the membrane potentials of neurons and glial cells might occur simultaneously in the same or in opposite directions. For example, neurons could be hyperpolarized during their 'positive' after-potential while glial cells are depolarized by K$^+$. The surface potential will then be the result of the algebraical summation of neuronal and glial currents. The extent of the glial contribution can only be assessed when length constants, input resistances and membrane potentials in both types of cell have been measured. Similar considerations might also apply to recordings made from peripheral unmyelinated nerves with the sucrose-gap techniques (Greengard and Straub 1958; Douglas and Ritchie 1962). The membrane potential of Schwann cells is not yet known, nor do we known whether they are electrically coupled to one another.

2. Cerebral cortex. The distribution of current flow that results in the cortex from glial depolarization will be more complex. Massive synchronous activity of groups of neighboring neurons is likely to depolarize glial cells in a circumscribed area. It has already been mentioned (Section IV, 5) that fused membrane contacts have been found between glial cells in the mammalian cortex, suggesting electrical coupling (cf. also Section VIII). Hence glial cells some distance from the depolarized region could act as sources of current and therefore cause changes which could be recorded with external electrodes. No specific analyses of these problems have yet been made in the brain because it has been difficult to identify reliably the glial cells.

Another point of interest is the spreading cortical depression of LEÃO [see BRINLEY (1963) for a recent review]. This is a local depolarization which can be initiated by tetanic excitation of transcallosal pathways (LEÃO and MORISON 1945) and also by a variety of chemical and electrical or mechanical stimuli. The depolarization spreads radially outward from a focus of stimulation at a slow speed of 2—3 mm/min. It is accompanied by release of K^+ from the affected area and the hypothesis has been advanced that it is K^+ accumulation around neurons which sets up the observed cortical depolarization. Intracellular recording from glial cells and neurons, coupled with applied known K^+ concentrations, could test this hypothesis.

A first step in exploring the contribution of glial cells to various forms of electrical activity in the brain will be their reliable identification with intracellular microelectrodes.

3. Glial cells and slow potentials in the retina. Some of the slow potentials recorded from single cells in the retina may be set up by glial cells. It is likely that these slow responses will eventually provide a basis for at least part of the well known electroretinogram (E.R.G.). In many species the E.R.G. has components which are slow and last for the duration of illumination (for a survey of the early literature see GRANIT 1947). Many attempts have been made to locate the cells which give rise to these components of the E.R.G. The most precise approach has been to record with finely tipped microelectrodes from single cells at different locations within the retina during light stimulation (TOMITA 1957; MACNICHOL and SVAETICHIN 1958; MITARAI 1960; BORTOFF 1965). There is now general agreement that at least two types of slow response can be recorded from individual cells within the retina. The responses are usually called slow (S) potentials and are subdivided into luminosity (L) and color (C) responses. The common characteristics of the potentials are that they are graded with the intensity of light stimulation and are maintained at a constant level for the duration of illumination. Most of the experiments have been done in a variety of fishes (for reviews see MACNICHOL and SVAETICHIN 1958; SVAETICHIN, LANGER, MITARAI, FATECHAND, VALLECALLE and VILLEGAS 1961; MOTOKAWA 1963; TOMITA 1963).

The luminosity (L) response is typically obtained after a cell with a resting potential of 10—40 mV has been entered. The effect of light, independent of wavelength, always causes a hyperpolarization of an additional 20—30 mV. The luminosity responses show the greatest sensitivity in the middle of the visible spectrum (500—600 mμ). Similar responses have also been seen in the cat by GRÜSSER (1957), MOTOKAWA, OIKAWA and TASAKI (1957) and BROWN and WIESEL (1959).

The color response differs from the luminosity response because it may be in the depolarizing or hyperpolarizing direction, depending on the wavelength of the light stimulus. Thus blue-green light increases and yellow-red decreases the cell's resting potential.

These phenomena are attributed by many workers to glial cells. Electro-phoretic deposition of dyes from microelectrodes indicates that the luminosity response originates in horizontal cells which lie in the outer plexiform layer between the primary receptors and bipolar cells. There is less agreement on the identity of the cells which give rise to the color response. They are located in a deeper portion of the retina (farther away from the receptors) in the inner nuclear and plexiform layer where the bipolar and amacrine cells are found; they have been assumed to be Müller cells. There is, however, some uncertainty about the location of the C-response; for example, SVAETICHIN et al. (1965), who have worked extensively in this field, stated quite recently that "...addi-tional microelectrode localization experiments are necessary to establish definitely the site of origin of the C-response."

There is little doubt that Müller cells are glial cells (see reviews noted above). The classification of horizontal cells is less certain. Electron microscopic studies by STELL (1965) in the goldfish indicate that the three layers of horizontal cells form synapses with the receptor cells. Other structural observations also make it likely that horizontal cells are neurons, perhaps somewhat atypical (GALLEGO 1964; DOWLING, personal communication; HAMA, personal com-munication). Other workers, however (see SVAETICHIN et al. 1965), believe that no synapses are formed on horizontal cells, more in keeping with the view that these structures are typical glial cells. Since CAJAL's early work (1933, p. 369) the horizontal cells in mammals have been generally considered to be neurons.

In the context of this review the various retinal responses are of great interest because it would be important to know by what mechanism they arise. One possibility is that they result from synaptic excitation, but no specific experiments have as yet tested this interpretation. It is, for instance, generally accepted that the luminosity and color responses are caused by stimulation of cones. If the responses do result from synaptic excitation one would expect permeability changes in the horizontal or in the other unidentified cells during light stimulation. The changes then could be compared with those seen in well studied excitatory and inhibitory synapses elsewhere.

If the luminosity and color responses are mediated by a non-synaptic mechanism a new explanation has to be found. At present only the hypothesis of SVAETICHIN and his co-workers (1965) is available. They present a complex scheme of glial-neuronal interaction and discuss the various metabolic pathways which may be involved during neuronal and glial activation. The experimental support for the various postulated steps has not yet advanced sufficiently for a detailed discussion. Nor can the results on the various retinal elements be usefully interpreted on the basis of the neuron-glial relationship in the leech, *Necturus* or frog.

The various cell types in the retina nevertheless provide a promising field for research. Many of them have a physiological behavior which is unusual as compared with cells studied elsewhere in the central nervous system. Whether they will eventually be classified as 'typical' glial or neuronal cells is at present of less importance than to obtain more accurate information about their physiological properties.

VIII. The "blood-brain barrier"

This review has already discussed the physiological evidence that intercellular spaces in the nervous system are open and serve as pathways for the rapid diffusion of ions and certain small molecules. We have not yet examined the physiological evidence which suggests that certain substances are kept out of the intercellular spaces or are delayed from entering them. These questions are usually referred to as the 'blood-brain barrier' problem. The experiments on the leech and the *Necturus* optic nerve are only relevant to one aspect of this problem since the circulation has been bypassed by applying substances to the bathing fluid. We have no information about the role of the endothelial cells that normally surround the nerve cord of the leech or about the permeability of the blood vessels in the optic nerve of amphibia (Fig. 8).

It has long been an attractive hypothesis that the brain is protected from the action of circulating substances, such as hormones, antibodies or metabolites (see FRIEDMANN 1942 for a review of the early literature). The brain has been assumed to be a special case as compared with other organs. It is also an old idea that the exchange between the C.S.F. and nerve cells meets fewer 'barriers'. There is now abundant evidence for the concept that many substances get into the 'brain' more slowly from capillaries than from the C.S.F. This has led some workers to assume that much, or even most, of the nutrition of the brain is derived from the ventricular system. This seems to us quite improbable for the reasons indicated by the following quotation (see also below) from a review by FRIEDMANN (1942): "it is utterly unlikely that in an organ of the vital importance of the brain the capillaries should lack their chief uses, namely, mediation of the exchange between blood and tissue and adaptation of blood supply to functional needs". Most neurons in the brain are no more than $50\,\mu$ away from a capillary (SCHARRER 1944; see also WOLFF and TSCHIRGI 1956) but are considerably farther from the C.S.F. For example in the basal ganglia the distance may be many millimeters. In the medulla of the puffer fish capillaries actually penetrate large neurons (NAKAJIMA, PAPPAS and BENNETT 1965). We will not quote the extensive literature in detail here because several comprehensive reviews are available (DAVSON 1956, 1963; R. EDSTRÖM 1958, 1964; TSCHIRGI 1960, 1962; RALL and ZUBROD 1962; BRINLEY 1963) and an excellent analysis of the experimental evidence has been provided by DOBBING (1961). We will, however, examine the basis on which some of the interpretations rest. It will become evident that on certain important specific points adequate experimental methods and information are lacking and therefore any interpretation remains questionable.

1. Structural considerations of the pathways into and out of the CNS. Before discussing physiological evidence for the blood-brain barrier it will be useful to examine several structural aspects of the pathways which are involved. There are two avenues into the extracellular space of the brain, i.e., into the intercellular clefts: first, from the capillaries and second, from the C.S.F. Once a substance has traversed the capillaries or ependyma there is a question of getting into the open intercellular spaces. Access to them may be closed at critical regions around the glial endfeet facing the capillaries and perhaps in the region facing ependyma.

a) The capillary-'brain' pathway. The lower section of Fig. 27 is a schematic presentation, based on electron microscopy of the cellular elements, that illustrates the capillary-glia-neuron relationship. To reach the neurons from the blood a substance has to pass through or between endothelial cells, next through the basement lamina and then through or in between the 'endfeet' of glial cells.

Concerning the structure of cerebral *capillaries*, they have a nonfenestrated endothelial lining and their perivascular space is restricted, usually to a thin layer of basement membrane (MAYNARD, SCHULTZ and PEASE 1957; BENNETT, LUFT and HAMPTON 1959; DONAHUE and PAPPAS 1961; WOLFF 1963). In areas which are stained by intravenous trypan blue they are fenestrated and have large perivascular spaces (Fig. 31) (PALAY 1957; FARQUHAR 1961).

With respect to the mechanism of passage across capillaries there are usually two suggestions (see MAJNO and PALADE 1961; PALADE 1961; JENNINGS, MARCHESI and FLOREY 1962; FLOREY 1964; FAWCETT 1965): 1. pinocytotic transport through the endothelial cells, 2. diffusion through the spaces between endothelial cells. These spaces are obliterated by tight junctions over much of their area and the question whether substances can use such pathways will be discussed below (MUIR and PETERS 1962). At present the pathways for the passage of materials through the capillary wall is uncertain not only in the brain but also in other organs of the body. It has not yet been established whether a special problem exists in the brain. The same applies to the properties of the basement lamina that surrounds capillaries. Many recent studies with electrondense markers suggest that the basement lamina does not act as a barrier to diffusion (FLOREY 1964).

b) The layer of glial endfeet as a possible barrier. The question is frequently asked whether 85 % (MAYNARD, SCHULTZ and PEASE 1957), 90 or 99 % (PETERS 1961; WOLFF 1963) of the capillary surface is covered by glial cells. If one considers the connective of the leech or the optic nerve of *Necturus* (Figs. 6, 9 and 10) it is clear that all but a very small fraction of the whole surface is covered by glia. The area of the clefts which communicate with the surface (assuming a cleft width of 150 Å) amounts to about 1—2 % of the total as measured in the leech connective (KUFFLER and NICHOLLS, unpublished). Yet

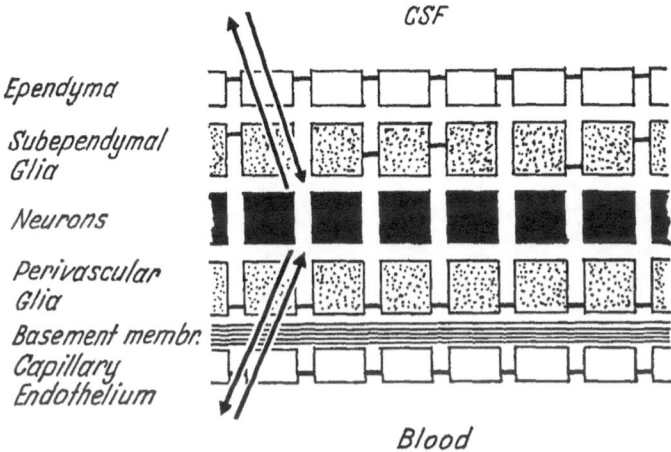

Fig. 27. Schematic presentation of the structures involved in the exchange between the blood and CSF and the neurons in vertebrate brains. The endothelial cells of capillaries, the glial cells and ependymal cells are assumed to be linked by tight junctions (horizontal bars). It is questionable whether tight junctions, or fused membranes, extend circumferentially around cells and close off intercellular channels

diffusion to the neurons occurs in a matter of seconds. Hence it does not seem important exactly how much of the capillary surface is surrounded by glia. Rather it is the nature of the seals between the endfeet of glial cells that will influence our thinking about the pathways.

There is now much evidence that the membranes of glial cells in mammalian brains are linked by tight junctions or membrane fusions similar to those described above for endothelial cells (GRAY 1961, 1964; PETERS 1962). Such areas presumably act as pathways for current to pass from a cell to its neighbor (see Section IV, 5). In addition, there is good evidence that tight junctions between epithelial or gland cells effectively prevent the movement of large molecules such as hemoglobin or ferritin through the intercellular spaces. In an organ, such as a gland, which may be secreting an acid protease, tight junctions appear to prevent diffusion from the lumen back into the tissue (MILLER 1960; PAPPAS and TENNYSON 1962; FARQUHAR and PALADE 1963). To be effective in such a manner, however, the membrane fusion would have to extend as a continuous belt around the cells.

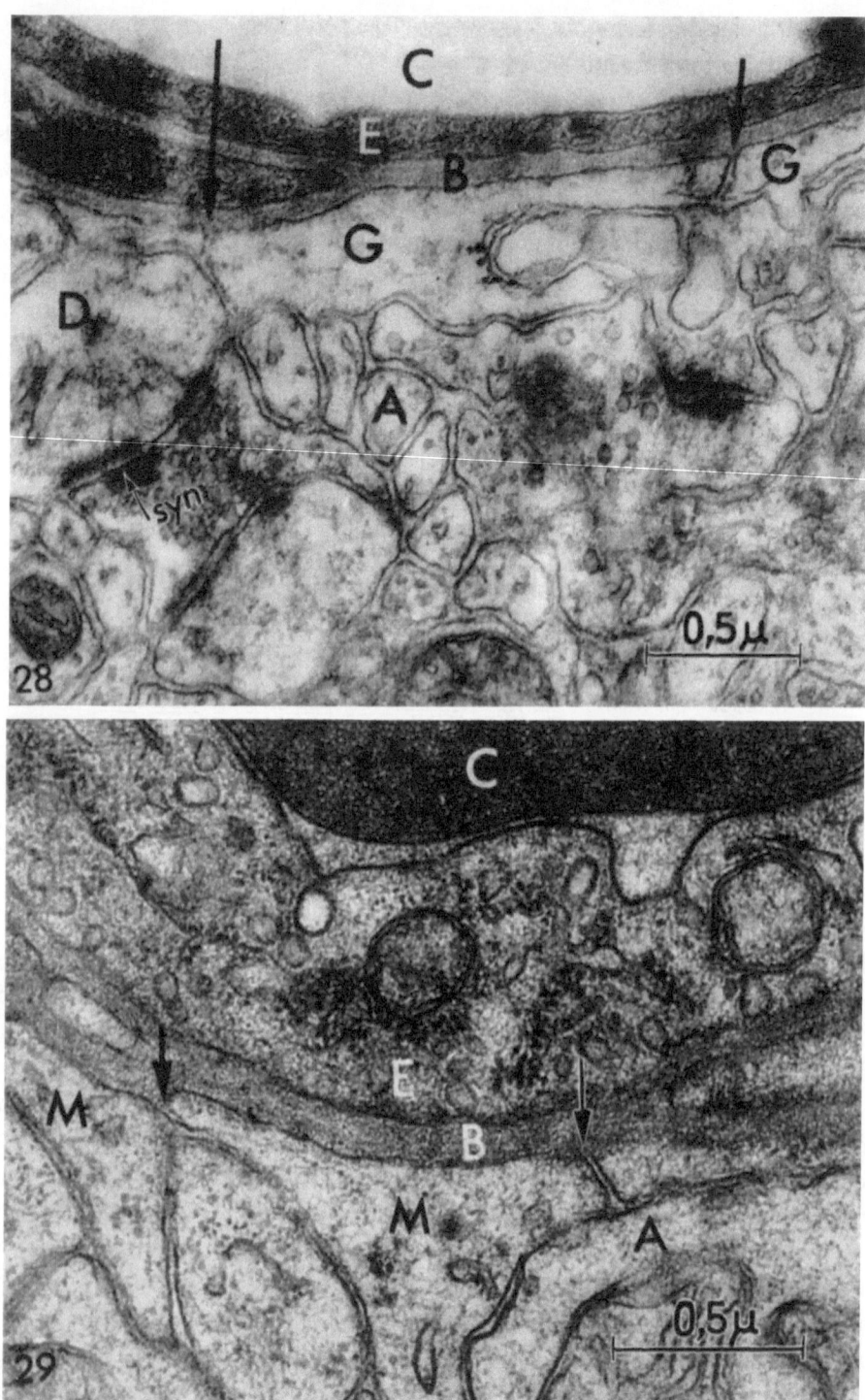

Fig. 28. *Electron micrograph from the superficial layer of the motor cortex of a rat. Unobstructed inter-cellular clefts between the endfeet of astrocytic glial cells (G) which surround a capillary (C) lead from the basement lamina (B) to dendrites (D) and axons (A). Top arrows mark two cleft openings. Syn = synapse; E = endothelial cell. Scale: 0.5 μ (kindly provided by Dr. G. D. PAPPAS, Columbia University)*

In relation to pathways along the clefts between glial cells, especially those around capillaries, a decisive question remains to be answered: are the peri-vascular glial cells joined together by occasional plaques or by membrane fusions that extend right around their circumference; or more graphically, are the intercellular clefts 'zippered up' or are they 'spot welded'? If glial cells are in fact 'zippered up' a very good case could be made for the assumption that the movement of substances would have to occur through the glial cell.

We have not been able to find publications in which a direct extracellular pathway had been pointed out in the mammalian brain leading from the base-ment membrane (lamina) to neurons or dendrites. We have, however, located several unpublished examples. Fig. 28, kindly supplied by Dr. GEORGE PAPPAS, is a section from the superficial layer of the rat motor cortex, showing a capil-lary (C), its basement lamina (B), and underneath several unobstructed inter-cellular clefts (arrows) between the endfeet of an astrocytic glial cell (G); the spaces can easily be followed until they merge with the widely ramifying system that separates all the cells. The clefts between the endfeet have the same dimension as other clefts at much greater distances from the capillary. Tight junctions, *presumably* maculae occludentes, i.e. fused spots, have been seen elsewhere in these preparations. Another relevant picture (Fig. 29) from the retina of a monkey has been obtained from Dr. TOISHIRO KUWABARA. It also presents several non-occluded clefts between pericapillary Müller cell processes (M), leading into the spaces around neuronal processes (LESSELL and KUWA-BARA 1963). Two of the open clefts are marked by arrows. This type of pathway precludes complete circumferential occlusion of the subcapillary region and therefore should permit diffusion to neurons or dendrites. We suspect that many similar examples have been observed by electron microscopists. It remains to be seen, however, how general this is.

c) The C.S.F.-'brain' pathway. In the upper section of Fig. 27 the pathway for the movement of substances from the C.S.F. to neurons is schematically shown to lead through or around ependymal cells that are linked by tight junctions (BRIGHTMAN and PALAY 1963) before reaching the subependymal glial cells which adjoin the neurons. Recently BRIGHTMAN (1965) injected ferritin into the ventricles of the rat and saw that the particles appeared in the intercellular clefts between ependymal cells on both sides of tight junctions. Since some ferritin enters ependymal cells in pinocytotic vesicles, several alternative explanations could account for its appearance beyond the tight junctions: either the seals are complete and ferritin is carried past them by intracellular transport (cf. PALAY and KARLIN 1959a, 1959b; PALAY and REVEL 1964; FAWCETT 1965) or else the tight junction is a 'spot' weld which

Fig. 29. Another example of perivascular glia with unobstructed clefts. Müller cell (M) in the retina of a monkey surrounding a capillary (C) with a red blood cell in its lumen. Direct extracellular pathways are seen between the glial processes (arrows), leading into the tortuous clefts system around axons (A) and dendrites. Scale: 0.5 μ (kindly provided by Dr. TOISHIRO KUWABARA, Harvard Medical School)

can be bypassed through the cleft. Or perhaps the tight junction, after all, does not occlude passage. The second possibility of getting around the 'fused' spot seems the more likely because the pictures show a high density of ferritin on either side of the tight junction with only a low concentration in the intra-cellular vesicles. Concerning the subependymal glial cells one may assume that they are joined by tight junctions. We know of no reports that these are circumferential around all cells, thereby closing off the intercellular spaces. In Fig. 27, therefore, only occasional tight junctions, irregularly placed, have been shown between subependymal glial cells.

To summarize the structural aspects of the pathway problem: it is not yet clear how much of the passage out of capillaries occurs through and how much between endothelial cells. In any event, this is a general unresolved problem and may not be specific for the brain. With respect to the relationship of glial endfeet around the capillary one needs more quantitative information about the extent of fused membranes. Having seen several examples as shown in Fig. 28 and 29 we are inclined to doubt the complete structural occlusion of intercellular clefts between subcapillary glial processes[1]. Concerning ependymal cells recent evidence indicates that an intercellular route allows substances to move from the C.S.F. to nerve cells. As with capillaries, the relative importance of intercellular and intracellular pathways is not known.

Although this discussion has emphasized the pathways into the brain, it should be understood that the questions raised refer equally to the movement of substances in the opposite direction (into capillaries or C.S.F.). In the case of protein it is not clear what the pathway could be since there are no lymphatics within the brain. Presumably, however effective the 'barrier' (see below), some protein must leak out of the capillaries and it will have to be reab-sorbed. Pinocytosis by glial cells might be important in this connection (see Section IX).

2. Physiological evidence for the blood-brain barrier. The term 'barrier' was useful while it was thought that certain substances such as dyes could be completely excluded from the central nervous system. Even then, however, the use of the term tended to cloud the issue. For example, 'blood-urine' barrier for glucose would give no insight into the mechanism by which the kidney keeps sugars out of the urine. A further drawback of the term is that it implies a homogeneity that is almost certainly an oversimplification. The capillaries in various regions of the brain differ structurally from each other. For instance in the neurohypophysis their endothelium is 'fenestrated' (PALAY 1957) and the same applies to the area postrema (Fig. 31). There is no reason to assume that capillaries or any other 'barriers' would have the same pro-perties throught the central nervous system (see Section IX).

---

[1] Since writing this review we have seen many further examples of unobstructed intercellular pathways leading from the basement lamina around cerebral capillaries to the intercellular spaces around neurons.

With the advent of radioactive tracers and more quantitative techniques the concept of an impermeable barrier had to be changed. For example, the evidence for the exclusion of dyestuffs from the brain has been shown to be quite unconvincing (see DAVSON 1956; DOBBING 1961). To establish a 'barrier', the former rigorous requirement that a substance be totally excluded is no longer necessary. Instead, the 'barrier' is now thought of as a structure or process that slows down rather than stops the entry of certain substances from the blood into the 'brain'. It supposedly exists for substances as diverse as Na^+ (DAVSON and POLLAY 1963), urea (DAVSON and LEVIN 1962; BRADBURY and DAVSON 1964) phosphate (BAKAY 1956), para-amino hippuric acid (DAVSON and SPAZIANI 1959), amino acids (LAJTHA 1964), adrenaline (WEIL-MALHERBE, WHITBY and AXELROD 1961) and a variety of drugs (RALL and ZUBROD 1962), to name only a few. It should be emphasized then that the critical test for any experiment on the blood-brain barrier is the *rate* at which any substance gets from blood plasma into the extracellular spaces of the brain, i.e., into the intercellular clefts which make up the environment of nerve cells. If we did speak of a structural barrier alone, disregarding active cell transport, it would follow that when a substance passes into the extracellular space, even if it does do so slowly, it is only a matter of time until the concentration within the extracellular space equilibrates with that in the plasma.

Experiments which are designed to prove the existence of a 'barrier' for a substance should, ideally, answer the following question: 1. at what rate does the substance pass from the blood plasma into the *extracellular spaces* within the brain? Its rate of entry should then be compared to that of various other substances. In other words, are some substances slowed more than others? Furthermore, similar measurements should be made in other organs with the same substances, otherwise it would not be certain that the results are peculiar or characteristic for the brain. If the rate of entry of diverse substances into the extracellular spaces of the brain did not differ from that into other organs, interest would turn to a component which is common to the tissues, most probably to the capillaries. 2. Having made the above tests one would like to determine whether the rates of penetration of these same substances into the intercellular spaces is different when they are applied to C.S.F. instead of blood. Differences between rates of penetration into 'brain' from blood plasma or from the C.S.F. would be important. Once established, they might indicate the structure on which to focus attention.

There are many difficulties of varying magnitude in the way of performing the experiments which are referred to above. The principal difficulty is to establish the extent and composition of the extracellular fluid. This arises partly from the relatively recent discovery that many substances that had been assumed to be purely extracellular in their distribution, are now known to be taken up into cells in significant amounts (see Section IX). The errors resulting from the intracellular uptake or binding are aggravated by the long periods (up to hours) that were usually employed for equilibrating and measuring procedures.

An extensive review critically analyzing the various sources of error in experimental procedures and the problems they raise has been published by Dobbing (1961). For example, penetration of solutions into the 'brain' from blood in intact animals has frequently been compared with penetration into isolated small pieces, cut out of the CNS, approximately 2.3 mm³ in size (Allen 1952; Davson and Spaziani 1959). Such a small piece of brain tissue, if it has survived the preceding anoxic period, must contain a large proportion of damaged cells; these will contribute to the extracellular' space which is measured. Another issue in many experiments is that penetration from blood into brain is slower when compared with penetration into muscle (see Dobbing 1961 for references). This comparison has been made so frequently because many of the techniques for measuring extracellular space were devised on muscle. Skeletal muscle is one of the few tissues between whose cells there are relatively large extracellular spaces. Although the total extracellular space of any tissue is not easily defined (see Dydýnska and Wilkie 1963) the procedure is simple in muscle as compared with brain. As far as intercellular spaces are concerned, tissues like epithelial or gland cells, would be more suitable for comparison with brain. In these the intercellular spaces seen in electronmicrographs are about 150 Å wide, and resemble those in the CNS (see Farquhar and Palade 1963), but it is not apparent that a study of the extracellular compartments of glands would be a fruitful pursuit.

It has already been mentioned that the situation of exchange between the C.S.F. and 'brain' has been accepted to be more rapid as compared with that between blood and brain. Even with much improved technical procedures of continuous perfusion of the ventricles, a non-uniform uptake of substances by the brain results (Feldberg and Fleischhauer 1960; Fleischhauer 1961, 1964; Feldberg 1963; Klatzo, Miquel, Ferris, Prockop and Smith 1964). This presumably reflects a lack of homogeneity within the brain or in ependyma and the underlying mechanisms have still to be clarified.

We have discussed in Section V, 5 some of the evidence that the H^+, HCO_3^-, and Cl^- concentrations in the intercellular fluid around respiratory neurons resemble those in C.S.F. rather than plasma (Pappenheimer, Fencl, Heisey and Held 1965; Fencl, Miller and Pappenheimer 1966). The principle of the experiments is to measure the alveolar ventilation in a goat whose ventricle is perfused with artificial C.S.F. The great advantage of the technique is that the activity of a group of neurons proves the basis for a sensitive bioassay which indicates the composition of the fluid around these neurons. In this way concentration is measured without the necessity of determining the volume of extracellular space. This considerably simplifies the interpretation. It was found that under a wide variety of steady-state conditions, such as alkalosis and acidosis, H^+ in the C.S.F. and not in the plasma regulates the rate of respiration. The simplest hypothesis to fit the observations was to suppose that the difference in H^+ concentration between blood plasma and C.S.F. is maintained by a pump situated in the region of the capillary wall; as a result there is a sharp discontinuity in the concentration profile between intercellular fluid and blood.

It is well known that the problem of the blood-brain barrier has been very extensively studied. In the context of this review we cannot critically examine this large field. The main difficulty lies in our inability to analyze many of the results which depend on a determination of extracellular spaces. As will be pointed out in Section IX, technical difficulties prevent a reasonable estimate of extracellular space. It has been shown that many hitherto unsuspected substances or metabolites do enter cells, especially when the cells have been exposed to test substances for long periods; this is the case in many studies on the blood-brain barrier in which 'equilibration' between blood, brain and C.S.F. is the aim. The extracellular spaces under such conditions can be determined only if one can distinguish the marker or test substance that has been retained within cells from that portion which remains in the intercellular clefts. The latter portion diffuses out rapidly after loading the preparation with a radioactive tracer.

In conclusion, it seems that in the case of H^+ and HCO_3^- the concentrations around respiratory neurons are different from those in the plasma under steady-state conditions. An explanation of the results of PAPPENHEIMER and his colleagues on normal goats requires the assumption not of a structural barrier, such as zonulae occludentes, but of a pumping mechanism around the capillaries. This would prevent escape of substances from capillaries into the extracellular spaces. We do see, however, a number of difficulties in extending this scheme because it forces us to postulate pumps for a diverse set of substances such as urea, Na^+, adrenaline, sucrose, inulin and other substances. These questions need to be resolved by specific experiments to determine whether such active transport can be attributed to glial cells, to endothelial cells or to both.

IX. The extracellular space of the brain

The estimate of the extracellular space of the brain ranges from about 4 % to 15 % (BARLOW, DOMEK, GOLDBERG and ROTH 1961; STREICHER 1961; VAN HARREVELD 1962; REED and WOODBURY 1963; RALL 1964) or more. The volume as such for the total brain might be of small physiological interest if, for instance, it were distributed quite unevenly in different areas. At present the main importance of determining extracellular space is its value for estimating the concentration and location of various test materials or drugs, thereby helping in the interpretation of the studies on the blood-brain barrier. In specific areas the amount of extracellular space may also be significant because it represents the number of intercellular channels which are available for exchange (see below).

1. The uptake of large molecules by glial cells and neurons. When a substance such as inulin or sucrose is used to measure the extracellular space the first assumption is that it does not enter cells. There is some evidence, however, that large (inert) molecules do in fact enter many types of cells, including renal

tubule cells (LEWIS 1931; PALAY and KARLIN 1959b; MILLER 1960; KAR-
NOVSKY 1962; FAWCETT 1965; MARSHALL and NACHMIAS 1965). Recently it has
also been shown that neurons, Schwann cells, glial cells and ependymal cells
can take up large molecules (KLATZO and MIQUEL 1960; DONAHUE and PAPPAS
1963; BONDAREFF 1964; ROSENBLUTH and WISSIG 1964; BRIGHTMAN 1965).

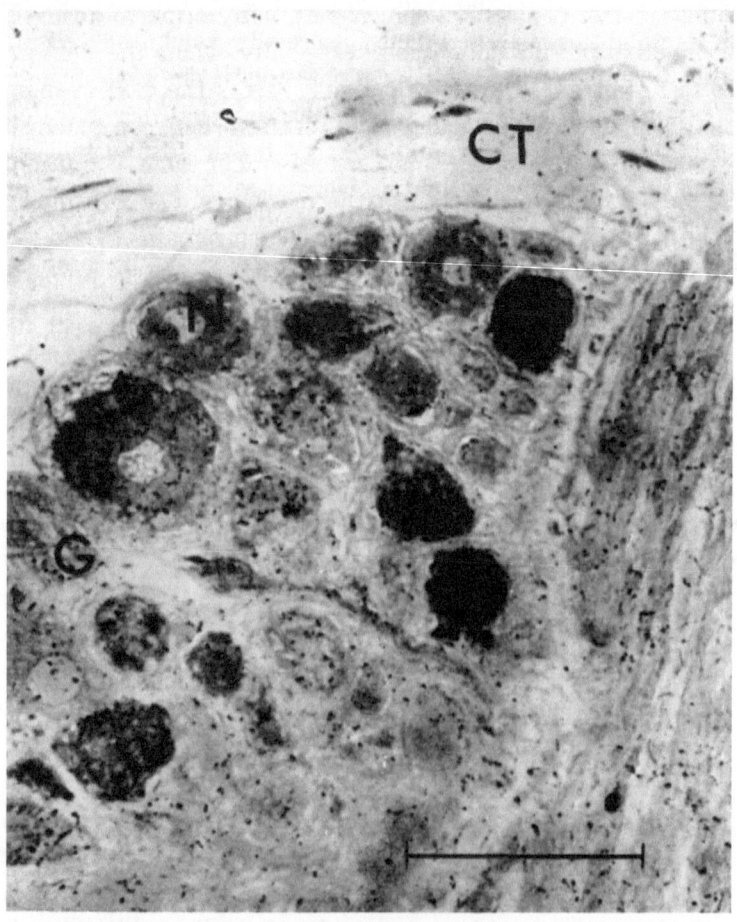

Fig. 30. Uptake of ¹⁴C-dextran into cells of the leech ganglion after loading for 45 minutes. Autoradiograph
made after the label in the extracellular spaces had been washed out for 31 minutes. Silver grains due to
radioactivity appear as black spots over neurons (N) and to lesser extent over glia (G). There are a few
grains over the connective tissue capsule (C.T.) and different amounts of label present in different neurons.
Scale: 50 μ (from NICHOLLS and WOLFE, in preparation)

Most of these studies have been made by electron microscopy with electrondense markers and have indicated that pinocytosis is the mechanism of uptake.
The functional significance of pinocytosis by glial cells and neurons is not clear;
possibly the uptake of proteins is important for both types of cell (see Sections XII, XIII, 2, 4). REED and WOODBURY (1962) have shown that the
sucrose space in the brain can reach 40 % if ¹⁴C sucrose is applied via the C.S.F.
and left for 48 hours. One can conclude that much of the sucrose must have
gone into cells. Sucrose is a normal constituent of squid axoplasm (DEFFNER

and HAFTER 1960) and there is evidence that it is metabolized by the central nervous system of the leech (NICHOLLS and WOLFE, in preparation).

Another approach has been to use ^{14}C-labelled sucrose, inulin and dextran, measuring the rates of tracer movement by conventional techniques and the distribution of label by autoradiography (NICHOLLS and WOLFE, in preparation). These experiments have shown that a significant part of the tracer that enters the nerve cord of the leech cannot be removed by washing for long periods. Autoradiography revealed that tracer was present inside neurons. Fig. 30 is an autoradiograph of part of a leech ganglion that had been soaked in ^{14}C-dextran (molecular weight 5000) for 45 min and then washed in inactive Ringer's fluid for 31 min. The half-time to clear the extracellular space of dextran in this preparation was about 90 sec. Numerous silver grains can be seen over the large neurons while the glial cells and connective tissue contain less activity. The extent of the uptake by cells was such that it would result in a serious overestimate (by a factor of 1.5—2) of the intercellular space between neurons and glial cells. These results emphasize the need for considering intracellular uptake of 'inert' materials in the vertebrate nervous system. One cannot determine extracellular space by experiments where the tracer content alone of a tissue is measured; intra- and extracellular tracer components must be distinguished by measuring the slow and fast phases of the efflux curve. It seems likely that different degrees of intracellular uptake may account for some of the wide discrepancies seen when various substances are used to measure extracellular space in the brain.

2. **Intercellular space in electron microscopy.** Until recently (see below) the most striking feature of electron micrography of central nervous tissue was the uniform appearance of extracellular space. Cell membranes appeared with a few exceptions to be separated by clefts about 150 Å wide. By the use of this figure the total extracellular space has been calculated to be about 5 % of the combined volume of neurons and glial cells (HORSTMANN and MEVES 1959; KUFFLER and POTTER 1964). In the optic nerve of *Necturus* where the axons are small, the extracellular space was about 9 % (WOLFE, in preparation). How reliable is the value of 150 Å for the width of clefts? As we have already seen, physiological experiments have shown that K$^+$ accumulates outside neurons after activity (FRANKENHAEUSER and HODGKIN, 1956; ORKAND, NICHOLLS and KUFFLER 1966). In squid axons an estimate of about 300 Å for the intercellular clefts has been made by these techniques (FRANKENHAEUSER and HODGKIN 1956). If the clefts were much larger than this, the dilution of the liberated K$^+$ would be too great to explain the results observed. In other studies the spacing of myelin has been found to be very similar when measurements were made *in vivo* by x-ray diffraction and by electron microscopy after fixation (FERNANDEZ-MORÁN and FINEAN 1957; FINEAN 1960). One can place a lower limit on the size of the spaces since they can be penetrated by ferritin

particles as large as 100 Å (ROSENBLUTH and WISSIG 1964; BRIGHTMAN 1965). On the other hand, it has been claimed that the spaces are considerably larger *in vivo* and that they shrink during the period between the arrest of the circulation and fixation (for review, see VAN HARREVELD 1962). The shrinkage is supposedly caused by the movement of water and electrolytes into glial cells, which swell and, therefore, compress the extracellular space, since the rigid skull prevents expansion in other directions. There are at least three lines of evidence to suggest that this is not the case in relevant tests on non-neural and nervous tissue: 1. Intercellular clefts of similar dimensions are seen in tissues of the body such as epithelia and glands (CURTIS 1962; FARQUHAR and PALADE 1963). 2. Electron micrographs of neurons and glial cells in tissue culture reveal contacts between cells, about 150 Å wide (BUNGE, BUNGE and PETERSON 1965; S. L. PALAY, personal communication; G. D. PAPPAS, personal communication). 3. Characteristic 150 Å gaps occur between the membranes of glial fragments and the membranes of neurons isolated from the nervous system and fixed *in vitro* (see Fig. 14). In these instances swelling of cells could not cause a diminution of the extracellular space. It is of course true that the size of the intercellular clefts in the central nervous system can be varied over a wide range by using various tonicities of fixative (KARLSSON and SCHULTZ 1965; SCHULTZ and KARLSSON 1965; TORACK, DUFFY and HAYNES 1965). While this emphasizes the need for caution in interpreting electron micrographs, it also shows that large extracellular spaces can be preserved by conventional fixation techniques (see also TORACK, TERRY and ZIMMERMANN 1960 for a discussion of brain edema).

A new method of fixation for electron microscopy has been introduced by VAN HARREVELD, CROWELL and MALHOTRA (1965). Instead of perfusing the brain with fixative the tissue is rapidly frozen and then subjected to substitution fixation in acetone containing osmium tetroxide. The procedure was to decapitate a mouse and expose its cerebellum. Blood and fluid on the cerebellum were removed by touching the surface with tissue paper moistened with Ringer's solution. The whole skull was then lowered at a controlled velocity of 30—40 cm/sec through cooled helium until it struck a silver plug coated by liquid nitrogen. The "cerebellum was flattened on contact with the silver surface but the cerebellar vermis and hemispheres could be recognized after fixation by the characteristic gyration" (VAN HARREVELD, CROWELL and MAL-HOTRA 1965). With this method the most superficial 10 μ of the surface were considered to be adequately fixed. Below this level ice crystals appeared. The tissue contained large amounts of extracellular space (23,6%), especially between the axons, when frozen in this way 30 sec after decapitation. Conventionally fixed tissue or tissue frozen 8 min after decapitation had the usual extracellular space of 5—6%. The interpretation is that water has moved into the cells during asphyxia.

It is beyond the scope of this review to comment on all the relative merits of the two techniques, fixation by perfusion and fixation by freeze substitution. It is, however, difficult to see why decapitation followed by rapid freezing should lead to less asphyxia than perfusion techniques; the circulation time is considerably less than 30 sec and most cells in the brain are probably within 50 μ of a cerebral capillary (SCHARRER 1944). As we have already stated, tissue

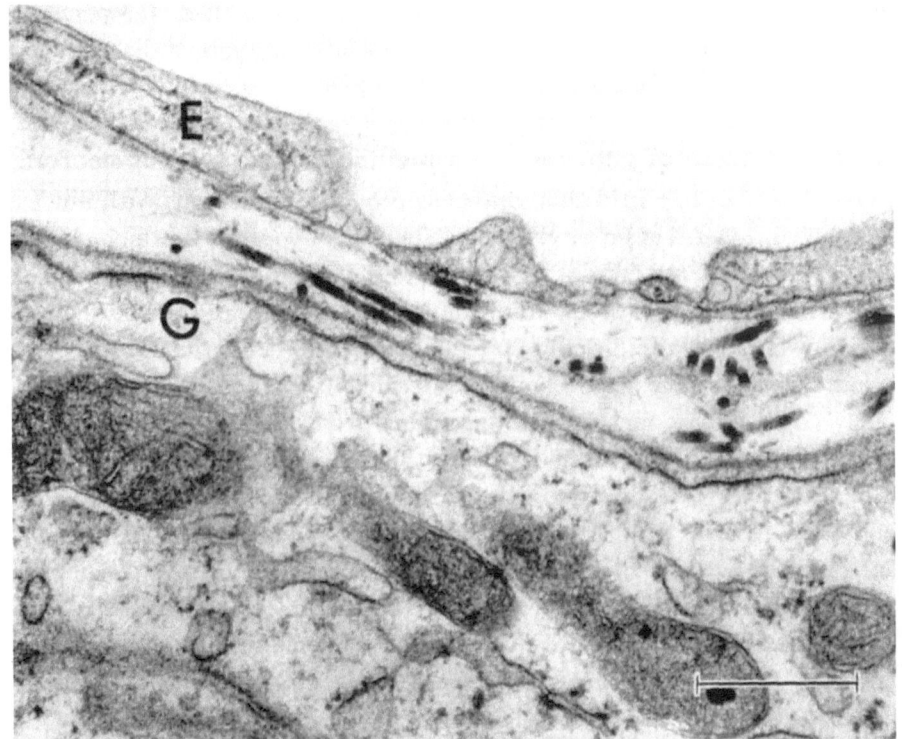

Fig. 31. Electron micrograph of the area postrema in the rat. Note the extensive perivascular space between the endothelium (E) and the neuroglia (G). Another difference from other areas is the fenestration of the capillaries. Scale: 0.5 μ (kindly provided by Dr. D. E. WOLFE, Harvard Medical School)

culture preparations fixed *in vitro* and the nervous system of the leech fixed *in vivo* both resemble the perfused brains in that there is little open space between axons. In these situations there is no question of asphyxia or compression, and fixation must have occurred within seconds.

At the present time it is not possible with electron microscopy to provide an exact figure for the total volume of the extracellular space of the central nervous system. There is, however, only limited physiological interest in this quantity, as its main value would be to help to define various 'compartments' of the brain, i.e, neurons, glia and extracellular space. There would be more interest in knowing the percent of extracellular space in specific regions of the brain, such as the area postrema or other areas where the "barrier" is lacking. Unfortunately, there is little published information about their electron microscopic appearance. Dr. D. E. WOLFE has made available to us an unpublished electron micrograph of part of the area postrema in the rat, shown in Fig. 31. The most striking feature is the large perivascular space, which is interposed between the glial cells and the endothelium. This is in contrast to the vast majority of perivascular spaces in the brain (Section VIII), as for example, those shown in Figs. 28 and 29. In addition, the capillaries in this area show fenestrations, as noted before. Thus in at least one of the regions of the brain where it is accepted that the "barrier" is absent or modified, the percentage of extracellular space around capillaries is considerably larger than elsewhere and the capillary structure is unusual.

X. The problem of glial cells as a metabolic reservoir for neurons

The view is widely held that glial cells provide the neurons with nutrients from their own stores of fat or glycogen. There are two ways in which glial cells could supply nutrients (WIGGLESWORTH 1960; LANDOLT 1965): 1. They could be the principal route of access from the blood if the rate of transport through glial cells were more rapid than extracellular diffusion or if the spaces between perivascular glial cells were completely sealed (see above Section VIII, 2). 2. They could release metabolites to neurons from their own stores but only under special conditions, e.g., if during long-maintained or high-frequency neuronal activity the supply via the clefts becomes inadequate; the neurons would then receive their supplies under normal conditions from the extra-cellular spaces without the intervention of glial cells, as already discussed for certain electrolytes, sucrose and choline. There is direct evidence that inter-cellular channels serve as pathways for the movement of nutrients such as glucose but their relative importance compared to that of transport across the glial cells is not known.

We shall confine the discussion to these two topics and shall simply mention that many biochemical studies have been made of the enzyme content and metabolism of neuroglial cells, Schwann cells and neurons. A wide variety of tissues has been investigated, for example single cells dissected from frozen dried sections (ROBERTS, COELHO, LOWRY and CRAWFORD 1958), isolated brain tissue (KOREY and ORCHEN 1959; TREHERNE 1962b), homogenates of different regions of the brain (ELLIOTT and HELLER 1957; HESS 1961; McILWAIN 1963), glial cells in tissue culture (see GEIGER 1963) and nervous tissue using histo-chemical techniques (POPE 1958; POPE and HESS 1958, POTANOS, WOLF and COWEN 1959; PIPA 1961; FRIEDE 1962; ROBERTS and BAXTER 1963). The work of HYDÉN and his colleagues on the metabolism of cells dissected from the brain will be discussed in Section XI.

In general, the enzyme and metabolic studies suggest a somewhat lower metabolic rate for glial cells than for neurons (see HESS 1961). The glial cells contain, as one might expect, a wide variety of respiratory enzymes; they also contain pseudocholinesterase (KOELLE 1955), but the physiological significance of this enzyme in glial cells is not yet clear.

1. Histological evidence for metabolic interactions. We have already mentioned some of the long standing anatomical considerations suggesting that glial cells provide the neurons with nutrients (see above, Section II). The most compelling feature has always been that astrocytes are interposed between neurons and capillaries, being intimately attached to both. It has also been observed that the number of glial cells around a neuronal cell body varies with the length of its axon (FRIEDE and VAN HOUTEN 1962; FRIEDE 1963). Several additional details of structure also suggest a nutrient role. For example, long processes of glial cells frequently invaginate the neuronal cell body. These glial fingers often contain fat and glycogen, and, purely on the assumption that they perform a nutritive function, they have been called the "trophospongium" (see HOLMGREN 1900). Recently, electron micrographic studies have revealed that the cytoplasm of the glial cells in the visceral ganglia of *Aplysia* is packed with glycogen granules and contains numerous lipid droplets (ROSENBLUTH 1963). The surface area of contact between the glial and neuronal membranes is greatly increased by the presence of many deep invaginations into the neuronal perikarya. The appearance clearly suggests that the glial cells are storing glycogen for use by the neurons.

In the nervous system of insects, the relation of glial cells to neurons also suggests a "trophic" role (for a review see SMITH and TREHERNE 1963). In the ant *(Formica lugubris Zett.)* the supraesophageal ganglion cells are seen in the electron microscope to be surrounded by glial cells that contain abundant glycogen and many ribosomes (LANDOLT 1965). The outermost processes of these glial cells end on tracheoles and on a layer of perilemmal cells that are also rich in glycogen. Over the surface of the neurons, however, there is only a thin film of glial cytoplasm that does not invaginate the surface.

The most suggestive experimental evidence so far for the transfer of nutrients from glial cells to neurons has been found in the cockroach, *Periplaneta americana*, by WIGGLESWORTH (1960), using light microscopy (see also SMITH and TREHERNE 1963). Triglycerides were identified after osmium tetroxide fixation, and glycogen was stained by the periodic acid-Schiff test. Neurons in the last abdominal ganglion are deeply invaginated by an extensive glial trophospongium. Most of the neuronal glycogen is concentrated in the axon hillock regions where the glial invaginations are most abundant. After starvation of the animal for 3—4 weeks, glycogen practically disappears from the ganglion "... but within 3 or 6 hours after the cockroach has been given a meal of honey the glycogen content of the ganglion has been restored and the deposits in the ganglion cells are again concentrated in the axon cone ..." (WIGGLESWORTH 1960). Small fat droplets which can be seen in the glial trophospongium are still present after starvation for 3—4 weeks, but are most abundant after an injection of sugar into the animal.

These elegant experiments by WIGGLESWORTH clearly demonstrate that the glycogen in neurons and glial cells is depleted in starvation, and is restored after feeding. While they suggest that the glial cells may be involved in the transfer process, they do not provide critical evidence about this point. If, for instance, the intercellular spaces in the cockroach were open for the movement of glucose from the blood, neurons could receive at least part of their supply directly. This possibility will be discussed in the next section. As for the invaginations, they need not have a trophic function and the numerous infoldings might merely increase the surface areas of both neurons and glial cells, thereby facilitating absorption from the intercellular space. In that case, one would also expect to observe the accumulation of metabolites around the trophospongium, near the site of uptake. Yet this alternate hypothesis is less attractive because it does not explain why glial cells contain the glycogen, unless it is for their own maintenance. On the other hand, we know neither the energy requirements of the glial cell nor those of the neurons for impulse generation; they may be adequately supplied by the direct route through the intercellular clefts. Similar considerations apply to other studies in which the glycogen content of glial cells has been shown to vary under different functional states of the animal (FRIEDE 1954; OKSCHE 1958, 1961).

2. **Conversion of glucose to glycogen by neurons and glial cells.** Recent experiments made on the central nervous system of the leech indicate that glucose, like sucrose, can reach the neurons rapidly by way of the intercellular spaces (WOLFE and NICHOLLS in preparation). The procedure combined conventional tracer techniques with autoradiography. First, the rate of incorporation of labelled glucose into glycogen was measured. Next, the distribution of radioactive glycogen between neurons and glial cells was determined by autoradiography.

Chains of ganglia were soaked in solutions containing radioactive glucose-6-^{3}H or ^{14}C-glucose for various periods of time. After removal from load solution, the preparations were washed in inactive Ringer's fluid until the extracellular space was cleared. The half-time for glucose efflux from the extracellular spaces is similar to that of sucrose, about 20 sec (NICHOLLS and KUFFLER 1964; NICHOLLS and WOLFE, in preparation). After the preparation has been washed for a few minutes, only the intracellular tracer and adsorbed tracer remains. Radioactivity can be detected in glycogen, amino acids, proteins, lipids and sugar phosphates. The rate of incorporation of tracer into glycogen in the leech nerve cord is rapid, as shown in Fig. 32. In this experiment, proteins and other large molecules were hydrolyzed by boiling the preparation in KOH, leaving only glycogen (solid circles) and "small molecules" which result from hydrolysis (open circles) (GOOD, KRAMER and SOMOGYI 1933). The figure shows that about 5 % of the total tissue radioactivity is in the form of glycogen after a load period of only one minute. The labelled glycogen after a load period of 30 min represents about 20 % of the total radioactivity and about 1 % of all the "cold" glycogen in the tissue.

These results show that glucose is rapidly converted to glycogen but do not indicate which cells are involved. The distribution of tracer in glial cells and neurons was therefore established by autoradiography of preparations soaked in radioactive glucose for different periods. During the process of fixation and dehydration, water-soluble molecules are lost and the only substances that are fixed are glycogen, proteins and lipids. Several lines of evidence suggest that clusters of silver grains seen over the tissue represent labelled glycogen. Autoradiographs obtained from a preparation soaked for 10 min in glucose-6-^{3}H are shown in Fig. 33. The section is through part of a leech ganglion and shows

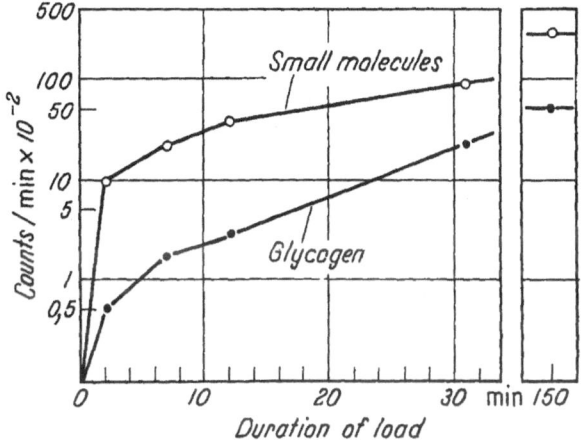

Fig. 32. Incorporation of tracer into cells of the leech ganglion after exposure to radioactive ^{14}C-glucose for various times. The abscissa is the duration of the load period and the ordinate is the radioactivity in the preparation in the form of glycogen (filled circles) or 'small molecules' (open circles). The small molecules result from hydrolysis of the preparation in boiling 30% KOH which preserves glycogen. Each pair of points represents the results obtained on a chain of 5 ganglia. Preparations were washed in inactive Ringer's fluid for 2 min before digesting in KOH (from WOLFE and NICHOLLS in preparation)

several neuronal cell bodies (N) and their axons. The neurons are heavily labelled, but the glial cell (G) that surrounds them is only lightly labelled. Neuropil (NP) in the lower right corner is also labelled but the neurons and glia cannot be distinguished. Autoradiographs made of preparations soaked for only one minute in radioactive glucose showed a less intensive label that was distributed in a similar manner.

The autoradiographs suggest that neurons as well as glial cells can convert glucose to glycogen. They do not, however, present direct evidence for this hypothesis because it could still be possible that the glial cell somehow transferred glycogen or some other intermediary product directly to the neuron.

The following experiment provides more direct evidence that neurons can take up glucose from the extracellular space and convert it to glycogen without the presence of glia. A single nerve cell without its glial investment was dissected out of a ganglion as described earlier (Figs. 12, 14). It had a resting potential of 40 mV with action potentials of 70 mV (as in Fig. 13). The cell was soaked in glucose-6-^{3}H for 30 min. Autoradiographs showed that it had converted glucose into glycogen. Thus a neuron without glia not only continues to give action potentials and maintains a resting potential, but it also metabolizes glucose taken up from the bathing fluid. One can, therefore, conclude that glial cells are not essential for the short-term metabolic needs of neurons. It also seems likely from these results that the neurons *in situ* normally receive at least part of their glucose by way of the intercellular spaces. The half-time

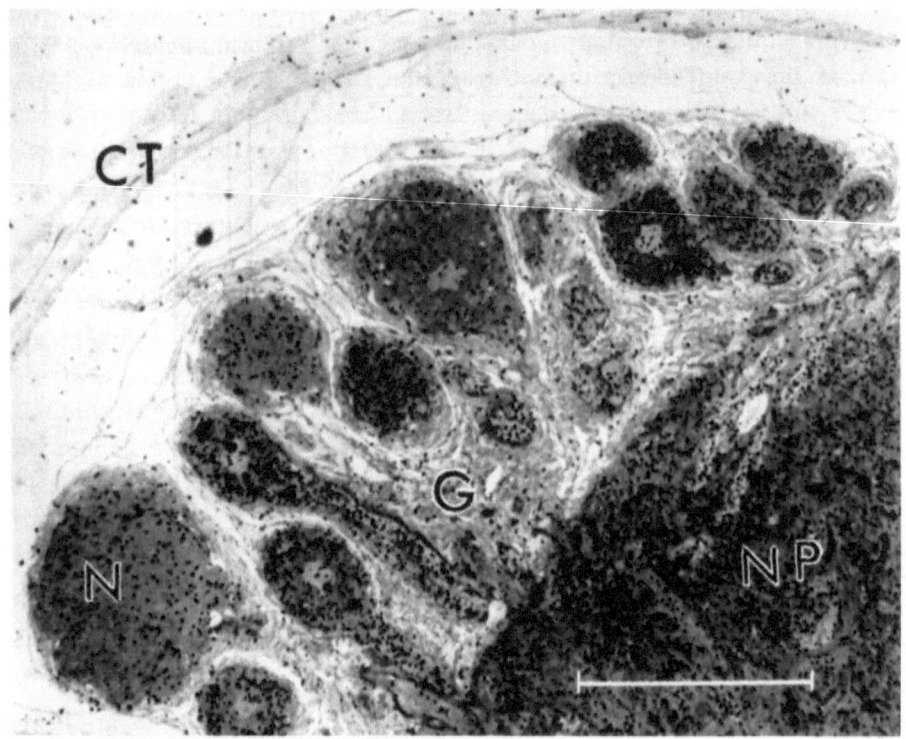

Fig. 33. Autoradiograph of a leech ganglion after exposure to glucose-6-³H for 10 minutes. Silver grains due to radioactivity appear as black spots and are mainly concentrated over neurons (N) which are, for the most part, more heavily labelled than the surrounding glia (G). Water soluble molecules are removed during fixation and dehydration and the remaining radioactivity is due to labelling present in glycogen and, to a lesser extent, protein. The connective tissue (CT) of the capsular material is faintly outlined above the nerve cells. The dark heavily labelled area (right lower corner) is the neuropil (NP). Scale: 100 μ
(from Wolfe and Nicholls in preparation)

for sucrose to travel through intercellular spaces is about 12 sec and glucose which has a similar coefficient of diffusion would probably move at the same rate. Unless, therefore, a specific system for glucose transport exists that is more rapid than intercellular diffusion one can conclude that neurons receive some glucose directly by way of intercellular channels. The relative importance of the glial cell as a means of transport of glucose through the nervous system has still to be established.

3. The fate of glycogen and fat stored in glial cells. The observations discussed above make it advisable to reconsider the questions of the "trophic" role of glial cells. The extracellular space can apparently supply many of the short-term metabolic requirements of neurons. There is no need to postulate that the glial cells act as a channel for rapid transit. This, therefore, focuses attention on the function of the glycogen and fat stored in glial cells. Are these substances required for the metabolism of the glial cell itself, or are they available for use by the neurons? Under what conditions would neurons need metabolites in addition to those that they are able to receive from the inter-cellular spaces? This could be tested directly by labeling glycogen or fat in the glial cell and demonstrating transfer to neurons. Such an experiment would be technically difficult and probably hard to interpret because 1. the penetration of glial cells with micropipettes large enough to inject such substances would almost certainly cause extensive damage, and 2. the appearance of radioactivity in neurons would not necessarily indicate that *net* transfer of material had occurred.

The existence of a trophospongium and the narrowness of the intercellular clefts would certainly facilitate transfer to neurons of materials stored in the glial cells. The diffusion time across a gap of 200 Å would be negligible compared to the time required for a molecule to escape or move away from the cell. However, until transfer is directly demonstrated, these ideas have to remain attractive hypotheses.

4. Lack of glia around some neurons. In some instances, glial cells could probably not act as efficient reservoirs for metabolites. The glial cell in the connectives of the leech is an example (see Fig. 6 and GRAY and GUILLERY 1963; COGGESHALL and FAWCETT 1964). It contains relatively few mitochondria, pinocytotic vesicles or ribosomes and little glycogen as compared to the glial cells in the ganglia. In the central nervous system of the lamprey larva, the glial cytoplasm is mostly reduced to thin sheets that are poor in organelles, glycogen and fat (SCHULTZ, BERKOWITZ and PEASE 1956; BERTOLINI 1964). In some animals the neuronal surfaces are only partially covered by glia, and in *Aplysia* one sees axons completely devoid of Schwann cells (ROSENBLUTH 1963). At the surface of the cerebellum, in the rat, there are axons in contact with the basement membrane under the pial covering (PALAY, personal communication); in the frog mesenteric ganglia, Schwann cells do not surround much of the surface of neuronal cell bodies (TAXI 1959). Thus, many types of neuron-glia associations exist, ranging from an elaborate trophospongium to complete absence of glia over the neuronal surface.

5. Conclusions regarding the transfer of materials. Recent studies on the leech nervous system have shown that glucose is rapidly converted into glycogen by isolated neurons deprived of their glia. Further, glucose can diffuse to the neurons from the bathing fluid via intercellular spaces. Therefore, if materials are transferred from glial cells to neurons it is likely to occur when the metabolic demands of the neurons cannot be met by the more direct supply routes through the intercellular spaces; this may occur during heavy neuronal activity or when for some other reason, such as starvation, the blood level is low in metabolites (see Wigglesworth 1960). According to such a hypothesis, the role of certain glial cells is to keep constant the level of meta-bolites in the clefts. Until one can account otherwise for the use of the metabolic stores such as glycogen or fat within glial cells, it will remain attractive to suppose that they serve as a reservoir for the use by neurons. The evidence for this widely held view is indirect.

XI. Changes in the biochemistry of neurons and glia during activity

The elegant procedures devised by J. E. Edström (1958, 1960), Hydén (1962) and their colleagues have provided new and quantitative chemical methods for the study of neurons and glial cells. In the studies of Hydén and his colleagues, rabbits and rats were subjected to a variety of experimental conditions; for example, they were rotated daily for various periods on a turntable or made to perform specific tasks. After the experiment the animals were killed and single nerve cells and glial material were dissected as quickly as possible from areas of the brain that were assumed to be particularly involved in the preceding activities. Various chemical analyses were then performed on such cells. Cells from similar regions in other animals or from elsewhere in the same brain were used as controls. When statistically sig-nificant biochemical changes occurred an attempt was made to correlate the altered metabolism of the cells with their previous functional state and also with the activity of the whole animal. For example if enzyme levels in cells of Deiters' nucleus changed after rotating the animal it was concluded that rotation had directly influenced the activity of the cells selected and had thereby changed their biochemistry. Further, the observation of reciprocal changes in neurons and glial material led to the suggestion that material was transferred from one cell to the other during activity. Conclusions have also been drawn regarding chemical changes which *characteristically* take place in cells during sleep, learning, transfer of handedness and other types of activity. A representative statement of their current views can be found in a recent article by Hydén and Lange (1965 b).

". . . the neuron and the glia form a functional unit with two parts which are eventually coupled and influence each other functionally. When functional demands increase precise changes in enzyme activity occur in the neurons and their glia within hours. The total protein contents per neuron and glia follow these enzyme changes. This predicts that a synthesis of enzyme protein occurs and not only changes in enzyme activities. Furthermore it seems significant that inverse RNA changes also occur concomitantly. A recent electrophoretic analysis of the RNA fractions which increased in the neuron and decreased in the glia showed an identical base ratio composition of the two RNA fractions.

This suggested that RNA molecules, alternatively nucleotides could be transferred from the glia to the neuron."

In the dissection procedure the animal is first killed by an air embolus and then exsanguinated to remove blood from the brain (see CUMMINS and HYDÉN 1962). The time to obtain isolated cells in this way is of the order of 10 min. Single neurons and clumps of glial cells are dissected free from a slice of brain immersed in 0.25 M sucrose solution. Several nerve cells can be removed from a single animal for analysis. The regions of brain that have been explored include Deiters' nucleus, the reticular formation, the cerebral cortex, the hypoglossal nucleus, the supraoptic nucleus and globus pallidus (HYDÉN and PIGON 1960; HYDÉN 1962a and b; HYDÉN and EGYHÁZI 1962; GOMIRATO and HYDÉN 1963; HAMBERGER and HYDÉN 1963). The parameters that have been measured include the volume, dry weight, protein content, RNA content, RNA base ratios, ATP-ase activity, ATP content, succinic oxidase activity, and cytochrome oxidase activity. One experiment consisted of rotating animals on a turntable for 25 min each day for 7 days. Neurons and glial clumps dissected from Deiters' nucleus were compared to those removed from control animals (HYDÉN and LANGE 1961, 1962). In other experiments animals were compelled to use the left forepaw instead of the right forepaw; cells were removed from the two sides of the cerebral cortex and compared (HYDÉN and EGYHÁZI 1965; HYDÉN and LANGE 1965a). In the same series of experiments, cells removed from rats that had learned to walk on a tight wire were analyzed and compared to cells taken from controls. Finally, cells have been removed from the reticular formation of sleeping and waking rabbits (HYDÉN and LANGE 1965b).

A general consideration of the experimental methods immediately raises the question of how individual cells that have comparable size, weight and function can be dissected out of different pieces of brain (see also KŘIVÁNEK 1965 for a recent review). For example, how can the forepaw area in a rat be located without careful electrical mapping of the motor cortex which was not performed in these experiments? Even if the forepaw areas were identified on the two sides of the brain how could cells that have similar functions be recognized? Another feature that makes it difficult to analyze critically the papers of HYDÉN and his colleagues is the lack of experimental detail presented. Thus in most cases it is not stated how many cells were taken from a single animal, how they varied from each other, which cells were rejected and for what reason. The standard deviation which is given does not in itself indicate the scatter unless the distribution of the population is known. The latter is stated in the detailed papers of EDSTRÖM (1958, 1960). In other words, many uncertainties in the interpretation of the results of HYDÉN and his co-workers are caused by the difficulty in evaluating the adequacy of the sampling technique. Somewhat different problems arise in the selection and chemical analysis of neurons and of glial tissue. We shall, therefore, discuss the results separately.

1. Evidence for changes in neurons. These studies need not be considered in detail except to comment on a few points in two sets of experiments that seem relevant for the interpretation. 1. For example, rotation of the animal led to statistically significant increases in the following constituents of neurons from Deiters' nucleus: dry weight (+10%), total proteins (+42%), RNA (+4%), cytochrome oxidase (+50%) and succinic oxidase (+200%) (see Hydén and Lange 1961, 1962). These values, surprisingly, are expressed not per weight of cell but simply per cell, assuming that the difference between dissected cells was negligible. Such results were then compared with results obtained on cells taken from control animals by the same technique and the values were expressed as percentages. 2. Another series of experiments concerns the enzyme activity of neurons from sleeping and awake rabbits (Hydén and Lange 1965b). An increase in succinic oxidase was seen during sleep in cells of the nucleus reticularis giganto-cellularis. In 29 neurons dissected out of rabbits that had been killed after 1.5 hours of sleep the mean succinic oxidase per cell was increased by about 140% as compared to the value found per cell in 24 cells dissected from control rabbits that had been kept awake for at least one hour before being killed. In this whole series 222 neurons were analyzed, taken from 165 animals (i.e. a mean of less than 2 cells per animal). Included were two groups of animals providing control observations on cells (that showed no change) from other parts of the brain.

The results described so far appear to show that such procedures as rotation of the animals or letting them sleep for one and a half hours leads to a change in the biochemistry of neurons which are removed at random from an area of brain containing many thousands of cells. Nevertheless the interpretation of the experiments is not clear. In no case is it known what the previous functional state of the nerve cells had been. Is the assumption that all the cells of Deiters' nucleus from animals which had been rotated were discharging at higher rates than cells from animals that had not been rotated? Or were some of the cells inhibited by the procedures? Were the 29 cells dissected out of the sleeping animals firing at higher or lower or unchanged frequencies as compared to the corresponding 24 cells isolated from the animals that were awake? The overall change in physiological activity of individual cells is not known. In pools of neurons that have been studied electrophysiologically a stimulus usually activates some cells and suppresses the firing of others (Hubel and Wiesel 1965). This principle, that a "response" may take the form of an increase or of a decrease in discharge, has been studied recently in great detail in sensory systems and also in the brain of waking and sleeping animals (Evarts 1965; Moruzzi 1963). In the experiments discussed here it is assumed that there is an "increased functional activity" in all the cells of the nuclear mass under study. While this is not impossible it would be surprising in view of what is known in systems that have been investigated neurophysiologically. A second doubtful point is the relation between a particular cell and the function under study. What assurance is there that one out of thousands of cells taken randomly from a portion of the reticular formation is specifically concerned with sleep (as is stated in the papers) unless other additional evidence of physiological activity is presented? The discharge pattern, at least of the cells that are selected, should be characteristically related to sleep. A similar line of reasoning applies to the cells from Deiters' nucleus during rotation, especially since the giant cells do not appear to have direct connections with the vestibular nerve (BRODAL 1960, p. 234).

An additional factor that makes the interpretation uncertain is that local circulatory changes accompany "activity" in the brain. Increased blood flow through the visual cortex has been demonstrated by autoradiography after shining light on the retina of the cat (SOKOLOFF 1961; see also SOKOLOFF and KETY 1960). Hence if RNA or enzymes change in neurons after some experimental procedure, one cannot at present say whether these changes are specifically linked with alterations in behavior (e.g. sleep). The biochemical changes may be a concomitant of decreased or increased firing by the cell or alternatively they might arise secondarily due to other changes in the brain such as variations in the local circulation. These uncertainties do not apply to the experiments by EDSTRÖM and GRAMPP (1965) on isolated stretch receptors. In cells which had been discharging for many hours they found no significant changes in the quantity of RNA or in the base ratios. When RNA synthesis was inhibited, an adenine-rich RNA fraction disappeared, without interfering with impulse activity or changing the membrane potential (cf. also GRAMPP and EDSTRÖM 1963).

A closely related question that naturally arises concerns the physiological condition of neurons isolated from the mammalian brain. Do they have resting and action potentials? HILLMAN and HYDÉN (1965) have stressed the importance of this question and have recently made recordings from isolated Deiters' cells with microelectrodes. It was found that the mean resting potential was about 40 mV provided that the glucose concentration in the Krebs' fluid was about 50 mM. Neurons immersed in the usual Krebs' fluid had resting potentials of only about 20—30 mV (HILLMAN and HYDÉN 1965, Table 2). The resting potential was reduced to about 15 mV in hypoxic media. A puzzling feature of the experiments, not seen in other cells that have been studied, was that the resting potentials were not abolished by 150 mM K^+ in the bathing fluid, a potential of about 15 mV remaining. It should be pointed out that a low resting potential in these cells need not detract from the interest of experiments in which enzyme activities or RNA are increased. Electron microscopic studies, however, of the cells which have been used, would seem quite desirable. For

instance, one would like to know how intact the cell surface is, whether
synapses and nerve terminals as well as glial remnants remain attached to the
dissected neurons (see a preliminary study by Roots and Johnston 1965).
Although fixation and dehydration of tissue introduces various artifacts
(Hydén 1961) these would not necessarily reduce the value of electron microscopic studies. The cells could still be compared with undissected neurons
which had been treated in a similar manner.

To conclude the discussion of results obtained on isolated neurons, one can
say that the cells appear to show differences after various experimental procedures. As yet, however, it seems doubtful that a correlation can be made with
any specific functional activity.

2. Measurements on isolated glial tissue. The experiments dealing with
clumps of glia are open to more serious objections. Hydén and Lange (1961)
describe the dissection thus:

"The sample of oligodendrocytes was obtained by taking those glial cells
surrounding the nerve cell body and the first part of the dendrites. The glial
cells situated further away among the dendrites were discarded as being
contaminated with too many fragments of nerve cell processes. As soon as
the glial cells are removed from the cell it is characteristic that they stick
together. They were trimmed down to approximately the same volume as that
of the nerve cell being surrounded. This was checked by forming the glial
clump into a sphereoid and determining the axes with a measuring ocular.
After a year of almost daily training visual estimation was sufficient to get a
sample of the right volume as also the variation coefficients of the numerical
results showed. An important question was the purity of the glial sample.
Pieces of nerve cell processes visible at 100 times magnification were easily
removed from the sample. The remaining nerve threads or parts of dendrites
left in this trimmed glial sample found by teasing out and staining is small
and can probably be overlooked as a source of error. From thirty-five to forty
oligodendrocytes were found to surround each Deiters' nerve cell. The proportion of astrocytes did not appear to exceed 10 % per sample." (see also Hamberger 1963).

In another section of the same article Hydén and Lange state:

"The CNS presents a typical example of a tissue which is inhomogeneous to
an extreme degree. Any attempt to isolate the components by homogenization
and differential centrifugation — proven so useful, for example in the case of
the liver — will cause the delicate membraneous (sic) processes of the neurons
and the glia to break into small fragments."

In view of the difficulties outlined in this second paragraph it seems rather
optimistic to assume that the glial samples obtained by gross dissection are
contaminated insignificantly or only about 5 % by nerve processes (Hamberger
1963). In electron micrographs it is quite clear that the shape of glial cells in

the CNS is extremely complex (see Fig. 1, 9 and 10). The perikarya of the glial cells are small; their processes are fine and inextricably interwoven with dendrites and nerve endings in a manner that would defy separation by dissection. In fact, the glial cells must have been removed in small fragments together with neuronal fragments judging from the complexity seen in electron microscopy. At present one cannot state with any assurance the source of the differences between the neurons and the fragmented glial cells contained in small clumps, because of the unknown extent of mixing within the tissue samples. The evidence that statistically significant changes were seen does not bear on this argument. If, for example, samples of cell bodies had been compared with samples of dendrites and nerve terminals, the results might also have been different.

Even if one assumed that measurements actually do represent reciprocal changes in glial cells and neurons, it still does not follow that neurons and their surrounding glial cells form a functional unit between which RNA and other substances do pass. The finding of reciprocal changes by themselves in the two types of cells would of course be of considerable interest. It would, however, be subject to the same difficulties in interpretation as we have mentioned above for the neurons, both in relation to sampling and also in relation to stimulation. It would finally have to be shown that interaction and/or exchange actually takes place. As a first step it seems essential to obtain more clear-cut controls of biochemical changes from neurons whose activity has been well monitored and from glial cells which are not fragmented and contaminated by neuronal material.

3. Learning, memory, and neuroglia. At this point one additional comment can be made regarding the role of glial cells in memory and learing. HYDÉN and his co-workers have reported experiments in which animals have been put through complex learning procedures (HYDÉN and EGYHÁZI 1963, 1965; HYDÉN and LANGE 1965 a). Afterwards samples of nerve cells or glial clumps have been taken for an analysis of their RNA content and for changes in the base ratio composition. During such learning procedures significant biochemical altera-tions were reported to occur.

In one set of experiments rats learned to balance on steel wires. The neurons and glial cells which were sampled were taken from Deiters' nucleus in the brain stem, because these neurons were "functionally involved in the establishment of this complex motor and sensory performance". In another set of experiments the learning procedure was transfer of handedness, such as learning to use the right hand (paw) rather than the left. Here it was supposed that there was an area in the 5th and 6th layers of the cortex which was critically involved and samples were taken from these regions.

A basic assumption in the above experiments is that learning and memory are functions of all the neurons or glial cells within a particular area. This area in turn was defined by somewhat arbitrary criteria (e.g. forepaw area of motor cortex) and its function was not verified by physiological techniques. The parts of the brain outside specific areas served as control, in most experiments. The first task, therefore, seems to establish the validity of the hypothesis that within a nuclear cell mass there occurs a diffuse molecular "storage" of memory and learning, presumably in RNA. If only a selected relatively small group of cells were involved in memory and learning, the limited sampling procedure would certainly be invalid.

The experiments of HYDÉN and his co-workers could be used to make their thesis more plausible, if it were possible to dissociate the biochemical changes due to non-specific activity from those due to learning. This task is more difficult than the linking of the activity of a specific set of neurons with the behavior of an animal, already discussed. The physiological correlates of learning or memory are still more obscure. To make a *reductio ad absurdum:* If protein synthesis were shown to be increased in muscles during the process of learning a particular exercise, this would not by itself imply that the muscles had learned because of the increased protein synthesis. Such a conclusion would require a whole series of specific controls to be performed at the same time, to exclude influences other than learning.

Unfortunately no theory of learning which can be tested by present methods seems to be on hand. One of the reasons is that we do not really know which neurons are involved, whether we should look for synaptic rearrangements and changed discharge patterns or for some other code which is not expressed in electrical signals. Detection of characteristic chemical changes may be particularly difficult in cells which are highly active. For example, if protein synthesis is involved, the increased or decreased chemical activity which is specific for "learning" may be small in relation to the total cellular metabolism. The question of glia is a further step removed because any theory involving glia has to assume hypothetical secondary changes resulting from neuronal influences which themselves are not yet known.

In summary, the experiments of HYDÉN and his colleagues are open to the serious criticisms that we have enumerated above. Nevertheless, it should be reiterated that a biochemical study of single cells in the central nervous system seems essential and there seems little doubt that such techniques will eventually find wide applications for the study both of neurons and of glial cells.

XII. Myelin formation and experimental allergic encephalomyelitis

Myelin formation by oligodendrocytes probably provides the best demonstration of functional neuron-glia interaction. To make the myelin sheath during normal development, these glial cells must wrap themselves around axons. In the process of forming the lamellae of the myelin sheath, the glial cytoplasm is gradually squeezed out and the cytoplasmic surfaces of the membrane become tightly apposed in a spiral (MATURANA 1960; PETERS 1960, 1964; BUNGE, BUNGE and RIS 1961; BUNGE, BUNGE and PAPPAS 1962; MUGNAINI and WALBERG 1964). The sequence of events is similar to that originally described in the peripheral axons where the Schwann cells form the myelin (GEREN 1954).

Recently a very valuable experimental approach to the study of neuron-glial interactions has been added by immunological techniques. The processes of demyelination and re-myelination can now be investigated in adult animals and tissue cultures. There is considerable evidence that antibodies are made to oligodendrocytes in experimentally induced demyelination of the central nervous system. The phenomenon of experimental allergic encephalomyelitis was discovered by RIVERS and SCHWENTKER (1935); for recent reviews see PETTE and BAUER (1964), WHIPPLE (1965). In general, either the C.S.F. is repeatedly removed and replaced or extracts of spinal cord are injected subcutaneously into animals which, after several days, develop neurological symptoms that are accompanied by characteristic pathological changes of demyelination in the central nervous system (BUNGE, BUNGE and RIS 1961; WAKSMAN and ADAMS 1962; BUBIS and LUSE 1964; GONOTAS, LEVINE and SHOULSON 1964; BUNGE and GLASS 1965). Mononuclear cells in localized regions of the brain insert themselves between the myelin and the axon cylinders (LAMPERT 1965; LAMPERT and CARPENTER 1965). The myelin lamellae separate first from each other and then from the axonal surface, thus leaving neurons completely free of myelin but otherwise normal in appearance. Myelin fragments can be seen inside phagocytes and also inside astrocytes (GONOTAS et al. 1964). After a variable period, oligodendroglial cells begin to wind around axons again, forming new layers of myelin. This recovery process appears to be a recapitulation of the original myelin formation in early development. A feature of great interest in allergic encephalomyelitis is that not all glial cells are affected and that little or no change is seen in the peripheral nervous system (see WAKSMAN and ADAMS 1959, 1962). If, however, peripheral nerve extracts or Schwann cells from tissue cultures are injected, instead of the usual central nervous tissue extract, the lesions appear in the peripheral nerves (WAKSMAN and ADAMS 1959; PALACIOS and PETTE 1964).

The evidence is now considerable that antibodies to glial cells are present in the circulation and in cerebrospinal fluid of animals with experimental allergic encephalomyelitis. Serum from such animals added to tissue cultures of rat cerebellum leads to demyelination (BORNSTEIN 1963; BORNSTEIN and APPEL 1965). The specificity of this reaction is shown by the use of fluorescent antibody techniques in which the globulins from the serum are conjugated to fluorescein. Their site of attachment can then be seen on the surface of glial cells. The attachment to neuroglial cells can be blocked by prior treatment of the tissue culture with antiglobulin. Conjugated globulin can also be seen to be attached to glial cells in sections of brain tissue (ALLERAND and YAHR 1964). Demyelination is not produced by normal serum in the absence of complement or by antisera specific for non-nervous tissue; conversely, serum from animals with allergic encephalomyelitis has no effect on cultures of fetal rat muscle, fibroblasts, or liver epithelium (BORNSTEIN and APPEL 1965).

It is interesting that the axons, although denuded of myelin may still conduct (BORNSTEIN and CRAIN 1965); but in any event they remain sufficiently intact for remyelination to occur (BORNSTEIN 1963; BORNSTEIN and APPEL 1965). The most striking feature of all these results is that the oligodendroglial cell can wrap itself around an axon during development or growth, can be removed experimentally, and then once again can go through the process of forming myelin. This indicates a specific chemical interaction between glial cells and neurons. It is quite likely that a comparable selective chemical mechanism operates to make specific connections in the case of astrocytes. Some of their processes end characteristically around capillaries and others end on neurons. The results described above are probably another indication that glial cells differ from each other in their chemistry as well as in their structure (see also SIDMAN, DICKIE and APPEL 1964). The Schwann cells and astrocytes are relatively unaffected by a factor that disrupts the organization of the oligodendroglial cell.

XIII. General conclusions regarding the functions of glial cells

The aim of most studies on neuroglial cells is to determine their role in the nervous system. It will be useful to recapitulate some of the reasons why, apart from general anatomical considerations, a functional relationship of neurons and glial cells has to be postulated. At the outset one must emphasize that the term "neuroglia" is used for cells that are probably specialized to perform a variety of functions. In view of the marked differences between glial cells and their varied associations with neurons (see Section X, 4), it seems unlikely that a single answer will be found to the question — "what do glial cells do?". The following section deals with several aspects of their function in the nervous system, some of which are well established and others remain speculative.

1. **Structural support.** Mechanical support was the first hypothesis proposed for glial function and this is probably a function that is shared by all glial cells. Some of them contain much fibrillar material and numerous desmosomes but are relatively poor in other organelles. A good example is provided by the glial cells in the leech connectives.

2. **Myelin formation.** This process provides one example in which a specific function can be attributed to a particular type of glial cell (see section XII). The myelin insulates long stretches of nerve fibers and speeds up conduction. During embryonic development oligodendrocytes attach themselves to axons

and, by a process which is not understood, wrap tightly packed lamellae around them, forming the myelin sheath. This faculty is retained, because the oligodendrocytes can perform the same task again in the adult during recovery from demyelination. In this instance, then, a glial cell supplies a specific neuron with a myelin covering. In at least one region of the brain glial cells isolate a part of a neuron electrically without forming myelin. This is seen in the Mauthner cell of the goldfish where a glial cap surrounds the axon hillock, apparently enabling "electrical" inhibition to occur (FURUKAWA and FURSHPAN 1963; ROBERTSON, BODENHEIMER and STAGE 1964).

3. Degeneration, regeneration, and growth. When axons are cut in mammals, the Schwann cells in the periphery or the glial cells in the central nervous system "react" visibly and progressively, starting during the first day. After a short time, glial cells replace those neurons which disappear. The classical two-colume treatise by CAJAL (English translation 1928) is one of the most valuable and comprehensive surveys of this subject (see also WINDLE 1956, and CLEMENTE 1964).

Some recent studies that deal with Schwann cells add indirectly to the evidence for an active relationship between glial cells and neurons. BIRKS, KATZ and MILEDI (1960) cut motor axons innervating frog muscles and observed the disintegration of their terminals, which were then replaced by Schwann cells. Much of the myelin and parts of the terminals became included in the Schwann cell cytoplasm. Although motor terminals were absent, the release of "packets" of acetylcholine from a source outside of the muscle was still observed. It looked as if Schwann cells had temporarily assumed part of the neuron's function or, in any event, that they could release acetylcholine. Unless one assumes that the release also occurs normally it appears that it has been induced by the degenerating axons. In sympathetic ganglia in which pre-synaptic neurons have been cut (HUNT and NELSON 1965), a similar replacement of nerve terminals occurs. However, no release of transmitter has been seen. In these experiments the Schwann cells are responding to the progressive disintegration of nerve fibers. It is interesting that the glial cells around moto-neurons have been observed to change after section of the ventral roots (ANDRES 1961). Thus glial cells, which are some distance from the site of injury have somehow been affected by chromatolysis in the motoneurons.

That Schwann cells act as "guides" for regenerating axons is well known (see CAJAL 1928). In the central nervous system of primitive vertebrates, ependymal and glial cells act in the same way during regeneration of the spinal cord (see, for example, KIRSCHE 1951; MARÓN 1959). After a transection or removal of a portion of the spinal cord, ependymal cells grow into the gap and around these a functioning segment of cord is later reconstituted.

4. Nutritive role. This aspect has been discussed in detail in Section X. The idea, formulated by GOLGI, has been widely accepted over the years and has been indirectly supported by electron microscopic evidence. Nevertheless, in our opinion no convincing demonstration has yet been made to show that substances are in fact exchanged between neurons and glia across the narrow intercellular clefts. Such an observation is a prerequisite for the theory that certain glial cells serve either as storage sites for the benefit of neurons or as distributing channels. Once this has been shown further possibilities can be explored, such as the glial cells supplying substances which the neurons need but cannot synthesize themselves.

5. Excretory role. At present one can only speculate that glial cells might be used for transport of materials into blood or cerebrospinal fluid (see LUGARO quoted by NAGEOTTE 1910). As we have already seen, glial cells are capable of both pinocytosis and phagocytosis and these processes could be involved in protein transport (see Section IX, 1; and KLATZO and MIQUEL 1960; GONOTAS et al. 1964). Though unsupported, this hypothesis is mentioned because we know of no way by which, in the absence of lymphatic vessels, protein could be returned to the circulation from the intercellular spaces of the brain.

6. Signalling between neurons and glia. An effect of nerves on glia by K^+ liberation appears well documented (Section VI), and it may be a mechanism by which neurons provoke glial cells to supply metabolites (Section X). But the response of glial cells to increasing K^+ concentrations in the clefts is not known. Finally, one may speculate along different lines about the meaning of the slow fluctuations in the level of the membrane potential of glial cells. By registering the changing K^+ concentrations in their environment glial cells indicate in a non-specific way the activity of groups of nervous elements, without differentiating the discharges in inhibitory and excitatory neurons. Whether the nervous system makes use of this 'information' is not known.

XIV. Summary

This review is concerned with recent studies on the physiology of glial cells and particular emphasis has been placed on the recent experimental work by the authors and their colleagues. Glial cells in the vertebrate and invertebrate brain are generally interposed between the neurons and other structures, such as the blood vessels, the brain surface and the ventricular system. There are many compelling reasons for the conclusion that they take part in the general functional activity of the central nervous system.

During recent years advances have been made in the study of neuroglia by electron microscopy and by newer histochemical, microchemical, immunological and tracer investigations. In addition, several relatively simple preparations have become available, such as tissue cultures and invertebrate and vertebrate nervous systems that can be analyzed by recording with intracellular micro-electrodes. As a result several widely held assumptions have had to be reconsidered.

The physiological properties of glial cells are in important respects different from those of neurons. Like neurons, they are surrounded by a high-resistance membrane, contain K^+ as their predominant intracellular cation and have high resting potentials. In contrast to neurons, their resting potential is at the K^+ equilibrium potential, and they do not generate propagated impulses; unlike most neurons they are linked to each other by special low resistance connections. All these differences are now established in the *Necturus*, frog and leech. In the latter, neurons also continue to maintain normal resting and action potentials and are still able to take up glucose and convert it to glycogen after removal of the glial cells.

A feature of the neuron-glial relationship is that the cell membranes are separated by narrow clefts approximately 150 Å wide. These channels serve as a pathway whereby certain substances (e.g. Na^+, K^-, sucrose, choline, dextran and inulin) move rapidly through the nervous system. The presence of an intercellular space containing fluid prevents electrical signals generated by neurons from spreading to adjacent glial cells. When neurons are activated, however, they exert an action on glial cells; they liberate K^+ and thereby increase the K^+ concentration in the intercellular clefts. The depolarization of glial cells produced in this way is assumed to contribute to an unknown extent to the slow potentials recorded with surface electrodes. In several important respects our knowledge about intercellular spaces is incomplete. For example, it is not yet known whether there is, as a rule, a continuous extracellular pathway between the inside of capillaries and the intercellular fluid in the brain. Such questions of structure need a solution for a better understanding of the nature of the "blood-brain barrier".

Concerning the role of neuroglia a variety of functions will probably emerge in addition to structural support. Oligodendrocytes are known to be concerned in myelin formation. They do this not only during development but also in tissue cultures, in adult animals and during recovery from experimentally produced demyelination. Glial cells also play a role during degeneration, regeneration, and growth of neurons, but these processes are little understood. For many years neuroglial cells have been considered to be a reservoir for metabolites and a source of supply of nutrient materials for neuronal activity. Direct evidence for this hypothesis is still lacking, but many of the observations make it clear that certain glial cells are highly active metabolically. What is most lacking at the moment is more precise information about the biochemical properties of various glial cells and about the nature of neuron-glia interaction.

Acknowledgement. We gratefully acknowledge the help of our colleagues, Drs. E. A. KRAVITZ, R. K. ORKAND, D. D. POTTER, and D. E. WOLFE, who have collaborated with us over the past few years, and with whom we have had many stimulating discussions. We also wish to thank Drs. R. E. COGGESHALL, D. W. FAWCETT, T. KUWABARA and S. L. PALAY of Harvard Medical School, and Dr. G. D. PAPPAS of Columbia University for kindly giving us unpublished electron micrographs.

Bibliography

Adrian, R. H.: The effect of internal and external potassium concentration on the membrane potential of frog muscle. J. Physiol. (Lond.) 133, 631—658 (1956).

Allen, J. N.: Extracellular space in the central nervous system. Arch. Neurol. Psychiat. (Chic.) 73, 241—248 (1955).

Allerand, C. D., and M. D. Yahr: Gamma globulin affinity for normal human tissue of the central nervous system. Science 144, 1141—1142 (1964).

Andres, K. H.: Untersuchungen über morphologische Veränderungen in Spinalganglien während der retrograden Degeneration. Z. Zellforsch. 55, 49 79 (1961).

Araki, T., and T. Otani: Response of single motoneurons to direct stimulation in toad's spinal cord. J. Neurophysiol. 18, 472 –485 (1955).

Bakay, L.: The blood-brain barrier. Springfield, Ill.: Ch. C. Thomas 1956.

Baker, P. F.: A method for the location of extracellular space in crab nerve. J. Physiol. (Lond.) 180, 439—447 (1965).

Barlow, C. F., N. S. Domek, M. A. Goldberg, and L. J. Roth: Extracellular brain space measured by S^{35} sulfate. Arch. Neurol. (Chic.) 5, 102—110 (1961).

Bennett, H. S., J. H. Luft, and J. C. Hampton: Morphological classification of vertebrate blood capillaries. Amer. J. Physiol. 196, 381- 390 (1959).

Bertolini, B.: Ultrastructure of the spinal cord of the lamprey. I. Ultrastruct. Res. 11, 1--24 (1964).

Birks, R., B. Katz, and R. Miledi: Physiological and structural changes at the amphibian neuromuscular junction, in the course of nerve degeneration. J. Physiol. (Lond.) 150, 145 –149 (1960).

Blackman, J. G., B. L. Ginsborg, and C. Ray: Some effects of changes in ionic concentration on the action potential of sympathetic ganglion cells in the frog. J. Physiol. (Lond.) 167, 374—388 (1963).

Bondareff, W.: Distribution of ferritin in the cerebral cortex of the mouse revealed by electron microscopy. Exp. Neurol. 10, 377—382 (1964).

Bornstein, M. B.: A tissue culture approach to demyelinative disorders. Nat. Cancer Inst. Monogr. 11, 197- 211 (1963).

—, and S. H. Appel: Tissue cuture studies of demyelination. In: Research in demyelinating diseases. Ann. N.Y. Acad. Sci. 122, 280- 286 (1965).

— , and S. M. Crain: Functional studies of cultured brain tissues as related to "demyelinative disorders". Science 148, 1242- 1244 (1965).

—, and M. R. Murray: Serial observations on patterns of growth, myelin formation, maintenance and degeneration in cultures of newborn rat and kitten cerebellum. J. biophys. biochem. Cytol. 4, 499- 504 (1958).

Bortoff, A.: Localization of slow potential responses in the Necturus retina. Vision Res. 4, 627--635 (1964).

Bradbury, M. W. B., and H. Davson: Transport of urea, creatinine and certain monosaccharides between blood and fluid perfusing cerebral ventricular system of rabbit. J. Physiol. (Lond.) 170, 195—211 (1964).

Brightman, M. W.: The distribution within the brain of ferritin injected into cerebrospinal fluid compartments. I. Ependymal distribution. J. Cell Biol. 26, 99 123 (1965).

—, and S. L. Palay: The fine structure of ependyma in the brain of the cat. J. Cell Biol. 19, 415— 439 (1963).

Brinley jr., F. J.: Ion fluxes in the central nervous system. Int. Rev. Neurobiol. 5, 185—242 (1963).

Brodal, A.: Fiber connections of the vestibular nuclei, p. 224—246. In: Neural mechanisms of the auditory and vestibular systems, edit. G. L. Rasmussen, and W. F. Windle. Springfield, Ill.: Ch. C. Thomas 1960.

Brown, K. T., and T. N. Wiesel: Intraretinal recording with micropipette electrodes in the intact cat eye. J. Physiol. (Lond.) 149, 537 562 (1959).

Bubis, J. J., and S. A. Luse: An electron microscopic study of experimental allergic encephalitis in the rat. Amer. J. Path. **44**, 299—318 (1964).

Bunge, M. B., R. P. Bunge, and G. D. Pappas: Electron microscopic demonstration of connections between glia and myelin sheaths in the developing mammalian central nervous system. J. Cell Biol. **12**, 448—454 (1962).

— — and H. Ris: Ultrastructural study of remyelination in an experimental lesion in adult spinal cord. J. biophys. biochem. Cytol. **10**, 67—94 (1961).

Bunge, R. P., M. B. Bunge, and E. R. Peterson: An electron microscope study of cultures of rat spinal cord. J. Cell Biol. **24**, 163—191 (1965).

—, and P. M. Glass: Some observations on myelin-glial relationships and on the etiology of the cerebrospinal fluid exchange lesion. In: Research in demyelinating diseases. Ann. N.Y. Acad. Sci. **122**, 15—28 (1965).

Cajal, S. Ramon Y.: Degeneration and regeneration in the nervous system. 2 vols., transl. from Spanish ed. of 1913. Oxford Univ. Press. London: Humphrey Milford 1928.

— Histology. Revised J. F. Tello-Munoz, Translated by M. Fernan-Nunez. Baltimore: William Wood & Co. 1933.

— Histologie du système nerveux de l'homme et des vertébrés. 2 vols. Madrid: Instituto Ramon Y Cajal 1952.

Cammermeyer, J.: Reappraisal of the perivascular distribution of oligodendrocytes. Amer. J. Anat. **106**, 197—231 (1960).

Chapman-Andresen, C.: Studies on pinocytosis in amoeba. C. R. Lab. Carlsberg **33**, 73—264 (1962).

Clemente, C. D.: Regeneration in the vertebrate central nervous system. Int. Rev. Neurobiol. **6**, 257—301 (1964).

Coggeshall, R. E., and D. W. Fawcett: The fine structure of the central nervous system of the leech, *Hirudo medicinalis*. J. Neurophysiol. **27**, 229—289 (1964).

Crain, S. M.: Resting and action potentials of cultured chick embryo spinal ganglion cells. J. comp. Neurol. **104**, 285—330 (1956).

—, and M. B. Bornstein: Bioelectric activity of neonatal mouse cerebral cortex during growth and differentiation in tissue culture. Exp. Neurol. **10**, 425—450 (1964).

Cummins, J. T., and H. Hydén: Adenosine phosphate levels and adenosine triphosphatases in neurons, glia and neuronal membranes of the vestibular nucleus. Biochim. biophys. Acta (Amst.) **60**, 271—283 (1962).

Curtis, A. S. G.: Cell contact and adhesion. Biol. Rev. **37**, 82—129 (1962).

Davson, H.: A comparative study of the aqueous humour and cerebrospinal fluid in the rabbit. J. Physiol. (Lond.) **129**, 111—133 (1955).

— Physiology of the ocular and cerebrospinal fluids. London: Churchill 1956.

— The cerebrospinal fluid. Ergebn. Physiol. **52**, 20—73 (1963).

—, and M. Pollay: Turnover of ²⁴Na in the cerebrospinal fluid and its bearing on the blood-brain barrier. J. Physiol. (Lond.) **167**, 247—255 (1963).

—, and E. Spaziani: The blood-brain barrier and the extracellular space of brain. J. Physiol. (Lond.) **149**, 135—143 (1959).

Deffner, G. G. J., and R. E. Hafter: Chemical investigations of the giant nerve fibres of the squid. III. Identification and quantitative estimation of free organic ninhydrin-negative constituents. Biochim. biophys. Acta (Amst.) **42**, 189—199 (1960).

De Robertis, E.: Some old and new concepts of brain structure. Wld. Neurol. **3**, 98—111 (1962).

—, and H. M. Gerschenfeld: Submicroscopic morphology and function of glial cells. Int. Rev. Neurobiol. **3**, 1—65 (1961).

Dobbing, J.: The blood-brain barrier. Physiol. Rev. **41**, 130—188 (1961).

Donahue, S., and G. D. Pappas: The fine structure of capillaries in the cerebral cortex of the rat at various stages of development. Amer. J. Anat. **108**, 331—347 (1961).

Douglas, W. W., and J. M. Ritchie: Mammalian nonmyelinated nerve fibers. Physiol. Rev. **42**, 297—334 (1962).

Dydýnska, M., and D. R. Wilkie: The osmotic properties of striated muscle fibres in hypertonic solutions. J. Physiol. (Lond.) 169, 312—329 (1963).

Eckert, R.: Electrical interaction of paired ganglion cells in the leech. J. gen. Physiol. 46, 573—587 (1963).

Edström, J. E.: Quantitative determination of ribonucleic acid in the microgram range. J. Neurochem. 3, 100—108 (1958).

— Extraction, hydrolysis and electrophoretic analysis of ribonucleic acid from microscopic tissue units (Microphoresis). J. biophys. biochem. Cytol. 8, 39—43 (1960).

—, and W. Grampp: Nervous activity and metabolism of ribonucleic acids in the crustacean stretch receptor neuron. J. Neurochem. 12, 735—741 (1965).

Edström, R.: An explanation of the blood-brain barrier phenomenon. Acta psychiat. scand. 33, 403—416 (1958).

— Recent developments of the blood-brain barrier concept. Int. Rev. Neurobiol. 7, 153—190 (1964).

Elliott, K. A. C., and I. H. Heller: Metabolism of neurons and glia, pp. 286—290. In: Metabolism of the nervous system, ed. D. Richter. London: Pergamon Press 1957.

Evarts, E. V.: Neuronal activity in visual and motor cortex during sleep and waking. In: Aspects anatomo-fonctionnels de la physiologie du sommeil, ed. Centre National de la Recherche Scientifique, pp. 189—212. Paris 1965.

Farquhar, M. G.: Fine structure and function in capillaries of the anterior pituitary gland. Angiology 12, 270—292 (1961).

—, and J. F. Hartman: Neuroglial structure and relationships as revealed by electron microscopy. J. Neuropath. 16, 18—39 (1957).

—, and G. E. Palade: Junctional complexes in various epithelia. J. Cell Biol. 17, 375—412 (1963).

Fawcett, D. W.: Surface specializations of absorbing cells. J. Histochem. Cytochem. 13, 75—91 (1965).

Feldberg, W.: A pharmacological approach to the brain from its inner and outer surface. London: Camelot Press 1963.

—, and K. Fleischhauer: Penetration of bromophenol blue from the perfused cerebral ventricles into the brain tissue. J. Physiol. (Lond.) 150, 451—462 (1960).

Fencl, V., T. B. Miller, and J. R. Pappenheimer: Studies on the respiratory response to disturbances of acid-base balance, with deductions concerning the ionic composition of cerebral interstitial fluid. Amer. J. Physiol. (1966) (in press).

Fernandez-Morán, H., and J. B. Finean: Electron microscope and low-angle X-ray diffraction studies of the nerve myelin sheath. J. biophys. biochem. Cytol. 3, 725—748 (1957).

Finean, J. B.: Electron microscope and X-ray diffraction studies of the effects of dehydration on the structure of myelin. I. Peripheral nerve. J. biophys. biochem. Cytol. 8, 13—29 (1960).

Fleischhauer, K.: Neuroglia. Dtsch. med. Wschr. 85, 2031—2035 (1960).

— Regional differences in the structure of the ependyma and subependymal layers of the cerebral ventricles of the cat, pp. 279—283 in: Regional neurochemistry, ed S. S. Kety and J. Elkes. London: Pergamon Press 1961.

— Fluoroscenzmikroskopische Untersuchungen über den Stofftransport zwischen Ventrikelliquor und Gehirn. Z. Zellforsch. 62, 639—654 (1964).

Florey, H. W.: The transport of materials across the capillary wall. Quart. J. exp. Physiol. 49, 117—129 (1964).

Frankenhaeuser, B., and A. L. Hodgkin: The after-effects of impulses in the giant nerve fibres of Loligo. J. Physiol. (Lond.) 131, 341—376 (1956).

Freygang jr., W. J., D. A. Goldstein, and D. C. Hellam: The after-potential that follows trains of impulses in frog muscle fibres. J. gen. Physiol. 47, 929—952 (1964).

Friede, R. L.: Der Kohlenhydratgehalt der Glia von Hirudo bei verschiedenen Funktionszuständen. Z. Zellforsch. 41, 509—520 (1955).

FRIEDE, R. L.: The cytochemistry of normal and reactive astrocytes. J. Neuropath. exp. Neurol. 21, 471—478 (1962).

— Relationship of body size, nerve cell size, axon length and glial density in the cerebellum. Proc. nat. Acad. Sci. (Wash.) 49, 187—193 (1963).

— Enzymatic response of astrocytes to various ions in vitro. J. Cell Biol. 20, 5—15 (1964).

—, and W. H. VAN HOUTEN: Neuronal extension and glial supply: Functional significance of glia. Proc. nat. Acad. Sci. (Wash.) 48, 817—821 (1962).

FRIEDMANN, U.: Blood-brain barrier. Physiol. Rev. 22, 125—245 (1942).

FURSHPAN, E. J.: "Electrical transmission" at an excitatory synapse in a vertebrate brain. Science 144, 878—880 (1964).

FURUKAWA, T., and E. J. FURSHPAN: Two inhibitory mechanisms in the Mauthner neurons of goldfish. J. Neurophysiol. 26, 140—176 (1963).

GALLEGO, A.: Déscription d'une nouvelle couche céllulaire dans la rétine des mammifères et son rôle functionnel possible. Bull. de l'Assoc. des Anatomists. XLIXᵉ Réunion (Madrid, 6—10 Septembre, 1964), pp. 624—631.

GEIGER, R. S.: The behavior of adult mammalian cells in tissue culture. Int. Rev. Neurobiol. 5, 1—52 (1963).

GEREN, B. B.: The formation from the Schwann cell surface of myelin in the peripheral nerves of chick embryos. Exp. Cell Res. 7, 558—562 (1954).

GERSCHENFELD, H. M., F. WALD, J. A. ZADUNAISKY, and E. DE ROBERTIS: Functions of astroglia in the water-ion metabolism of the central nervous system. Neurology (Minneap.) 9, 412—425 (1959).

GLEES, P.: Neuroglia morphology and function. Oxford: Blackwell Sci. Publ. 1955.

GOLGI, C.: Opera Omnia, vol. 1, p. 40. Milano: U. Hoepli 1903a.

— Opera Omnia, vol. 2, p. 460. Milano: U. Hoepli 1903b.

GOMIRATO, G., and H. HYDÉN: A biochemical glia error in Parkinson's disease. Brain 86, 773—780 (1963).

GONATAS, N. K., S. LEVINE, and R. SHOULSON: Phagocytosis and regeneration of myelin in an experimental leukoencephalopathy. An electron microscopic study. Amer. J. Path. 44, 565—584 (1964).

GOOD, C. A., H. KRAMER, and M. SOMOGYI: The determination of glycogen. J. biol. Chem. 100, 485—491 (1933).

GRAMPP, W., and J. E. EDSTRÖM: The effect of nervous activity on ribonucleic acid of the crustacean stretch receptor neuron. J. Neurochem. 10, 725—732 (1963).

GRANIT, R.: Sensory mechanisms of the retina. London: Oxford University Press 1947.

GRAY, E. G.: In: Electron microscopy in Anatomy, edit. J. D. BOYD, pp. 54—61. London: Arnold 1961.

— Tissue of the central nervous system. In: Electron microscopic anatomy, ed. S. M. KURTZ, pp. 369—471. New York: Academic Press, Inc. 1964.

—, and R. W. GUILLERY: An electron microscopical study of the ventral nerve cord of the leech. Z. Zellforsch. 60, 826—849 (1963).

GREENGARD, P., and R. W. STRAUB: After-potentials in mammalian non-myelinated nerve fibres. J. Physiol. (Lond.) 144, 442—462 (1958).

GRÜSSER, O. J.: Rezeptorpotentiale einzelner retinaler Zapfen der Katze. Naturwissenschaften 44, 522 (1957).

HAGIWARA, S., and H. MORITA: Electrotonic transmission between two nerve cells in leech ganglion. J. Neurophysiol. 25, 721—731 (1962).

HAMBERGER, A.: Difference between isolated neuronal and vascular glia with respect to respiratory activity. Acta physiol. scand. 58, Suppl. 203, 1—52 (1963).

—, and H. HYDÉN: Inverse enzymatic changes in neurons and glia during increased function and hypoxia. J. Cell Biol. 16, 521—526 (1963).

—, and H. RÖCKERT: Intracellular potassium in isolated nerve cells and glial cells. J. Neurochem. 11, 757—760 (1964).

HARREVELD, A. VAN: Water and electrolyte distribution in central nervous tissue. Fed. Proc. **21**, 659—664 (1962).

— J. CROWELL, and S. K. MALHOTRA: A study of extracellular space in cortical nervous tissue by freeze substitution. J. Cell Biol. **25**, 117—137 (1965).

HERTZ, L.: Possible role of neuroglia: A potassium-mediated neuronal-neuroglial-neuronal impulse transmission system. Nature (Lond.) **206**, 1091—1094 (1965).

HESS, A.: The ground substance of the central nervous system and its relation to the blood-brain barrier. Wld Neurol. **3**, 118—124 (1962).

HESS, H. H.: The rates of respiration of neurons and neuroglia in human cerebrum. In: Regional neurochemistry, edit. S. S. KETY and J. ELKES, pp. 200—212. London: Pergamon Press 1961.

HILD, W.: Observations on neurons and neuroglia from the area of the mesencephalic fifth nucleus of the cat *in vitro*. Z. Zellforsch. **47**, 127—146 (1957).

—, and I. TASAKI: Morphological and physiological properties of neurones and glial cells in tissue culture. J. Neurophysiol. **25**, 277—304 (1962).

HILL, A. V.: The diffusion of oxygen and lactic acid through tissues. Proc. roy. Soc. B **104**, 39—96 (1928).

HILLMAN, H., and H. HYDÉN: Membrane potentials in isolated neurones *in vitro* from Deiters' nucleus of rabbit. J. Physiol. (Lond.) **177**, 398—410 (1965).

HITCHCOCK, D. I.: In: Physical chemistry of cells and tissues, edit. HÖBER. Philadelphia: Blakiston 1945.

HODGKIN, A. L.: Ionic movements and electrical activity in giant nerve fibres. Proc. roy. Soc. B **148**, 1—37 (1957).

HOFFMAN, H. J., and J. OLSZEWSKI: Spread of sodium fluorescein in normal brain tissue. A study of the mechanism of the blood-brain barrier. Neurology (Minneap.) **11**, 1081—1085 (1961).

HOLMGREN, E.: Weitere Mitteilungen über ''Saftkanälchen'' der Nervenzellen. Anat. Anz. **18**, 290—296 (1900).

HORSTMANN, E.: Zur Frage des extracellulären Raumes im Zentralnervensystem. Verh. Anat. Ges. (Jena), 1959, Suppl. to Anat. Anz. **105**, 100—107 (1958).

— Die Neuroglia und ihre physiologische Bedeutung. Verh. Anat. Ges. (Jena), 1962, Suppl. to Anat. Anz. **109**, 196—203 (1960—1961).

—, and H. MEVES: Die Feinstruktur des molekulären Rindengraues und ihre physiologische Bedeutung. Z. Zellforsch. **49**, 569—604 (1959).

HOSOKAWA, H., and H. MANNEN: Some aspects of the histology of neuroglia. In: Morphology of neuroglia, edit. J. NAKAI, pp. 1—52. Springfield, Ill.: Ch. C. Thomas 1963.

HUBEL, D. H., and T. N. WIESEL: Receptive fields and functional architecture in two non-striate visual areas (18 and 19) of the cat. J. Neurophysiol. **28**, 229—289 (1965).

HUNT, C. C., and P. G. NELSON: Structural and functional changes in the frog sympathetic ganglion following cutting of the presynaptic fibers. J. Physiol. (Lond.) **177**, 1—20 (1965).

HYDÉN, H.: The neuron. In: The cell, edit. J. BRACHET and A. MIRSKY, vol. IV, p. 215. New York: Academic Press, Inc. 1961.

— A molecular basis of neuron-glia interaction. In: Macromolecules and biological memory, edit. F. O. SCHMITT, pp. 55—69. Cambridge, Mass.: M.I.T. Press (1962 a).

— The neuron and its glia — a biochemical and functional unit. Endeavour **21**, 144—155 (1962 b).

—, and E. EGYHÁZI: Changes in the base composition of nuclear ribonucleic acid of neurons during a short period of enhanced protein production. J. Cell Biol. **15**, 37—44 (1962).

— — Glial RNA changes during a learning experiment in rats. Proc. nat. Acad. Sci. (Wash.) **49**, 618—624 (1963).

— — Changes in RNA content and base composition in cortical neurons of rats in a learning experiment involving transfer of handedness. Proc. nat. Acad. Sci. (Wash.) **52**, 1030—1035 (1965).

HYDÉN, H., and P. W. LANGE: Differences in the metabolism of oligodendroglia and nerve cells in the vestibular area. In: Regional neurochemistry, edit. S. S. KETY and J. ELKES, pp. 190—199. London: Pergamon Press 1961.

— — Kinetic study of neurone-glia relationship. J. Cell Biol. 13, 233—237 (1962).

— — A differentiation in RNA response in neurons early and late during learning. Proc. nat. Acad. Sci. (Wash.) 53, 946—952 (1965a).

— — Rhythmic enzyme changes in neurons and glia during sleep. Science 149, 654—656 (1965b).

—, and A. PIGON: A cytophysiological study of the functional relationship between oligodendroglial cells and nerve cells of Deiter's nucleus. J. Neurochem. 6, 57—72 (1960).

ITO, T.: Zytologische Untersuchungen über die Ganglienzellen des japanischen medizinischen Blutegels, Hirudo nipponica, mit besonderer Berücksichtigung auf die „dunkle Ganglienzelle". Okajimas Folia anat. jap. 14, 111—170 (1936).

JENNINGS, M. A., V. T. MARCHESI, and H. FLOREY: Transport of particles across the walls of small blood vessels. Proc. roy. Soc. B 156, 14—19 (1962).

KARLSSON, U., and R. L. SCHULTZ: Fixation of the central nervous system for electron microscopy by aldehyde perfusion. I. Preservation with aldehyde perfusates versus direct perfusion with osmium tetroxide with special reference to membranes and the extracellular space. J. Ultrastruct. Res. 12, 160—186 (1965).

KARNOVSKY, M. L.: Metabolic basis of phagocytic activity. Physiol. Rev. 42, 143—168 (1962).

KATZ, B., and R. MILEDI: A study of spontaneous miniature potentials in spinal motoneurones. J. Physiol. (Lond.) 168, 389—422 (1963).

KATZMAN, R.: Electrolyte distribution in mammalian central nervous system. Are glia high sodium cells? Neurology (Minneap.) 11, 27—36 (1961).

KAYE, G. I., S. DONAHUE, and G. D. PAPPAS: Electron microscopical evidence for the uptake of colloidal particles by Schwann cells in situ. J. de Microscop. 2, 605—612 (1963).

—, and G. D. PAPPAS: Studies on the cornea. I. The fine structure of the rabbit cornea and the uptake and transport of colloidal particles by the cornea in vivo. J. Cell Biol. 12, 457—479 (1962).

KEYNES, R. D., and J. M. RITCHIE: The movement of labelled ions in mammalian nonmyelinated nerve fibers. J. Physiol. (Lond.) 179, 333—367 (1965).

KIRSCHE, W.: Die regenerativen Vorgänge am Rückenmark erwachsener Teleostier nach operativer Kontinuitätstrennung. Z. mikr.-anat. Forsch. 56, 190—265 (1951).

KLATZO, I., and J. MIQUEL: Observations on pinocytosis in nervous tissue. J. Neuropath. exp. Neurol. 19, 475—487 (1960).

— — P. J. FERRIS, L. D. PROCKOP, and D. E. SMITH: Observations on the passage of fluorescein labelled serum proteins (FLSP) from the cerebrospinal fluid. J. Neuropath. exp. Neurol. 23, 18—35 (1964).

KLEEMAN, C. R., H. DAVSON, and E. LEVIN: Urea transport in the central nervous system. Amer. J. Physiol. 203, 739—747 (1962).

KOCH, A., J. B. RANCK jr., and B. L. NEWMAN: Ionic content of the neuroglia. Exp. Neurol. 6, 186—200 (1962).

KOELLE, G. B.: The histochemical identification of acetyl-cholinesterase in cholinergic, adrenergic and sensory neurons. J. Pharmacol. exp. Ther. 114, 167—184 (1955).

KONIGSMARK, B. W., and R. L. SIDMAN: Origin of brain macrophages in the mouse. J. Neuropath. exp. Neurol. 22, 643—676 (1963).

KOREY, S. R., and M. ORCHEN: Relative respiration of neuronal and glial cells. J. Neurochem. 3, 277—285 (1959).

KRIVÁNEK, J.: Quantitative histochemistry of central nervous system. Fed. Proc. Trans. Suppl. 24, 786—798 (1965).

KUFFLER, S. W., J. G. NICHOLLS, and R. ORKAND: Physiological properties of glial cells in the central nervous system of amphibia. J. Neurophysiol. 29, July 1966.

KUFFLER, S. W., and D. D. POTTER: Glia in the leech central nervous system. Physiological properties and neuron-glia relationship. J. Neurophysiol. **27**, 290—320 (1964).

LAJTHA, A.: Protein metabolism of the nervous system. Int. Rev. Neurobiol. **6**, 1—98 (1964).

LAMPERT, P. W.: Demyelination and remyelination in experimental allergic encephalomyelitis. J. Neuropath. exp. Neurol. **24**, 371—385 (1965).

—, and S. CARPENTER: Electron microscopic studies on the vascular permeability and the mechanism of demyelination in experimental allergic encephalomyelitis. J. Neuropath. exp. Neurol. **24**, 11—24 (1965).

LANDOLT, A. M.: Elektronmikroskopische Untersuchungen an der Perikaryenschichte der Corpora pedunculata von Waldameisen (*Formica lugubris* ZETT.) mit besonderer Berücksichtigung der Neuron-Glia-Beziehung. Z. Zellforsch. **66**, 701—736 (1965).

LASANSKY, A., and F. WALD: The extracellular space in the toad retina as defined by the distribution of ferrocanide. A light and electronmicroscopic study. J. Cell Biol. **15**, 463—479 (1962).

LEÃO, A. A. P., and R. S. MORISON: Propagation of spreading cortical depression. J. Neurophysiol. **8**, 33—45 (1945).

LESSELL, S., and T. KUWABARA: Retinal neuroglia. Arch. Ophthal. **70**, 671—678 (1963).

LEWIS, W. H.: Pinocytosis. Bull. Johns Hopk. Hosp. **49**, 17—23 (1931).

LITTLE, M. S., and J. MORRIS: Glia bibliography 1960—1964. Neurosci. Res. Program Bull. **2**, Suppl. (1965).

LOEWENSTEIN, W. R., and Y. KANNO: Studies on an epithelial (gland) cell junction. I. Modifications of surface membrane permeability. J. Cell Biol. **22**, 565—586 (1964).

— S. J. SOCOLAR, S. HIGASHINO, Y. KANNO, and N. DAVIDSON: Intercellular communication: renal, urinary bladder, sensory, and salivary gland cells. Science **149**, 295—298 (1965).

LUMSDEN, C. E.: Histological and histochemical aspects of normal neuroglial cells. In: Biology of neuroglia, edit. W. F. WINDLE, pp. 141—161. Sprinfield, Ill.: Ch. C. Thomas 1958.

—, and C. M. POMERAT: Normal oligodendrocytes in tissue culture. Exp. Cell Res. **2**, 103—114 (1951).

LUSE, S. A.: Ultrastructure of the brain and its relation to transport of metabolites. Res. Publ. Ass. nerv. ment. Dis. **40**, 1—26 (1962).

MACNICHOL, E. F., and G. SVAETICHIN: Electric responses from the isolated retinas of fishes. Amer. J. Ophthal. **46**, 26—46 (1958).

MAJNO, G., and G. E. PALADE: Studies on inflammation. I. The effect of histamine and serotonin on vascular permeability: An electron microscopic study. J. biophys. biochem. Cytol. **11**, 571—606 (1961).

MARÓN, K.: Regeneration capacity of the spinal cord in *Lampetra fluviatilis* larvae. Folia biol. (Warsaw) **7**, 179—189 (1959).

MARSHALL, J. M., and V. T. NACHMIAS: Cell surface and pinocytosis. J. Histochem. Cytochem. **13**, 92—104 (1965).

MATURANA, H. R.: The fine anatomy of the optic nerve of Anurans — an electron microscope study. J. biophys. biochem. Cytol. **7**, 107—120 (1960).

MAYNARD, E. A., R. L. SCHULTZ, and D. C. PEASE: Electron microscopy of the vascular bed of rat cerebral cortex. Amer. J. Anat. **100**, 409—433 (1957).

MCILWAIN, H.: Chemical exploration of the brain. London: Elsevier Publ. Co. 1963.

MILLER, F.: Hemoglobin absorption by the cells of the proximal convoluted tubule in mouse kidney. J. biophys. biochem. Cytol. **8**, 689—718 (1960).

MITARAI, G.: Determination of ultramicroelectrode tip position in the retina in relation to S potential. J. gen. Physiol. **43**, Suppl. 95—100 (1960).

MORUZZI, G.: Active processes in the brain stem during sleeping. Harvey Lect. Series **58**, 233—297 (1963).

MOTOKAWA, K.: Mechanism for the transfer of information along the visual pathways. Int. Rev. Neurobiol. **5**, 212—181 (1963).

MOTOKAWA, K., T. OIKAWA, and K. TASAKI: Receptor potential of vertebrate retina. J. Neurophysiol. **20**, 186—199 (1957).

MUGNAINI, E., and F. WALBERG: Ultrastructure of neuroglia. Ergebn. Anat. Entwickl.-Gesch. **37**, 193—236 (1964).

MUIR, A. R., and A. PETERS: Quintuple-layered membrane junctions at terminal bars between endothelial cells. J. Cell Biol. **12**, 443—448 (1962).

NAGEOTTE, J.: Phénomènes de sécrétion dans la protoplasma des cellules nérvogliques de la substance grise. C. R. Soc. Biol. (Paris) **68**, 1068—1069 (1910).

NAKAI, J., edit.: Morphology of neuroglia. Springfield, Ill.: Ch. C. Thomas 1963.

NAKAJIMA, V., J. D. PAPPAS, and M. V. L. BENNETT: The fine structure of the supramedullary neurons of the Puffer with special reference to endocellular and pericellular capillaries. Amer. J. Anat. **116**, 471—492 (1965).

NICHOLLS, J. G., and D. E. WOLFE: The distribution of ^{14}C-labelled sucrose, inulin and dextran in extracellular space and in cells of the central nervous system of the leech (in preparation).

—, and S. W. KUFFLER: Extracellular space as a pathway for exchange between blood and neurons in central nervous system of leech: The ionic composition of glial cells and neurons. J. Neurophysiol. **27**, 645—673 (1964).

— — Na and K content of glial cells and neurons determined by flame photometry in the central nervous system of the leech. J. Neurophysiol. **28**, 519—525 (1965).

NURNBERGER, J. I., and M. W. GORDON: The cell density of neural tissues: Direct counting method and possible applications as a biologic referent. In: Ultrastructure and cellular chemistry of neural tissue, edit. H. WAELSCH. NewYork: Hoeber 1957.

OKSCHE, A.: Histologische Untersuchungen über die Bedeutung des Ependyms, der Glia und der Plexus Choroidei für den Kohlenhydratstoffwechsel des ZNS. Z. Zellforsch. **48**, 74—129 (1958).

— Der histochemisch nachweisbare Glykogenaufbau und -Abbau in den Astrocyten und Ependymzellen als Beispiel einer funktionsabhängigen Stoffwechselaktivität der Neuroglia. Z. Zellforsch. **54**, 307—361 (1961).

ORKAND, R. K., J. G. NICHOLLS, and S. W. KUFFLER: The effect of nerve impulses on the membrane potential of glial cells in the central nervous system of amphibia. J. Neurophysiol. **29**, July 1966.

PALACIOS, O., and G. E. PETTE: Zur Frage der Erzeugung einer „Allergischen Polyneuritis" in Kaninchen mit Schwannschem Zellgewebekultur-Antigen. Z. Immunitäts-u. Allergieforsch. **126**, 122—124 (1964).

PALADE, G. E.: Blood capillaries of the heart and other organs. Circulation 24, 368—384 (1961).

PALAY, S. L.: Synapses in the central nervous system. J. biophys. biochem. Cytol. **2** (Suppl.), 193—201 (1956).

— The fine structure of the neurohypophysis. In: Progress in neurobiology. II. Ultrastructure and cellular chemistry of neural tissue, edit. H. WAELSCH, pp. 31—44. NewYork: Hoeber-Harper 1957.

— An electron microscopical study of neuroglia. In: Biology of neuroglia, edit. W. F. WINDLE, pp. 24—38. Springfield, Ill.: Ch. C. Thomas 1958a.

— The morphology of synapses in the central nervous system. Exp. Cell Res., Suppl. **5**, 275—293 (1958b).

—, and L. J. KARLIN: An electron microscopic study of the intestinal villus. I. The fasting animal. J. biophys. biochem. Cytol. **5**, 363—371 (1959a).

— — An electron microscopic study of the intestinal villus. II. The pathway of fat absorption. J. biophys. biochem. Cytol. **5**, 372—383 (1959b).

—, and J. P. REVEL: The morphology of fat absorption. In: Lipid transport, edit. H. C. MENG, pp. 1—11. Springfield, Ill.: Ch. C. Thomas 1964.

PAPPAS, G. D., and V. M. TENNYSON: An electron microscopic study of the passage of colloidal particles from the blood vessels of the ciliary processes and choroid plexus of the rabbit. J. Cell Biol. **15**, 227—239 (1962).

88 S. W. KUFFLER and J. G. NICHOLLS: The physiology of neuroglial cells

PAPPENHEIMER, J. R.: Passage of molecules through capillary walls. Physiol. Rev. **33**, 387—423 (1953).
— V. FENCL, S. R. HEISEY, and D. HELD: Role of cerebral fluids in control of respiration as studied in unanesthetized goats. Amer. J. Physiol. **208**, 436—450 (1965).
PENFIELD, W., edit.: Cytology and cellular pathology of the nervous system, vol. 2. New York: Paul B. Hoeber 1932.
PETERS, A.: The formation and structure of myelin sheaths in the central nervous system. J. biophys. biochem. Cytol. **8**, 431—446 (1960).
— Anatomical considerations of the site of the blood-brain barrier. J. Anat. (Lond.) **95**, Suppl, 20—22 (1961).
— Plasma membrane contacts in the central nervous system. J. Anat. (Lond.) **96**, 237—248 (1962).
— Observations on the connexions between myelin sheaths and glial cells in the optic nerves of young rats. J. Anat. (Lond.) **98**, 125—134 (1964).
—, and S. L. PALAY: An electron microscope study of the distribution and patterns of astroglial processes in the central nervous system. J. Anat. (Lond.) **99**, 419 (1965).
PETTE, E., and H. BAUER, edits.: Experimental contributions to the pathogenesis of the demyelinating encephalomyelitides. Z. Immunitäts- u. Allergieforsch. **126**, 1—248 (1964).
PIPA, R. L.: Studies on the hexapod nervous system. III. Histology and histochemistry of cockroach neuroglia. J. comp. Neurol. **116**, 15—26 (1961).
POMERAT, C. M.: Dynamic neurogliology. Tex. Rep. Biol. Med. **10**, 883—913 (1952).
— Cinematographic analysis of cell dynamics. Fed. Proc. **17**, 975—984 (1958).
— Functional concepts based on tissue culture studies of neuroglial cells. In: Biology of neuroglia, edit. W. F. WINDLE, pp. 162—180. Springfield, Ill.: Ch. C. Thomas 1958.
POPE, A.: Implication of histochemical studies for metabolism of the neuroglia In: Biology of neuroglia, edit. W. F. WINDLE, pp. 211—233. Springfield, Ill.: Ch. C. Thomas 1958.
—, and H. H. HESS: Cytochemistry of neurones and neuroglia, pp. 72—82. In: Metabolism of the nervous system, edit. D. RICHTER. London: Pergamon Press 1957.
POTANOS, J. N., A. WOLF, and D. COWEN: Cytochemical localization of oxidative enzymes in human nerve cells and neuroglia. J. Neuropath. exp. Neurol. **18**, 627—635 (1959).
RALL, D. P.: The structure and function of the cerebrospinal fluid, pp. 269—282. In: The cellular functions of membrane transport, edit. J. F. HOFFMAN. Englewood Cliffs, N. J.: Prentice-Hall 1964.
—, and C. G. ZUBROD: Mechanisms of drug absorption and excretion. Ann. Rev. Pharmacol. **2**, 109—128 (1962).
REED, D. J., and D. M. WOODBURY: Kinetics of movement of iodide, sucrose, inulin and radio-iodinated serum albumin (RISA) in the central nervous system and cerebrospinal fluid of the rat. J. Physiol. (Lond.) **169**, 816—850 (1963).
RÍO HORTEGA, P. DEL: Tercera aportacion al conocimiento morfologico y interpretacion functional de la oligodendroglia. Mem. Real. Soc. Esp. Hist. Nat. **14**, 1—122 (1928).
— Microglia. In: Cytology and cellular pathology of the nervous system, vol. II, edit. W. PENFIELD, pp. 483—543. New York: Paul B. Hoeber Inc. 1932.
RIVERS, T. M., and F. F. SCHWENTKER: Encephalomyelitis accompanied by myelin destruction experimentally produced in monkeys. J. exp. Med. **61**, 689—702 (1935).
ROBERTS, E., and C. F. BAXTER: Neurochemistry. Ann. Rev. Biochem. **32**, 513—552 (1963).
ROBERTS, N. R., R. R. COELHO, O. H. LOWRY, and E. J. CRAWFORD: Enzyme activities of giant squid axoplasm and axon sheath. J. Neurochem. **3**, 109—116 (1958).
ROBERTSON, J. D., T. S. BODENHEIMER, and D. E. STAGE: The ultrastructure of Mauthner cell synapses and nodes in goldfish brains. J. Cell Biol. **19**, 157—199 (1964).
ROOTS, B. I., and P. V. JOHNSTON: Isolated rabbit neurons: electron microscopical observations. Nature (Lond.) **207**, 315—316 (1965).

ROSENBLUTH, J.: The visceral ganglion of *Aplysia californica*. Z. Zellforsch. **60**, 213—236 (1963).
—, and S. L. WISSIG: The distribution of exogenous ferritin in toad spinal ganglia and the mechanism of its uptake by neurons. J. Cell Biol. **23**, 307—325 (1964).
RUGH, R.: Vertebrate embryology, p. 437. New York: Harcourt, Brace and World, Inc., 1964.
RYSER, H. J. P.: The measurement of I^{131}-serum albumin uptake by tumor cells in tissue culture. Lab. Invest. **12**, 1009—1017 (1963).
SCHARRER, E.: The blood vessels of the nervous tissue. Quart. Rev. Biol. **19**, 308—318 (1944).
SCHULTZ, R. L.: Macroglial identification in electron micrographs. J. comp. Neurol. **122**, 281—296 (1964).
SCHULTZ, R., E. C. BERKOWITZ, and D. C. PEASE: The electron microscopy of the lamprey spinal cord. J. Morph. **98**, 251—274 (1956).
SCHULTZ, R. L., and U. KARLSSON: Fixation of the central nervous system for electron microscopy by aldehyde perfusion. II. Effect of osmolarity, pH of perfusate, and fixative concentration. J. Ultrastruct. Res. **12**, 187—206 (1965).
SCHULTZ, R. L., E. A. MAYNARD, and D. C. PEASE: Electron microscopy of neurons and neuroglia of cerebral cortex and corpus callosum. Amer. J. Anat. **100**, 369—407 (1957).
SIDMAN, R. L., M. M. DICKIE, and S. APPEL: Mutant mice (QUAKING and JIMPY) with deficient myelination in the central nervous system. Science **144**, 309—311 (1964).
SJÖSTRAND, F. S.: Topographic relationship between neurons, synapses and glial cells. In: The visual system: Neurophysiology and psychophysics, edit. R. JUNG and H. KORNHUBER, pp. 13—24. Berlin: Springer 1961.
SMITH, D. S., and J. E. TREHERNE: Functional aspects of the organization of the insect nervous system. In: Advances in insect physiology, vol. I, ed. J. W. L. BEAMONT, J. E. TREHERNE and V. B. WIGGLESWORTH. New York: Academic Press, Inc. 1963.
SOKOLOFF, L.: Local cerebral circulation at rest and during altered cerebral activity induced by anaesthesia or visual stimulation. In: Regional neurochemistry, ed. S. S. KETY and J. ELKES, pp. 107—117. Oxford: Pergamon Press 1961.
—, and S. S. KETY: Regulation of cerebral circulation. Physiol. Rev., Suppl. **4**, 38—44 (1960).
STELL, W. K.: Correlation of retinal cytoarchitecture and ultrastructure in Golgi preparations. Anat. Rec. **153**, 389—397 (1965).
STREICHER, E.: Thiocyanate space of rat brain. Amer. J. Physiol. **201**, 334—336 (1961).
SVAETICHIN, G., M. LANGER, G. MITARAI, R. FATEHCHAND, E. VALLECALLE, and J. VILLEGAS: Glial control of neuronal networks and receptors. In: The visual system: Neurophysiology and psychophysics, ed. R. JUNG and H. KORNHUBER, pp. 445—463. Berlin: Springer 1961.
—, and E. F. MacNICHOL jr.: Retinal mechanisms for chromatic and achromatic vision. Ann. N. Y. Acad. Sci. **74**, 385—404 (1958).
— K. NEGISHI, R. FATEHCHAND, B. D. DRUJAN, and A. SELVIN DE TESTA: Nervous function based on interactions between neuronal and non-neuronal elements. In: Progress in brain research. Biology of neuroglia, vol. 15, ed. E. DE ROBERTIS and R. CARREA, pp. 1513—1535. Amsterdam: Elsevier Publ. Co. 1965.
TASAKI, I., and J. J. CHANG: Electric response of glia cells in cat brain. Science **128**, 1209—1210 (1958).
TAXI, J.: Sur la structure des travées du plexus d'Auerbach: confrontation des données fournies par le microscope électronique. Ann. Sci. Nat. Zool., Ser. XII (1959).
TOMITA, T.: A study on the origin of intraretinal action potential of the cyprinid fish by means of a pencil-type microelectrode. Jap. J. Physiol. **7**, 80—85 (1957).
TORACK, R. M., M. L. DUFFY, and J. M. HAYNES: The effect of anisotonic media upon cellular ultrastructure in fish and fixed rat brain. Z. Zellforsch. **66**, 690—700 (1965).
— R. D. TERRY, and H. M. ZIMMERMANN: The fine structure of cerebral fluid accumulation. Amer. J. Path. **36**, 273—288 (1960).

TREHERNE, J. E.: The distribution and exchange of some ions and molecules in the central nervous system of *Periplaneta americana* L. J. exp. Biol. **39**, 193—217 (1962a).
— Transfer of substances between the blood and central nervous system in vertebrate and invertebrate animals. Nature (Lond.) **196**, 1181—1183 (1962b).
TSCHIRGI, R. D.: Chemical environment of the central nervous system. In: Neurophysiology III. Handbook of physiology, edit. J. FIELD. Washington, D. C.: Amer. Physiol. Soc. 1960.
— Blood-brain barrier: fact or fancy? Fed. Proc. **21**, 665—671 (1962).
VILLEGAS, G. M., and R. VILLEGAS: Extracellular pathways in the peripheral nerve fibres: Schwann-cell-layer permeability to thorium dioxide. Biochim. biophys. Acta (Amst.) **88**, 231—233 (1964).
VILLEGAS, R., L. VILLEGAS, M. GIMENEZ, and G. M. VILLEGAS: Schwann cell and axon electrical potential differences: Squid nerve structure and excitable membrane location. J. gen. Physiol. **46**, 1047—1064 (1963).
VIRCHOW, R.: Cellular pathology as based upon physiological and pathological histology. Translated by F. CHANCE from 2nd edit. of R. VIRCHOWs Cellularpathologie. Berlin: Hirschwald 1859.
WAKSMAN, B. H., and R. D. ADAMS: A histologic study of the early lesion in experimental allergic encephalomyelitis in the guinea pig and rabbit. Amer. J. Path. **41**, 135—162 (1962).
WARDELL, W. M.: "Dielectric breakdown" as a second mechanism of the electrical response of neuroglia. J. Physiol. (Lond.) **175**, 52—54 P (1964).
WEIL-MALHERBE, H., G. WHITBY, and J. AXELROD: The blood-brain barrier for catecholamines in different regions of the brain. In: Regional neurochemistry, ed. S. S. KETY and J. ELKES. London: Pergamon Press 1961.
WHIPPLE, H. E., ed.: Research in demyelinating diseases. Ann. N.Y. Acad. Sci. **122**, 1—570 (1965).
WIGGLESWORTH, V. B.: The nutrition of the central nervous system in the cockroach *Periplaneta americana* L. The role of the perineurium and glial cells in the mobilization of reserves. J. exp. Biol. **37**, 500—512 (1960).
WINDLE, W. F.: Regeneration of axons in the vertebrate central nervous system. Physiol. Rev. **36**, 427—440 (1956).
— edit.: Biology of neuroglia. Springfield, Ill.: Ch. C. Thomas 1958.
WOLFE, D. E.: Electron microscopic observations on the optic nerve of *Necturus*. (In preparation.)
—, and J. G. NICHOLLS: The uptake of radioactive glucose and its conversion to glycogen by neurons and glial cells in the central nervous system of the leech (in preparation).
WOLFF, J.: Beiträge zur Ultrastruktur der Kapillaren in der normalen Großhirnrinde. Z. Zellforsch. **60**, 409—431 (1963).
— Elektronmikroskopische Untersuchungen über Struktur und Gestalt von Astrozytenfortsätzen. Z. Zellforsch. **66**, 811—828 (1965).
WOLFF, P. H., and R. D. TSCHIRGI: Inability of cerebrospinal fluid to nourish the spinal cord. Amer. J. Physiol. **184**, 220—222 (1956).
WYCKOFF, R. W. G., and J. Z. YOUNG: The motorneuron surface. Proc. roy. Soc. B **144**, 440—450 (1956).

The Influences of Insulin on the Hepatic Metabolism of Glucose[*]

By

ROBERT STEELE[**]

With 3 Figures

Table of Contents

[*] This review is dedicated to the memory of my mentor and colleague, Professor R. C. DE BODO, who died on April 22, 1965.

[**] Biology Department, Brookhaven National Laboratory, Upton, New York 11973, U.S.A. (Supported by the U.S. Atomic Energy Commission.)

Introduction

The study of the hormonal control of metabolism is fascinating at a purely intellectual level. Also it is given practical support, and to some considerable extent it is undertaken by scientists, because of faith in its practical applicability to the relief of human disorders. It is hoped that the present review will contribute to each of these interests, although the relevance to practical matters of the knowledge which is reviewed here is usually only imperfectly understood.

A deficiency in insulin action sets off responses which are spread out over a vast time scale. Some of the eventualities take years to develop and require that the deficiency in insulin action be of a marginal nature so that earlier death does not intervene. The earliest known effects of insulin are observable in 2 minutes.

The recognition of earlier and later consequences of the presence of insulin has led many workers in the field to apply value judgement to the terms "insulin action" and "insulin effects". In the extreme application of this viewpoint even the earliest manifestation of the influence of insulin may be down-graded to the status of an "effect" by the discovery not only of some earlier effect but also of a later effect which does not require the intermediation of the earliest manifestation known.

The working concept of a single initial action of a hormone is not the only possible one even for those interested in elucidating primary actions. Another has been expressed in connection with the concept that induced changes in protein conformation ("allosteric transitions") are responsible for regulation of metabolic activity (MONOD, CHANGEUX and JACOB 1963): ". . . all in fact of these physiologically essential and chemically bewildering properties could be accounted for by the assumption that hormones in general (but not necessarily in all of their manifestations) act as allosteric effectors, *each* of them able specifically to trigger allosteric transitions *in a variety of different proteins*."

It is important to state clearly the implications of this alternative point of view. It does not deny that a later developing manifestation of hormone action may have the same primary cause as an earlier manifestation; it does deny that this should be assumed to be probable. Causative relationships between the sequentially appearing effects have to be demonstrated.

At the outset an attempt was made to prepare this review in accordance with the concept of multiple primary mechanisms of insulin action. Eventually it became apparent that the prejudices of the reviewer make it impossible for him to maintain this point of view.

This is not to say that the highest value is put on knowing the ultimate molecular mechanism of insulin action. Rather than this, a high value is put on all of the knowledge which is gained in exploring the territory toward this goal. Very likely if the molecular action of insulin were discovered by accident, it would require as much effort to move from this toward the various ultimate expressions of insulin action as is now required to move from the consequences toward the molecular action. When all is understood, the knowledge of the molecular mechanism will be but a trivial part of the knowledge of the physiological significance of insulin action.

I. The induction of liver enzymes for glucose metabolism

A mechanism of hormone action which is of current interest postulates the selectively increased rate of synthesis of particular protein enzymes, mediated by events which include the prior synthesis of specific messenger RNA molecules.

In mammalian tissues the intervention of messenger RNA synthesis and new protein synthesis in bringing about a change in tissue enzyme content has often been inferred from the sensitivity of the observed change to actinomycin and puromycin respectively, rather than to actual observation of the synthesis of messenger RNA and the increased synthesis of a specific enzyme protein. In the cases of liver enzymes the action of puromycin rapidly to cause hepatic glycogenolysis (HOFERT, GORSKI, MUELLER and BOUTWELL 1962) points to a danger in the uncritical acceptance of this kind of inference. In general terms the danger is that these agents may have pharmacological actions which are not presently understood, and that these actions may affect the amount of a particular tissue enzyme in ways other than by direct interference with messenger RNA or protein synthesis in the tissue cell.

Perhaps the most complete example of "induction" by a hormone is the variety of liver enzymes which can be increased in amount by the administration of glucocorticoids (ROSEN and NICHOL 1963). These increases in enzymes concerned with gluconeogenesis are discussed further in section II, B 3). The pattern suggests a purposeful modification of consecutive enzymatic reactions; in this respect it seems similar to the induction in unicellular organisms of a whole set of enzymes acting in sequence in the synthesis (for example) of a particular amino acid.

Jacob and Monod (1961), in their original enunciation of a hypothesis for enzyme induction and repression, introduced the concept of the "operon". "The synthesis of the messenger by the structural gene . . . can be initiated only at certain points on the DNA strand, and the cytoplasmic transcription of several, linked, structural genes may depend upon a single initiating point or *operator.*" The formation of messenger RNA by the whole "operon" is thought to be blocked by a reversible combination of the "operator" gene with a "repressor". In the original hypothesis the "repressor" was visualized as a special RNA synthesized by a "regulator" gene; in a revised version (Monod, Changeux and Jacob 1963) of the hypothesis, the "repressor" was visualized as an "allosteric" protein. In the newer version, the "repressor", still held to be formed as the result of the operation of a "repressor gene", if combined with an "inducer" (e.g. a potential metabolite whose utilization depends on an enzymatic pathway to be created by the "operon") is no longer able to combine effectively with the "operator" gene and this leads to activation of the "operon", to the resulting synthesis of the necessary messenger RNA, and finally to the synthesis of the enzymes required for the metabolism of the "inducer".

A potentially interesting feature of the concept just described, insofar as endocrinology is concerned, lies in the possible modification of "genetic repressors" by hormones. Modification of a repressor to allow activation of an "operon" might furnish a new kind of explanation for the integrated, purposeful, metabolic changes, involving many enzymes, which hormones tend to bring about. If, however, a sequence of de-repressions is required, involving many "operons" and many substrate concentration changes for the expression of the action of a hormone, the hoped-for integrating feature of the concept is not realized.

A. Low affinity glucokinase

In 1960 di Pietro and Weinhouse reported a lowered soluble hexokinase activity in livers from 48-hour-fasted rats and alloxan-diabetic rats. Their assay for hexokinase activity utilized an unusually high glucose concentration (0.1 M or 1800 mg/100 ml); the significance of this was made clear later. In a comparison of the glucose concentration dependence of the liver hexokinase system with hexokinase from yeast and adipose tissue, published in May of 1962, di Pietro, Sharma and Weinhouse reported a continued rapid rise in reaction rate with the liver system as glucose concentration was increased in the range from 0.05 M to 0.10 M. The hexokinases derived from other sources, in keeping with their previously described high affinities for glucose, were operating at near their maximal rates at 0.002 M to 0.005 M glucose concentration (36—90 mg/100 ml).

It is of some interest that WEINHOUSE and his co-workers were stimulated to perform their experiments by previous observations of others 1. that glucose uptake by liver slices continued to increase with increasing glucose concentration at high medium glucose concentrations, 2. that glucose transport across the hepatic cell membrane did not appear to be the rate-limiting factor, and 3. that a defect in glucose uptake was apparent in liver slices removed from fasted or alloxan-diabetic rats (CAHILL, ASHMORE, RENOLD and HASTINGS 1959, CHERNICK and CHAIKOFF 1951, SPIRO, ASHMORE and HASTINGS 1958). The discovery of the low affinity (often called "high K_m") hexokinase of liver turned out to depend, also, upon the fortunate use of the proper animal species. BALLARD and OLIVER (1964) demonstrated that the adult sheep liver, in contrast with the adult rat liver, does not contain any low-affinity hexokinase; nevertheless liver slices from adult sheep do increase their glucose uptake rapidly in response to increases in glucose concentration at high medium glucose concentrations. They do so because they contain a large amount of glucose dehydrogenase of low glucose affinity, which leads glucose to glucose 6-phosphate over the pathway:

$$\text{glucose} \rightarrow \text{gluconate} \rightarrow \text{gluconate-6-P} \rightarrow \text{glucose-6-P}.$$

This pathway was explored earlier in adult rat liver by WEINHOUSE and his co-workers (1962) and was stated not to be able to account for the results in this species because adult rat liver was found not to contain a large enough amount of glucose dehydrogenase. METZGER, WILCOX and WICK (1964) recently have reported further investigations on glucose dehydrogenase in rat liver.

WALKER in July of 1962, in a preliminary note, noted the presence of two hexokinases in dialyzed supernatant solution from adult guinea pig liver homogenate. One of these, with a high glucose affinity, was similar to the hexokinase of foetal guinea pig liver; the other, judged to be a specific gluco-kinase, had a lower affinity for glucose. In another preliminary note in April, 1963, WALKER and RAO reported on the liver hexokinase of rats. They introduced the concept that the specific (low affinity) glucokinase could be measured in a mixture with high affinity hexokinase by comparing glucose phosphorylation at 0.1 M glucose (total glucose phosphorylating activity) with the mean of the phosphorylation rates at 0.001 M and 0.0002 M glucose (high affinity hexokinase phosphorylating activity). Using this procedure they demonstrated that 48-hour-fasted and alloxan-diabetic adult rats had decreased specific glucokinase activities but unchanged nonspecific hexokinase activities. A full account of the early work of WALKER was published late in 1963.

In September, 1962, in a preliminary communication, NIEMEYER, PÉREZ, GARCÉS and VERGARA reported that ethionine administration prevented the recovery of hepatic hexokinase activity which takes place on refeeding 48-hour-fasted rats; on this basis they concluded that the restoration of enzyme activity was the consequence of an increased rate of synthesis of enzyme protein and suggested that the increased synthesis was mediated by enzyme induction of the kind seen in microorganisms. In their assay procedure 0.009 M glucose was employed (Niemeyer, Clark-Turri, Garcés and Vergara 1962), which indicates that all the high affinity hexokinase activity and little of the potential low affinity glucokinase activity was being measured; under these circumstances they were fortunate to obtain positive findings. In a further preliminary communication in June, 1963, Niemeyer, Clark-Turri and Rabajille extended these observations on ethionine suppression to recovery of hexokinase activity provoked in rats by carbohydrate feeding after subsistence on a high fat diet.

In March, 1963, in a preliminary communication, Viñuela, Salas and Sols reported the separation of rat liver hexokinase from glucokinase by fractional ammonium sulphate precipitation. The glucokinase was not inhibited by glucose-6-P, had a low affinity for glucose and was affected differently from the hexokinase by inhibitors. It was found to phosphorylate mannose as well as glucose. Changes in rat liver glucokinase-plus-hexokinase activity as the result of fasting or alloxan diabetes were shown to occur only in the glucokinase enzyme by an assay in which glucose phosphorylation at 0.1 M and 0.0005 M glucose concentration were compared.

In November and December of 1963 Salas, Viñuela and Sols and Sharma, Manjeshwar and Weinhouse, respectively, reported that the low-affinity glucokinase of rat liver is an inducible enzyme, dependent for its restoration, after depletion by fasting or by alloxan diabetes, on both RNA synthesis (actinomycin sensitivity) and protein synthesis (puromycin or amino acid analogue sensitivity). Salas, Viñuela and Sols (1963) found a barely noticeable effect of insulin to increase the low glucokinase level of 4-day alloxan-diabetic rats in 3 hours after its administration *in vivo*; full restoration to the normal level required more than 24 hours. During fasting the liver glucokinase level was significantly decreased in 43 hours, but not in 23 hours; again there was a barely noticeable increase in 3 hours when 80-hour-fasted rats were refed on stock diet; at 6 hours the increase was quite large. In view of the low endogenous secretion rate of insulin in the fasted animal, Salas et al. noted that the availability of insulin was the common denominator in the induction of glucokinase. They remarked also on the sluggish appearance and disappearance of glucokinase; in view of this they suggested that the uptake of glucose by the liver during hyperglycemia is in part regulated by the nature of the enzyme itself, which increases in activity as a result of increases in blood glucose concentration which are within the physiological range. They also suggested that changes in glycogen synthetase activity may play a part in changes in hepatic glucose uptake in the normal fed animal.

SHARMA, MANJESHWAR and WEINHOUSE (1963) found hepatic glucokinase decreased in amount 24 hours after withdrawal of insulin from alloxan-diabetic rats maintained on insulin; further decreases continued up to 96 hours. On administration of glucose to 72-hour-fasted normal rats, an increase in glucokinase level to near normal was seen in 4 hours; shorter intervals were not studied. SHARMA et al. mentioned preliminary evidence that adrenalectomy did not alter the recovery of hepatic glucokinase on refeeding, nor did cortisol enhance the process. Insulin administration to the alloxan-diabetic rat increased glucokinase to normal only after 16 to 24 hours, whereas glucose feeding in the normal fasted rat brought about the change in 4 hours. From this, SHARMA et al. argued that insulin may not be the sole requirement for induction and that insulin may be "... required in order for the liver cell to assume full responsiveness to glucose in its capability for glucokinase synthesis."

B. Glycogen synthetase

STEINER and WILLIAMS, in 1959, reported the glucose-6-phosphate concentration of livers of fed rats, fasted rats and alloxan-diabetic rats. They noted that diabetic rats given insulin showed a marked increase in liver glycogen content in 7 hours whereas the glucose-6-phosphate concentration showed a small decrease and did not increase above the low level characteristic of the diabetic animal until 48 hours after the initiation of the insulin regimen. They reasoned from this that an early increase in the level or activity of glycogen synthetase might have taken place after insulin administration. In 1961, STEINER, RAUDA and WILLIAMS assayed liver homogenates (freed from nuclei and unbroken cells by centrifuging at $700 \times g$ for 10 minutes) prepared in 0.3 M sucrose for their ability to incorporate the C^{14} of UDP-glucose-C^{14} into glycogen in the presence of added rat liver glycogen. The activity was measured both in the presence and absence of glucose-6-phosphate at 6.6 mM concentration. In the absence of added glucose-6-phosphate the C^{14} incorporation rate was lower in fasted than in normal animals and was even lower in alloxan-diabetic rats; however, when glucose-6-phosphate was added, the C^{14} incorporation rate was higher in fasted than in normal rats and was markedly higher in the alloxan-diabetic animals. Administration of insulin (4 U) to diabetic rats 2 hours before sacrifice resulted in a 6- to 7-fold increase in C^{14} incorporation over the diabetic control in the absence of added glucose-6-phosphate and in only a 60% increase in the presence of glucose-6-phosphate. In 1964, STEINER and KING, using a similar but slightly modified assay system, followed the time course of the insulin-induced response in alloxan-diabetic rats. They reported an increase in C^{14} incorporation into glycogen from UDP-glucose-C^{14} when rats were killed for assay only 60 minutes after intraperitoneal injection of 5 U of insulin; this increase was evident both in the presence and absence of glucose-6-phosphate in the assay medium. When insulinization of the rats was continued up to 48 hours, the response was found to reach a maximum at 7 hours and to fall back to the control diabetic level at 24 hours. The response was prevented by puromycin and by an amount of actinomycin D which had no effect on the incorporation of leucine-1-C^{14} into total liver protein. From these results the authors concluded that hepatic glycogen synthetase is induced by a process involving messenger RNA and the synthesis of new protein on a template which has a short lifetime. It was recognized that insulin *per se* is ruled out as the inducer substance because the diabetic animals have high rather than low "synthetase" activity in their livers. The authors suggested that insulin could a) affect the concentration of an inducer metabolite present in higher than normal concentration in the liver of the diabetic animal, or b) augment or supplement the effect of this metabolite at the specific site where the synthesis of messenger RNA is initiated. Presumably when this metabolite is reduced in concentration as the result of continuing insulin action, insulin is no longer able to act upon (or together with) the metabolite to continue the induction process.

Clearly the above observations demand clarification as to a) the validity of regarding the assay in the crude homogenate as being a measure of glycogen synthetase, b) the possible relationship of the activity changes observed to the glycogen synthetase activation process proposed by LARNER (1964), and c) the significance of the actinomycin D and puromycin inhibitions.

HILZ, TARNOWSKI and AREND (1963) have studied changes in glycogen synthetase activity at early times after cortisol acetate injection in adrenal-ectomized rats. They observed a 50% increase at about 4 hours after a cortisol acetate suspension (2.5 mg/100 g) was injected intraperitoneally, a time at which the first significant increase in liver glycogen level was also observed. These authors suggest an elevated glucose 6-phosphate concentration in the liver cell as the inducer of glycogen synthetase. Unfortunately, in this preliminary communication, assurance was not given that the method of analyzing for glucose 6-phosphate (analysis was done on $HClO_4$ homogenates of "rapidly removed" liver) measured the true level of this intermediate in the living tissue.

C. Relationship to other hepatic enzymes

A challenging evaluation of the earlier work having to do with hormonal effects on enzyme-forming systems is given by TEPPERMAN and TEPPERMAN (1960). They point out the characteristic sluggishness of these changes as compared with other changes produced by the same hormones, and they say: "The fact that these effects are secondary does not mean that they are of secondary interest."

With regard to the induction of the hepatic enzymes glucokinase and glycogen synthetase, two much earlier responses to insulin by the liver have been described and will be reviewed here in later sections. One is decreased K+ efflux which occurs in a few minutes (see section III, C 1) and the other is restraint of glucose release, which occurs in 5 to 10 minutes (see section III, B 4). It seems to have been unfortunate, in view of these developments, for STEINER and KING (1964) to have said that the singular characteristics of the time course of response of glycogen synthetase (as compared with glucokinase, and the two dehydrogenases) seems to rule out the possibility that all these enzymes could be assigned to a single unit of genetic expression or "operon", and that the levels of these more slowly responding enzymes are controlled by changes in the concentration of certain key metabolites, which serve as inducers or repressors. Their argument might be used just as well to relegate the induction of glycogen synthetase to the status of a secondary effect.

FITCH and CHAIKOFF (1960), in discussing the disparate patterns of hepatic enzymes which are produced in the rat by the feeding of a diet high in fructose or glucose, respectively, after a period of maintenance on a stock diet, concluded that the hepatic enzymes increase in amount according to the use which is made of the pathways utilizing them. If induction of more rapid enzyme synthesis is the mechanism involved here, then substrates might be thought to be the obvious inducers, each substrate for its own enzyme or enzymes. On the other hand, there is another way in which a substrate might increase enzyme content. That is, by the preservation of preformed enzyme from degradation, for example, by maintenance of the enzyme in the form of a less easily degraded enzyme-substrate complex. A mechanism of this kind is suggested by recent findings of HIATT and BOJARSKI (1961), of SCHIMKE (1963), and of KNOX (1964). An increase in enzyme content brought about in this way would be dependent upon continued protein synthesis, hence would be "puromycin sensitive". Provided the destruction of the RNA template were rapid enough (TRAKATELLIS et al. 1964a and b), the increase in enzyme content would also be dependent upon continued messenger RNA synthesis, hence would be "actinomycin sensitive".

The central position which might be occupied by genetic induction as the mediator of hormonal effects on metabolism is threatened if in many instances "induction" by substrate of an increase in enzyme content as a consequence of hormone action is brought about in the way just decribed. Also, as noted previously, the attractive possibility that genetic induction or repression acts as an integrating mechanism for the metabolic effects of hormones is weakened if not a single "operon" in a tissue is affected, but a series of "operons" is affected in sequence, each by a change in the concentration of a particular metabolite.

The sections of this review which follow support the concept that most or all of the changes in hepatic enzyme content which insulin brings about are secondary in nature.

II. Effects of insulin on liver enzymatic functions related to carbohydrate metabolism

A. Hepatic lipogenesis

In 1950, Chernick, Chaikoff, Masoro and Isaeff reported that the conversion of glucose-U-C^{14} to fatty acids was impaired in liver slices prepared from alloxan-diabetic rats, and Chernick and Chaikoff (1950) reported that pretreatment of alloxan-diabetic rats with insulin restored the capacity of slices from their livers to carry out this conversion. In the same year Brady and Gurin (1950b) reported that acetate-C^{14} conversion to fatty acids was impaired in liver slices from alloxan-diabetic rats and pancreatectomized cats. In 1951, Brady, Lukens and Gurin (1951a and b) found that liver slices from hypophysectomized-pancreatectomized cats converted acetate-C^{14} to fatty acids at a normal rate; growth hormone treatment of such an animal reversed the effect of hypophysectomy, that is, it reinstituted the impairment of lipogenesis from acetate seen in the pancreatectomized cat before hypophysectomy. Adrenalectomy of the pancreatectomized cat restored to some extent the ability of liver slices from such animals to incorporate acetate-C^{14} into fatty acids.

Following these initial observations, which are discussed further in section II, A 2 below, hundreds of publications have appeared in which liver slices incubated *in vitro* have been used to study hormonal influences on hepatic lipogenesis.

1. Enzymatic processes in hepatic lipogenesis. The term lipogenesis has been used to mean the accumulation of isotope from tagged precursors into long-chain (usually triglyceride) fatty acids.

The specific enzymatic pathways of fatty acid synthesis are under active investigation. These have been reviewed recently by Wakil (1962) and by Winegrad (1964).

The mechanism whereby increased amounts of free fatty acids in the hepatic cell rapidly inhibit fatty acid synthesis is now understood, at least in part. The fact that they do has received ample documentation since the earlier evidence was reviewed by Langdon (1960) and by Fritz (1961).

Bortz and Lynen (1963) have observed competitive inhibition of purified acetyl coenzyme A carboxylase (see Fig. 1 b) *in vitro* by long-chain acyl coenzyme A derivatives, the free fatty acids themselves being without such an effect. However, the conversion of the free fatty acids to the acyl coenzyme A derivatives in the intact liver cell is rapid, so that increased nonesterified fatty acid flux from adipose tissue to the liver would automatically result in an increased formation of hepatic intracellular acyl coenzyme A derivatives.

Hill, Webster, Linazasoro and Chaikoff (1960) demonstrated a pronounced decrease in fatty acid synthesis from acetate by liver slices within

1 hour after oral administration of fat to the rat. BORTZ, ABRAHAM and CHAI-KOFF (1962) studied particle free supernates from sucrose homogenates of fat-fed rats and found an inhibition of fatty acid synthesis (less than the inhibition seen in slices) centered in the acetyl coenzyme A carboxylase step, and visible in 2 hours after fat feeding. WIELAND, WEISS, EGER-NEUFELDT and MÜLLER (1963) demonstrated decreased conversion of acetate to fatty acids by rat liver slices after 30 minutes of infusion of autologous chylomicrons in normal animals. These authors found a correlation between the total esterified

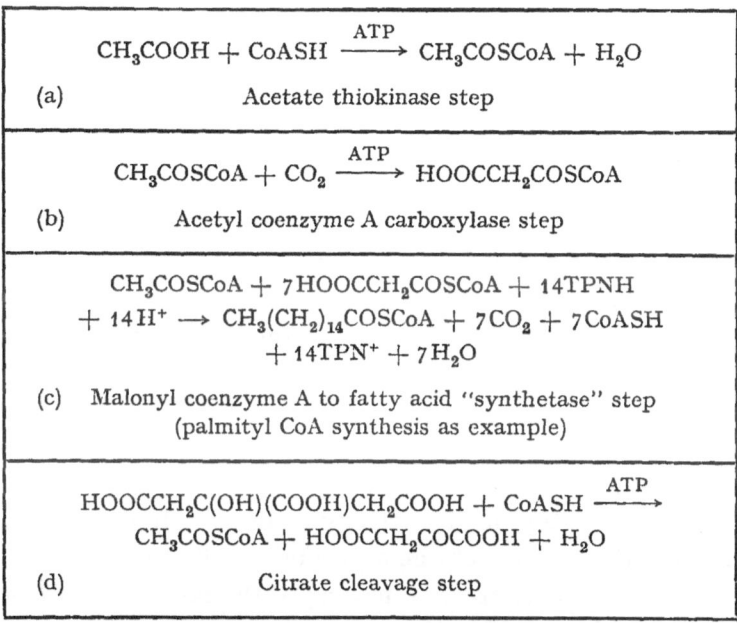

Fig. 1. Some enzymatic steps in fatty acid synthesis

fatty acid content of liver (which was increased during the chylomicron infusion) and the inhibition of fatty acid synthesis. In view of the rapid lipolysis and re-esterification of fatty acids seen in hepatic cells, and involving the triglycerides carried there as chylomicrons (see below), it is quite likely that liver free fatty acid and fatty acyl coenzyme A concentrations, as well as triglyceride concentrations, were increased during the chylomicron infusion.

GIBSON, ALLMANN and ASHMORE, as mentioned in a preliminary report (GIBSON and ALLMANN 1963), have found that acetate-C^{14} incorporation into fatty acids by liver slices from rats treated shortly before sacrifice with anti-insulin serum is greatly impaired, whereas the cell-free fatty acid synthesizing system prepared from the same liver tissue is fully competent. In these experiments, non-esterified fatty acids transferred in increased amounts from adipose tissue to liver during the *in vivo* portion of the experiment may have been responsible for decreased lipogenesis in the slices and hence the impaired acetate incorporation.

Changes in specific hepatic enzyme activities concerned with lipogenesis, which are probably due to changes in enzyme amounts but which usually have not been proven to be so, have been observed in situations in which prolonged increases in the flux of fatty acids to the liver are known to occur.

NUMA, MATSUHASHI and LYNEN (1961) have studied the activities of enzymes present in particle-free supernatant solutions of sucrose homogenates of livers of 2-day-fasted rats, and WIELAND, NEUFELDT, NUMA and LYNEN (1963) have studied alloxan-diabetic rats in a similar way. In the supernates so prepared from livers of fasted animals, the enzymes acetate thiokinase, malonyl coenzyme A to fatty acid "synthetase", and isocitrate dehydrogenase were found lower than normal in activity but not rate-limiting, whereas acetyl coenzyme A carboxylase was extremely low and was proven to be the rate-limiting enzyme for the overall synthesis of fatty acid from acetate. The steps in fatty acid synthesis just referred to are shown in Fig. 1. In the supernates from the livers of diabetic animals, also, acetyl coenzyme A carboxylase activity was low, and the addition of purified enzyme to the supernates restored full synthesis of fatty acid from acetate.

SPENCER and LOWENSTEIN (1962) have postulated that citrate, formed intra-mitochondrially from either carbohydrate (*via* pyruvate) precursors or two-carbon precursors of fatty acids, is transported outside the mitochondria where it is converted to acetyl coenzyme A and oxaloacetate by the operation of the citrate cleavage enzyme (see Fig. 1 d). KORNACKER and LOWENSTEIN (1964) found a marked reduction in the activity of the citrate cleavage enzyme in supernatant fractions of sucrose homogenates of alloxan-diabetic rat livers; either insulin treatment or fructose feeding of the rat brought about a large increase in citrate cleavage enzyme activity.

The inducer (or inducers) for increased acetyl coenzyme A carboxylase and citrate cleavage enzyme activities is unknown. KORNACKER and LOWEN-STEIN suggest that the inducer is hepatic cell α-glycerophosphate. Since α-glycerophosphate is utilized in the conversion of fatty acids to triglycerides, fructose (or insulin) might increase the concentration of this metabolite in the hepatic cell by reducing the flux of non-esterified fatty acids from adipose tissue to the liver for re-esterification.

At one time it was believed that the limited availability of the reduced form of triphosphopyridine nucleotide (TPNH or NADPH) was the controlling factor for lipogenesis in liver in both diabetic and fasted animals. (For recent evidence contra-indicating this see GORDON 1963.) This belief stimulated interest in the hepatic levels of glucose-6-phosphate dehydrogenase and 6-phosphogluconate dehydrogenase (both of which produce TPNH from TPN^+) under conditions in which altered rates of lipogenesis were occurring, since it was evident that high levels of these enzymes were usually associated with high rates of lipogenesis. The evidence that enzyme induction is involved in these changes in enzyme levels has been weighed by TEPPERMAN and TEPPERMAN (1963) who considered that their own data and the data published by POTTER and ONO (1961), and by WEBER, BANERJEE and BRONSTEIN (1962), support the tentative conclusion that enzyme induction is involved. The changes in the dehydrogenase levels in these experiments were brought about by changes in dietary intake, and the specific identity of the inducer is not known; the experiments of TEPPERMAN and TEPPERMAN (1963) suggest that induction is not caused by a high level of hepatic glucose 6-phosphate.

COHN and JOSEPH (1959) showed that a high carbohydrate diet when force-fed (in two feedings per day) resulted in higher levels of these dehydrogenases than the same amount of the same diet fed *ad libitum;* the difference in the challenge to the two groups lay in the fact that the *ad libitum* fed animals nibbled at the diet whereas the force-fed animals received the diet in two definite meals. It is certainly not to be expected that endogenous insulin secretion in the "nibblers" was inadequate, nor even that total daily endogenous insulin secretion in the "nibblers" was less than in the "meal eaters". As in the case of low affinity glucokinase discussed above (see section I, A), it may be inferred that insulin is probably not the sole requirement for the induction of these dehydrogenases, although the presence of insulin may be necessary in order for the liver cell to assume responsiveness to the inducer or inducers.

2. **Hormonal control of free fatty acid release by adipose tissue, and its relation to hepatic lipogenic capacity.** If the idea be accepted that hepatic enzymes may be immediately lowered in activity and eventually decreased in amount as a result of increased amounts of fatty acid substrate presented to the liver, the relationship between the hormonal control of fatty acid release by adipose tissue (or, in older terminology, of "fat mobilization") and the problem of hormonal effects on hepatic lipogenesis becomes apparent.

In 1963, JUNGAS and BALL demonstrated a new action of insulin on the release of non-esterified fatty acids by rat epididymal adipose tissue. By the older known mechanism, insulin is thought to promote re-esterification of the fatty acids released by lipolysis; it does this by making α-glycerophosphate available as a result of increased uptake of precursor glucose in the presence of insulin; the glycerol produced by lipolysis is unavailable, presumably because it cannot be phosphorylated in adipose tissue, and the glycerol escapes to the blood. The experiments establishing these relationships have been reviewed by FRITZ (1961). JUNGAS and BALL observed that *in vitro* in the absence of glucose in the medium (absence of glucose in the absence of insulin has little effect on the lipolytic action of epinephrine) insulin has a powerful anti-lipolytic effect opposing the release from the tissue of glycerol and non-esterified fatty acids otherwise stimulated by epinephrine or glucagon. These findings are in line with the observations of PERRY and BOWEN published in 1962, that growth hormone-, adrenocorticotropin-, or epinephrine-stimulated increased nonesterified fatty acid production by adipose tissue *in vitro* is inhibited by the presence of insulin in the medium in the absence of glucose. The involvement of cyclic 3',5'-adenosine monophosphate in epinephrine- and adrenocortico-tropin-induced lipolysis in adipose tissue is suggested by the findings of RIZACK (1964) and of BUTCHER, HO, MENG and SUTHERLAND (1964).

Soon after the significance of the non-esterified fatty acids (NEFA) of plasma was discovered, it became evident that insulin is a most important hormone in inhibiting their release by adipose tissue. BIERMAN, DOLE and ROBERTS proposed in 1958 that a major mechanism in precipitating diabetic acidosis in human subjects is "failure in the control of NEFA release from adipose tissue, leading to a marked rise in plasma level, an overloading of liver with fatty acid substrate, and excessive production of ketone acids".

TARRANT, MAHLER and ASHMORE (1964) have studied adipose tissue taken from mice made acutely insulin-deficient by the injection of anti-insulin serum. Insulin added *in vitro* to such adipose tissue reversed the high rate of non-esterified fatty acid release brought about by the anti-insulin serum. Since the high rate of non-esterified fatty acid release was not accompanied by decreased glucose uptake by the adipose tissue *in vitro*, and since insulin added *in vitro* significantly reduced the rate of glycerol release by the adipose tissue, it was concluded that the free fatty acid mobilization in acute insulin deficiency was due to the absence of the normal restraining action of insulin on hormones promoting adipose tissue lipolysis, rather than to decreased glucose uptake by adipose tissue. In agreement with the thought that a factor other than deficient glucose uptake by adipose tissue is the cause of the increased non-esterified fatty acid release by adipose tissue in diabetes, CARLSON and ÖSTMAN (1963) found that when subcutaneous adipose tissue from human normal and diabetic subjects was incubated in each subject's own plasma, the diabetic fat showed a significantly higher rate of glycerol release in spite of a higher glucose uptake at the elevated (diabetic) plasma glucose concentration. When incubated in buffered solution, containing 20 mg human albumin and 1 mg glucose per milliliter, the diabetic fat, although not significantly different from normal fat in glucose uptake, showed a significantly higher rate of non-esterified fatty acid release.

DE BODO and ALTSZULER (1957), LANGDON (1960), WERTHEIMER and SHAFRIR (1960) and WINEGRAD (1962) have reviewed the evidence for the participation of growth hormone, thyroid hormone, the adrenal glucocorticoids, epinephrine, norepinephrine and glucagon in promoting fat mobilization.

SCOW and CHERNICK proposed in 1960 that the development of ketosis, fatty liver and hyperlipemia in the diabetic rat is the result of a direct action of the glucocorticoids on adipose tissue. JEANRENAUD and RENOLD (1960) reported in the same year that cortisol or corticosterone when present in the incubation medium increased the release of non-esterified fatty acids by rat adipose tissue *in vitro*. MUNCK reported (1961 a) that cortisol injection in adrenalectomized rats led within 1.5 hr to reduced glucose uptake by adipose tissue removed from the rats and incubated *in vitro*. The same author demonstrated (MUNCK 1961 b) that cortisol or corticosterone, added *in vitro*, decreased glucose uptake by adipose tissue from adrenalectomized rats over the 0 to 2.5 hr interval of incubation and decreased it further on longer incubation; he suggested that the decrease in glucose uptake might affect other metabolic pathways, such as those leading to the increase in the release of non-esterified fatty acids. In 1962 MUNCK reported that adipose tissue from normal and from hypophysectomized-adrenalectomized rats, as well as from adrenalectomized rats, showed decreased glucose uptake *in vitro* if cortisol or corticosterone had been injected *in vivo* 30 minutes or more prior to sacrifice; the same steroids, when added to the incubation medium at as low as 10^{-7} M concentration, decreased glucose uptake by adipose tissue taken from alloxan-diabetic adrenalectomized rats as well as by adipose tissue from adrenalectomized rats. In the same year LEBOEUF, RENOLD and CAHILL (1962) reported that glucose uptake, and the contribution of medium glucose-C^{14} to CO_2, glyceride glycerol, tissue fatty acids and glycogen were all diminished in normal rat adipose tissue incubated for 3 hrs *in vitro* when cortisol (30 μg/ml) was present in the medium. FAIN, SCOW and CHERNICK in 1963 correlated the effects of a synthetic glucocorticoid, dexamethasone (9 α-fluoro-11 β, 17α, 21-trihydroxy-16α-methyl-1,4-pregnadiene-3,20-dione), added *in vitro* at 4×10^{-8} M concentration (0.016 μg/ml), on glucose uptake and fatty acid release by normal rat adipose tissue. In spite of some discrepancies (for example when dexamethasone was present only during hours 2 to 4 of incubation it increased fatty acid release without significantly decreasing glucose uptake) the authors concluded that the results in general supported the concept that increased fatty acid release by adipose tissue is secondary to decreased glucose metabolism brought about by the glucocorticoids. The reverse of this, i.e. that a decrease in glucose uptake is a result of increased free fatty acid concentration, has been found in the case of muscle tissue (see section IV, A). This relationship is not the governing one in adipose tissue, apparently, inasmuch as the other agents which increase free fatty acid release by adipose tissue *in vitro*, such as epinephrine or ACTH, cause increased rather than decreased glucose uptake. Part of the hormonal control of fatty acid release by adipose tissue is exerted through a mechanism other than fatty acid re-esterification promoted by glucose uptake, as has been emphasized by the work of JUNGAS and BALL (1963) reviewed earlier in this section. For this reason the way in which the glucocorticoids cause increased fatty acid release seems to be open to reinterpretation after further investigation.

Earlier evidence regarding a permissive action of the glucocorticoids, an action sensitizing adipose tissue to the fat-mobilizing action of epinephrine has been reviewed by Ramey and Goldstein (1957). This evidence has since been supplemented by the work of Shafrir and Steinberg (1960) and Reshef and Shapiro (1960).

Fain and Wilhelmi (1962) recently have studied fatty acid synthesis in the whole rat by measuring the incorporation of tritium from tagged body water into fatty acids. In these experiments it was found that a single injection of growth hormone in normal or hypophysectomized rats inhibited tritium incorporation into fatty acids during the 6- to 14-hour period; however, when growth hormone was given for 4 days prior to the incorporation study it had no effect on tritium incorporation. In view of the inhibition of hepatic fatty acid synthesis produced by an increase in plasma non-esterified fatty acid concentration, these results may be related to the fact that in the rat (Franklin and Knobil 1961), as in the dog (Winkler, Steele, Altszuler and de Bodo 1964), elevated plasma non-esterified fatty acid levels are seen early after growth hormone injection, but disappear after a few days.

Goodman and Knobil (1959) have presented evidence for the importance of thyroid hormones in sensitizing adipose tissue to the fat-mobilizing action of epinephrine. Hypophysectomized monkeys fail to show elevated plasma non-esterified fatty acid levels in response to epinephrine injection. Pretreatment with cortisol or prolactin did not restore the response, but pretreatment with either thyrotropin or triiodothyronine did. Pretreatment with monkey growth hormone (1 mg/kg per day for 4 days) had a small effect which was attributed to its known contamination with thyrotropin.

The response of fat mobilization to growth hormone and the glucocorticoids in an animal with an intact pancreas are confused by increases simultaneously induced in the rate of endogenous insulin secretion. Growth hormone injection increases plasma insulin levels as shown directly by Campbell and Rastogi (1964), and as inferred from indirect evidence by de Bodo et al. (1963 b). The evidence for the influence of the glucocorticoids to increase insulin secretion has been summarized by Fajans (1961) and Conn and Fajans (1956). Additional evidence in this regard has been furnished by Hausberger (cf. Hausberger and Ramsay 1959). Hypophysectomy, conversely, leads to a decreased rate of endogenous insulin secretion which may cancel out, in part, the decreased fat mobilization which the absences of adrenocorticotropin, thyrotropin and growth hormone are tending to bring about. Rats kept on a high carbohydrate diet, in the experiments of Fain and Wilhelmi (1962) reported above, were susceptible for a longer time to the inhibitory action of growth hormone on tritium incorporation into fatty acids. This might be interpreted as due to an increased demand for insulin caused by the consumption of the high carbohydrate diet, a demand sufficient to exhaust the ability of an overtaxed pancreas to secrete enough insulin to counter the effect of growth hormone in the between-feeding periods.

In 1951 CHERNICK and CHAIKOFF reported that liver slices prepared from alloxan-diabetic rats are impaired in their ability to convert glucose-C^{14} to $C^{14}O_2$ whereas they are not impaired in their ability to convert fructose-C^{14} to $C^{14}O_2$. The impairment in glucose metabolism may now possibly be explained on the basis of the decreased content of low-affinity glucokinase in the liver of the diabetic rat. The second impairment in carbohydrate utilization discerned by CHERNICK and CHAIKOFF in the livers of alloxan-diabetic rats was in the conversion of hexose-C^{14} to fatty acids; separate studies showed that lipogenesis from other precursors was also impaired. BAKER, CHAIKOFF and SCHUSDEK showed, in 1952, that fructose feeding of the diabetic animal led to a normal rate of fatty acid synthesis from acetate-C^{14} or lactate-C^{14} by slices prepared from its liver, without correcting the impairment in conversion of glucose-C^{14} to $C^{14}O_2$ or fatty acids.

Extensive investigations of the changes in hepatic enzyme activities which result from the maintenance of rats on high glucose diets and high fructose diets, respectively, have been summarized by FITCH and CHAIKOFF (1960).

A relationship between the above observations and the rate of release of non-esterified fatty acids from adipose tissue may be suspected. Fructose is utilized well by adipose tissue (FROESCH and GINSBERG 1962), and is as effective as glucose in inhibiting the release of non-esterified fatty acids from adipose tissue *in vitro* (PERRY and TJADEN 1962). Infused intravenously, fructose lowers the elevated plasma non-esterified fatty acid level of a diabetic human when glucose infused in like amount fails to do so (GORDON 1958). Thus there is an alternative to the thought that it is entirely the uptake of fructose by hepatic cells which corrects the defective hepatic lipogenesis of the alloxan-diabetic rat.

There is a troublesome point in the usual explanation of how insulin administration, like fructose feeding, corrects the defect in hepatic lipogenesis. This explanation is that insulin, by increasing glucose uptake by the liver, supplies the same intracellular metabolite which fructose supplies in the absence of insulin. The troublesome point is that the liver of the diabetic animal, although it may have a lowered steady-state glucose 6-phosphate concentration (STEINER and WILLIAMS 1959), has a supernormal rate of production of glucose 6-phosphate, which is also the first product of glucose phosphorylation. It is therefore not clear how insulin-induced increased hepatic glucose uptake could be responsible, in the *in vivo* situation, for the production of a metabolite, low or missing in the diabetic state, which is required for hepatic lipogenic enzyme induction.

The situation would be clarified if both fructose feeding and insulin were to be found to correct the defect in hepatic lipogenic capacity by decreasing the flux of fatty acids from adipose tissue to the liver.

The changes in hepatic lipogenic capacity demonstrated by BRADY, LUKENS and GURIN (1951a and b) by the use of liver slices of untreated and growth hormone-treated hypophysectomized-pancreatectomized cats and of non-adrenalectomized and adrenalectomized depancreatized cats are perhaps to be understood on the basis of the effects of growth hormone and the gluco-corticoids to condition adipose tissue for increased lipolysis and the effect of insulin to oppose lipolysis and promote fatty acid re-esterification. Later experiments by others along these lines, using rats, gave results which were not always as clear cut. Some of the difficulties may be attributed to the compensatory changes in insulin secretion brought about by an excess or deficit of growth hormone in the animal with an intact pancreas. Perry and Bowen (1955), and Allen, Medes and Weinhouse (1956) found only a slight impairment in lipogenesis from acetate by liver slices from normal rats given growth hormone; Greenbaum and Glascock (1957) obtained greater effects.

Hill, Bauman and Chaikoff (1955) found that maintenance of hypophysectomized rats on a 60% starch or glucose diet corrected what Baruch and Chaikoff (1955) had found to be a deficient, rather than an increased, rate of lipogenesis from acetate by liver slices from such rats; however, lipogenesis from glucose or fructose was still impaired. Later, Nejad, Chaikoff and Hill (1962) reported that adrenocorticotropin or thyroxine, or prolactin-plus-growth hormone could correct the latter defect. The overlapping of these effects suggests a common mechanism, i.e. the restoration of sufficient endogenous insulin secretion. Fain and Wilhelmi (1962) found that fatty acid synthesis *in vivo* was also depressed in hypophysectomized rats and could be restored to normal by thyroxine treatment.

Bauman, Hill and Chaikoff (1957) found that hypophysectomy of the alloxan-diabetic rat increased lipogenesis from acetate by liver slices only if the animals were kept on a 60% glucose diet, and Spiro (1958) reported that, in the alloxan-diabetic rat kept on a stock diet, hypophysectomy caused only a small increase in lipogenesis from fructose by liver slices. Aside from the species difference, these animals differed from the surgically depancreatized cats of Brady, Lukens and Gurin (1951b) in having a continued supply of pancreatic glucagon.

3. **Rapid effects of insulin on hepatic lipogenesis.** Whether or not liver, besides adipose tissue, demonstrates the antagonistic action of insulin and lipolytic agents on triglyceride breakdown is not yet known. The rapid breakdown of plasma triglycerides taken up by liver, and the re-esterification of the free fatty acids formed, has been shown by Stein and Shapiro (1960), Olivecrona (1962) and Chernick and Scow (1964).

If liver does respond in this way, light may be thrown on early experiments in which insulin added *in vitro* was shown to promote lipogenesis in liver slices, since it is now clear that an increased amount of hepatic intracellular free fatty acid has a prompt inhibitory action on lipogenesis.

BLOCH and KRAMER, in 1948, reported that insulin increased acetate-1-C^{14} incorporation into fatty acids by liver slices from normal rats incubated with unlabeled pyruvate as substrate. BRADY and GURIN (1950a) found insulin active on such slices when acetate-C^{14} was the sole added substrate. MASRI, LYON and CHAIKOFF, in 1952, reported that insulin, in the presence of glucose substrate, increased lipogenesis from acetate-C^{14} by liver slices from 18-hour-fasted rats.

BRADY, LUKENS and GURIN (1951a and b) found that liver slices from hypophysectomized-pancreatectomized cats, in contrast with pancreatectomized cats, converted acetate to fatty acids at a normal rate; growth hormone treatment of the hypophysectomized-pancreatectomized cat reversed the effect of hypophysectomy. Insulin added *in vitro* was effective in increasing acetate incorporation in liver slices from the hypophysectomized-pancreatectomized cat, but not in liver slices from such animals after they were previously treated with growth hormone. Adrenalectomy of the pancreatectomized cat restored to some extent the ability of liver slices to incorporate acetate into fatty acids; insulin added *in vitro* was ineffective in increasing acetate incorporation in these slices. Liver slices from hypophysectomized rats incorporated acetate into fatty acids at a greater than normal rate; insulin did not increase the incorporation further in such slices.

In 1953, HAUGAARD and STADIE, using liver slices from fed normal rats, reported that glucagon-free insulin added *in vitro* stimulated the incorporation of acetate-C^{14}, present as the sole substrate, into fatty acids; ordinary insulin, which contained glucagon, was not effective, and glucagon or epinephrine added *in vitro* was shown to inhibit acetate incorporation in this system. HAUGAARD and HAUGAARD, in 1954, extended these observations to include an inhibitory effect of glucagon on incorporation of the C^{14} of glucose-U-C^{14} and fructose-U-C^{14} into fatty acids; the net ketone body release to the medium by similar slices, in the absence of substrate, was increased by glucagon added *in vitro*.

HAFT and MILLER (1958 a and b), using perfused livers from fasted (48-hour) normal and fed alloxan-diabetic rats in which insulin treatment had been withheld for 40 hours, found that very high concentrations of glucose or fructose in the perfusion medium, which increased the total net sugar uptake by the perfused livers, did not increase fatty acid synthesis from acetate-1-C^{14}. In contrast, the inclusion of insulin in the perfusing medium stimulated fatty acid synthesis from acetate-1-C^{14} in livers from both normal and nonketotic diabetic donors. In these experiments the perfusing medium both in control and in added insulin trials contained about 350 mg glucose per 100 ml. These effects were not obtained when excessive trauma accompanied the removal of livers for perfusion.

WILLIAMS, HILL and CHAIKOFF, in 1960, infused large amounts of a) glucose alone or b) glucose and insulin into the portal veins (duodenal portion of small intestine exteriorized under anesthesia) of alloxan-diabetic rats which had been deprived of maintenance insulin for 14 hours. Liver slices prepared from these rats after 60 minutes or more of glucose and insulin infusion showed higher rates of *in vitro* incorporation of acetate-1-C^{14} into fatty acids than liver slices from rats perfused with glucose alone, in which blood sugar levels were higher. In general, this outcome is in accord with the findings of Haft and Miller given above, but the interpretation of the finding is complicated by the possibility that repression by insulin of non-esterified fatty acid release by adipose tissue may have played a part in stimulating lipogenesis in the experiments in which insulin was infused in the intact rat.

4. Summary. Rapid changes in hepatic lipogenesis are brought about by changes in the amount of non-esterified fatty acid or triglyceride transported to the hepatic cells. An enzyme of importance in lipogenesis is inhibited by the long-chain acyl coenzyme A derivatives which increase in amount in the hepatic cell when the amount of non-esterified fatty acids in the liver is increased. The early effects of acute insulin deprivation, and of fat administration, to inhibit hepatic lipogenesis cannot be explained on the basis of a loss of hepatic enzymes for lipogenesis.

In the intact animal in the postabsorptive state, adipose tissue is the principal source of the non-esterified fatty acids which are transported to the liver. A number of hormones promote lipolysis in adipose tissue, whereas insulin opposes lipolysis. Insulin opposes lipolysis directly in a way not dependent upon increased glucose uptake. In addition, insulin promotes the re-esterification of free fatty acids in adipose tissue with a consequent reduction in the amount of non-esterified fatty acids (but not of glycerol) released to the blood. In the animal with an intact pancreas, many of the hormones which enhance the release of non-esterified fatty acids by adipose tissue also cause an increase in the rate of insulin secretion, which eventually tends to counter adipose tissue lipolysis.

In addition to its rapid effects to enhance ketone body formation and inhibit fatty acid synthesis in the liver, an increased flux of non-esterified fatty acids to the liver has a long-term effect to decrease the amounts of enzymes concerned with fatty acid synthesis. The induction of these enzymes as a result of insulin action may be attributed to one or more hepatic cell metabolites which are increased in concentration as a result of a lessened flux of non-esterified fatty acids to the liver; α-glycerophosphate, which is used up in triglyceride synthesis, has been suggested in this connection. Fructose feeding, which results in the induction of hepatic enzymes for lipogenesis in the absence of insulin, may do so indirectly by promoting the re-esterification of fatty acids in adipose tissue and so preventing their release for transport to the liver. The decreased capacity of liver tissue for lipogenesis from glucose (after fructose feeding of the diabetic animal has corrected the deficient lipogenesis from other precursors) may be ascribed to the still lowered level of low-affinity glucokinase.

Hormones such as growth hormone which promote the release of non-esterified fatty acids from adipose tissue and also eventually cause an increased rate of insulin secretion have an uncertain effect on the lipogenic capacity of the liver; in such cases the effects of these hormones to give rise to an impaired hepatic lipogenic capacity may be reversed.

Older findings of a direct effect of insulin added *in vitro* (an effect antagonized by glucagon or epinephrine added *in vitro*) to increase acetate incorporation into fatty acids by liver slices incubated in the absence of glucose may take on new significance in the light of the findings in adipose tissue of the antagonistic actions of insulin on the one hand, and glucagon and epinephrine on the other, on the process of lipolysis in the absence of glucose. If insulin and the lipolytic agents have the same effects in the liver cell as in adipose tissue, the intracellular concentration of non-esterified fatty acids (and of acyl coenzyme A derivatives) in the liver cell, as determined by the rate of breakdown of hepatic cell triglycerides, may be responsible for these *in vitro* effects.

The major impact of the newer concepts just summarized is to offer alternatives to the thought that insulin promotes hepatic lipogenesis solely as a secondary consequence of its effect to increase the uptake and utilization of blood glucose.

B. Hepatic gluconeogenesis

The liver of the diabetic animal synthesizes and releases a greater than normal amount of free glucose. It also converts a greater than normal amount of amino acid nitrogen to urea nitrogen. When insulin decreases the amounts of the hepatic enzymes concerned with these processes, its action must be mediated by "repressors" or by a decrease in the amounts of "inducers".

1. Enzymatic processes in gluconeogenesis. The pathways involved in gluconeogenesis have been described recently by KREBS (1963) who points out that, aside from transaminative and deaminative steps, these pathways are identical for gluconeogenesis from amino acids and from substances such as lactate, pyruvate, and the intermediates of the tricarboxylic acid cycle, and involve critical steps which are not simple reversals of glycolysis.

In the fasting animal an important source of carbon for a long-term increase in hepatic gluconeogenesis is undoubtedly the glucogenic amino acids. On the other hand, an increase in hepatic gluconeogenesis may also result from the diversion toward gluconeogenesis of non-nitrogenous intermediates, such as lactate, pyruvate, or glycerol, which would otherwise suffer a different metabolic fate after being transported to the liver.

As emphasized by KREBS, the direct net conversion of pyruvate to phosphopyruvate does not occur in the cell; two pathways are known for the indirect conversion, both involving CO_2 fixation and both having oxaloacetate as an intermediate. The first pathway involves the malic enzyme; the malate formed by this enzyme from pyruvate and CO_2 is converted to oxaloacetate by a dehydrogenase (see Fig. 2a). The second pathway involves pyruvate carboxylase, which forms oxaloacetate directly from pyruvate and CO_2 (see Fig. 2b). In both pathways the oxaloacetate formed is converted to phospho-pyruvate and CO_2 by the enzyme phosphopyruvate carboxykinase (see Fig. 2c).

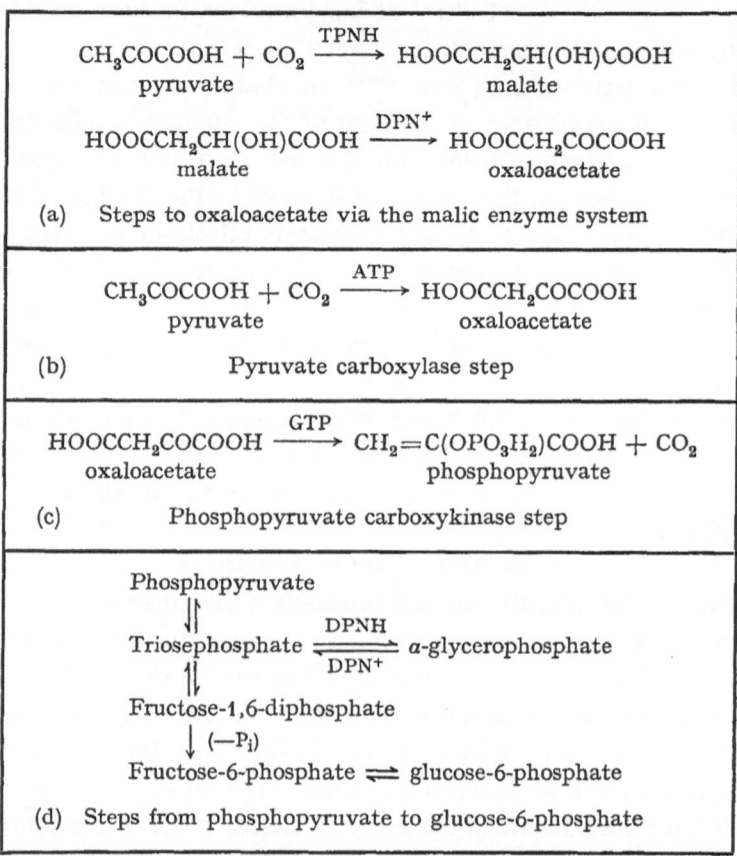

$$CH_3COCOOH + CO_2 \xrightarrow{\text{TPNH}} HOOCCH_2CH(OH)COOH$$
pyruvate   malate

$$HOOCCH_2CH(OH)COOH \xrightarrow{\text{DPN}^+} HOOCCH_2COCOOH$$
malate   oxaloacetate

(a) Steps to oxaloacetate via the malic enzyme system

$$CH_3COCOOH + CO_2 \xrightarrow{\text{ATP}} HOOCCH_2COCOOH$$
pyruvate   oxaloacetate

(b) Pyruvate carboxylase step

$$HOOCCH_2COCOOH \xrightarrow{\text{GTP}} CH_2{=}C(OPO_3H_2)COOH + CO_2$$
oxaloacetate   phosphopyruvate

(c) Phosphopyruvate carboxykinase step

Phosphopyruvate
$\updownarrow$
Triosephosphate $\underset{\text{DPN}^+}{\overset{\text{DPNH}}{\rightleftharpoons}}$ α-glycerophosphate
$\updownarrow$
Fructose-1,6-diphosphate
$\downarrow$ $(-P_i)$
Fructose-6-phosphate $\rightleftharpoons$ glucose-6-phosphate

(d) Steps from phosphopyruvate to glucose-6-phosphate

Fig. 2. Some enzymatic steps in gluconeogenesis

Krebs, Dierks and Gascoyne (1964) have recently described a cell-free preparation from pigeon liver which is capable of forming carbohydrate from L-lactate. Krebs, Newsholme, Speake, Gascoyne and Lund (1964) have discussed the utilization of this preparation for the study of the control of hepatic gluconeogenesis. The activity of the hepatic enzyme fructose, 1,6-diphosphatase is lowered by adenosine monophosphate and also by a high concentration of its own substrate and this limits the production of fructose 6-phosphate in gluconeogenesis. Precursors of gluconeogenesis, such as lactate, malate, succinate and the α-keto acids arising from amino acids, besides sup-plying carbon for gluconeogenesis are also readily oxidized and so do two things: a) they supply reducing power for the conversion of triosephosphates to α-glycerophosphate, and so remove the inhibiting excess of fructose, 1,6-

diphosphate, since this compound is in reversible equilibrium with the triose-phosphates (see Fig. 2d, and b) they result in the conversion of adenosine monophosphate to adenosine triphosphate, which further unblocks the fructose 1,6-diphosphatase reaction and allows gluconeogenesis to proceed from the gluconeogenic precursors.

In addition, UTTER and KEECH (1960) have shown that pyruvate carboxylase has an absolute requirement for acetyl coenzyme A for the expression of its enzymatic activity. UTTER, KEECH and SCRUTTON (1964) have recently summarized their studies of the mechanism of this activation. WIELAND and WEISS (1963) have noted that the acetyl coenzyme A content of liver is increased in alloxan-diabetic rats from which insulin maintenance has been withdrawn for 24 to 72 hours. Thus an increased flux of fatty acids to the liver, resulting in higher hepatic acetyl coenzyme A concentration, in addition to inhibiting lipogenesis as discussed above, also may activate gluconeogenesis without contributing carbon to gluconeogenesis.

Very recently HAYNES (1964) has presented evidence for the stimulating effect of a synthetic glucocorticoid (triamcinolone: 1,4-pregnadiene-9α-fluoro-11β, 16α, 17α, 21-tetrol-3,20-dione) added in vitro on glucose production from pyruvate by liver slices from fasted adrenalectomized rats. This work evolved from earlier work of the same author which had shown that in the presence of alanine at substrate concentrations, triamcinolone, and natural glucocorticoids as well, stimulated glucose synthesis by such slices. The successful demonstration of increased glucose synthesis with pyruvate as substrate was found to be possible only at a low concentration of either pyruvate or bicarbonate, which was explained by the author as being due to the sensitivity of pyruvate carboxylase and not of the malic enzyme to the steroid action. When both bicarbonate and pyruvate were present in excess, the malic enzyme (low substrate affinity) catalyzed reaction was supposed to be driven fast enough to obscure any steroid effect on the pyruvate carboxylase reaction. Either reaction may intermediate in the synthesis of glucose from pyruvate by way of oxaloacetate as described above.

HENNING, HUTH and SEUBERT (1964) have also reported an in vitro activation of pyruvate carboxylase by a glucocorticoid. In their work, rat kidnex cortex slices were incubated for 1 hour with cortisol. Pyruvate carboxylase activity was increased by about 30% and glucose synthesis from added pyruvate by about 70%.

The above findings of an effect of glucocorticoids on pyruvate carboxylase activity are too recent to be properly evaluated. In particular it is not known whether or not an increase in acetyl coenzyme A concentration precedes the increase in pyruvate carboxylase activity.

2. Hormonal control of the flow of amino acids to the liver. Insulin, added in vitro under conditions which make it very unlikely that the effect is

secondary to increased glucose uptake, stimulates the incorporation of tagged amino acids into protein by diaphragm muscle, perfused heart muscle and bone marrow. For a review of this matter and a discussion of its relationship to the situation in other tissues in which insulin will not produce this effect except in the presence of oxidizable substrate see WOOL (1964). WOOL and WEINSHELBAUM (1959 and 1960) reported that adrenalectomy of the rat prior to removal of diaphragm tissue for study *in vitro* increased the incorporation of amino acids into protein. Cortisone treatment of either the normal or the adrenalectomized rat reduced the amino acid incorporation of the diaphragm studied later *in vitro*. SHIMIZU and KAPLAN (1964) have recently confirmed the inhibition of amino acid incorporation brought about by the prior administration of cortisone to the adrenalectomized rat.

MANCHESTER and YOUNG (1961), who together with WOOL developed the earlier observations of KRAHL (1952) and of SINEX, MACMULLEN and HASTINGS (1952) on the effect of insulin on amino acid incorporation in diaphragm, have reviewed the hormonal control of protein metabolism in the whole animal from the point of view that insulin, by promoting protein synthesis, limits the supply of amino acids available for gluconeogenesis. Growth hormone is postulated to fit into the picture by increasing the sensitivity of the tissues to the action of insulin to promote protein synthesis and possibly also by increasing insulin secretion. The glucocorticoids are postulated to fit into the picture by decreasing the basal level of protein synthesis and depressing the action of insulin to further protein synthesis. Thus the continuing insulin supply of the normal animal, by promoting protein synthesis, prevents the excessive nitrogen loss seen in the diabetic animal.

It is to be noted that the increased protein synthesis postulated to prevent nitrogen loss is considered to occur mostly in the extrahepatic tissues. As free amino acids appear in extrahepatic tissues as the result of the natural rate of protein turnover there, these are considered to be resynthesized to a greater extent into protein *in situ*, under the action of insulin, rather than being released from the cells in which they arise into the blood for transport. The adrenal glucocorticoids are considered to antagonize this action of insulin by interfering with the process of protein resynthesis, rather than by increasing the rate of protein degradation in the extrahepatic tissues. A somewhat similar viewpoint was taken by GLENN, MILLER and SCHLAGEL (1963), with the large difference that these authors attributed the inhibitory action of the glucocorticoids on protein synthesis to an inhibition of glucose utilization.

This picture does not seem to be complete in view of some older findings. For example, DOUGHERTY and WHITE demonstrated, in 1945, that the dissolution of lymphocytes begins within an hour after the administration of glucocorticoids to mice or rabbits. As reported by BLECHER and WHITE (1959), this effect was not correlated with a specific inhibition by these steroids, as distinct from other steroids, of protein synthesis in lymphocytes studied *in vitro*, nor were the inhibitory effects of the steroids on lymphocyte metabolism confined to inhibition of protein synthesis.

In 1962, SACHS, DE DUVE, DVORKIN and WHITE reported promising results in correlating thymus involution after cortisol treatment of rats *in vivo* with the increase in thymus tissue of cathepsin, aryl-sulfatase and β-glucuronidase activities. This was postulated to represent the intracellular release of these enzymes from the lysosomes of the lymphocytes, together with the selective loss of nonenzymic protein as a consequence of the intracellular action of the released lysosomal enzymes. However, in a later discussion, WHITE (1964) stated that the observed increases in enzyme activities had subsequently been found to be due to an influx of phagocytic cells into the thymus to clean up cellular debris left there after thymus involution had already taken place, rather than being a cause of the thymus involution.

ENGEL showed, in 1952, that epinephrine injection in cortisone-maintained adrenalectomized rats increases urea production in a situation in which no increased amount of glucocorticoid occurs. Epinephrine was known not to be the only factor which could operate in this way, since INGLE and his collaborators had shown in 1947 that stress could cause nitrogen loss in the adrenalectomized rat maintained on adrenal cortical extract in a situation in which the main source of epinephrine was absent. ENGEL postulated that increased amounts of glucocorticoids led to increased sensitivity of the rat to the protein catabolic influence of factors such as epinephrine, so that a minor stress such as fasting would lead to excessive nitrogen loss in the fasting cortisone-treated animal as compared with the untreated animal, thus giving the impression that the injected cortisone was directly causing increased protein catabolism. Comments are currently made (LEVINE 1963, MANCHESTER and YOUNG 1961), on the basis of this and related evidence (see ELLIS 1956), that epinephrine increases protein catabolism.

After several publications had appeared indicating that glucagon increases protein catabolism, SALTER, DAVIDSON and BEST, in 1957, reported that rats force-fed a high carbohydrate diet showed continuous glucosuria over a 5-day-period, and a doubled urinary nitrogen excretion, when given 1.2 mg glucagon per day by subcutaneous injections spaced at 6-hour intervals. Similar treatment, with glucagon, of fasted rats resulted in no glucosuria or hyperglycemia, but urinary nitrogen excretion was about $1^{1}/_{2}$ times that of untreated fasted controls. In adrenalectomized rats, also, glucagon injection by a similar schedule for 5 days resulted in about $1^{1}/_{2}$ times as much urinary nitrogen excretion as seen in untreated adrenalectomized rats whose food intake was limited to the same amount consumed by the glucagon-treated animals.

After further studies SALTER (1960) reported that rats given 40 µg glucagon every 8 hours for 22 days had less body protein and fat than pairfed controls.

Hyperglycemia was not present in the glucagon-treated animals just prior to a glucagon injection (i.e. 8 hours after the previous injection) but the hyperglycemic response 30 minutes after a glucagon injection was seen to increase over the first 3 days of the glucagon regimen and to stay at this higher level thereafter. The production of permanent diabetes, with glycogen infiltration of the β-cells, in some glucagon-treated rabbits by Logotheto-poulos, Sharma, Salter and Best (1959), whereas other rabbits in the series evidenced hypertrophy and hyperplasia of the pancreatic β-cells, suggests that the administration of excessive amounts of glucagon to the fed animal results in an increased demand for insulin secretion which is extensive enough to damage the pancreatic β-cells.

Davidson, Salter and Best (1960), in a further study of glucagon, reported that oxygen consumption in rats was elevated about 45 % above the basal level about one hour after subcutaneous injection of 1 mg glucagon per rat. This effect was the largest which could be produced by glucagon; larger doses did not increase it. A similar maximal effect was produced by epinephrine given at 0.1 mg/rat in oil. However, when both hormones were given together, oxygen consumption was further elevated so that it became more than 70 % higher than the basal rate. The separate effects of both glucagon and epinephrine on oxygen consumption were strikingly diminished by prior adrenalectomy and were restored by cortisone treatment of the adrenalectomized rat (2.5 mg/rat per day). The effect of glucagon on oxygen consumption was also strikingly diminished by prior thyroidectomy, and 5 to 10 µg of thyroxin per day restored the response to glucagon. In the case of epinephrine the dependence of the induced increase in oxygen consumption on the level of circulating thyroxin had already been established.

Izzo and Glasser, in 1961, compared the effects of hydrocortisone (0.5 mg/rat per day) and glucagon (1.5 mg/rat per day) on the urinary nitrogen excretion of the rat during a 5-day fast. The hydrocortisone-treated animals excreted little more urinary nitrogen than the controls during the first and second days, but thereafter excreted much more. In contrast, the largest increase over controls in urinary nitrogen excretion in the glucagon-treated rats was during the first day, and this declined therafter. On the basis of this and the fact that at the end of the 5-day fast the glucagon-treated rats had lowered blood α-amino nitrogen values whereas the hydrocortisone-treated rats had elevated values, the authors concluded that the two hormones increased amino acid catabolism by two separate mechanisms. In 1962, the same authors Glasser and Izzo confirmed the fact that adrenalectomy fails to abolish the effect of glucagon to increase urinary nitrogen excretion in the fasted rat. The same authors Izzo and Glasser, in 1963, found that glucagon had no effect on the already greatly elevated nitrogen excretion of the hypophysectomized rat during a 3-day fasting period starting 36 to 48 hours after hypophysectomy. When this nitrogen loss was brought somewhat below that seen in normal fasting control rats by bovine growth hormone given at 1.5 mg/rat per day, glucagon was again rendered effective in raising urinary nitrogen excretion.

Glucagon was shown by BONDY and CARDILLO (1956) to cause a fall in blood amino acids shortly after its administration to human subjects; this fall was stated to be comparable with that caused by a similar hyperglycemia brought about by glucose administration. A similar action of epinephrine has been reviewed by RUSSELL (1955) and by ELLIS (1956). WEINGES (1959) confirmed this amino-acid-lowering effect of glucagon in the dog. SHOEMAKER and VAN ITALLIE (1960) showed that the uptake of α-amino acids by the liver increased about $2^1/_2$-fold at 10 minutes after the intravenous injection of glucagon (20 μg/kg) in the dog; at 30 minutes and 60 minutes after glucagon injection this increased uptake was still seen.

Thus glucagon in large doses has the overall effect of increasing urinary nitrogen excretion. It is not known what its contribution is, relative to that of epinephrine, to the enhancement of protein catabolism in the normal animal. Also, glucagon has not been studied fully with regard to the possibility that small doses of glucocorticoids in the adrenalectomized animal, or small doses of thyroxin in the thyroidectomized animal, may sensitize the tissues to its protein catabolic action, although the potentiating action of these hormones on the increased oxygen consumption induced in the whole animal by both glucagon and epinephrine has been clearly shown.

Leaving aside the question of how the adrenal glucocorticoids, epinephrine and glucagon cause the arrival in the liver cell of more amino acids derived from extrahepatic tissues, and how insulin antagonizes this end result, it remains clear that, by present concepts of protein metabolism, a long-term increased rate of nitrogen loss together with the development of atrophic changes in many tissues, as seen after glucocorticoid treatment (see BAKER 1952), implies increased transport of free amino acids from extrahepatic tissues to the liver. The hormonal control of blood free amino acid metabolism has been reviewed by RUSSELL (1955), by ELLIS (1956) and by DE BODO and ALTSZULER (1957). NOALL, RIGGS, WALKER and CHRISTENSEN (1957), on the basis of studies with the nonutilizable amino acid analogue, α-amino-iso-butyric acid, have postulated that cortisol enhances the ability of hepatic cells to take up amino acids from the plasma for catabolism, an effect which would tend to obliterate any increase in blood amino acid concentration which would otherwise result from increased protein breakdown in the peripheral tissues.

Unfortunately, because of the technical difficulties involved, the flux of amino acids from extrahepatic tissues to the liver has not been measured by tracer techniques. Transhepatic catheterization in the unanesthetized dog, as utilized by Shoemaker and his collaborators in the particular instance of glucagon administration (see above), is a promising approach. The lack of such a demonstration, in the general case, encourages skepticism about the accuracy of the current belief that protein in quantity moves through the blood from other tissues to the liver only after first being broken down to free amino acids. The opposite point of view survives (see Roberts and Kelley 1956), and was expressed with great clarity by Roberts (in discussion, p. 313, of a contribution by Engel 1951), in connection with glucocorticoid-induced protein catabolism.

Without regard to how increased amounts of amino acids arrive in the liver under the influence of several hormones in opposition to the effect of insulin, the fact that they do arrive probably plays an important part in the induction of hepatic enzymes for gluconeogenesis from protein. The fact that insulin opposes their arrival in the liver cell, as well as the arrival of free fatty acids, glycerol and pyruvate, seems adequate for an understanding of the induction of enzymes for gluconeogenesis which is seen in the diabetic animal.

3. The induction of hepatic enzymes concerned with gluconeogenesis. Rosen and Nichol (1963) have reviewed recently the relationships between the glucocorticoids and the free amino acids in the induction of hepatic enzymes. Tryptophan pyrrolase and tyrosine transaminase reach a peak of activity within 4 to 6 hours after the injection of cortisol *in vivo* in the rat, whereas alanine transaminase, glucose 6-phosphatase, the urea cycle enzymes, and others, increase gradually over a period of 48 hours. In the case of tyrosine transaminase and tryptophan pyrrolase, the isolated perfused liver has been used to demonstrate induction *in vitro* in 4 hours and 2 hours, respectively. In the case of tyrosine transaminase, studied by Goldstein, Stella and Knox (1962), and by Barnabei and Sereni (1962), the increase *in vitro* depended upon the simultaneous presence of both cortisol and tyrosine in the perfusing medium.

Ewald, Hübener and Wiedemann (1963) have correlated the induction of tryptophan pyrrolase and tyrosine transaminase in the 24-hour-fasted rat with the increase in liver glycogen content in the 2- to 8-hour period after oral administration of cortisol. They found a peak in tryptophan pyrrolase at 4 hours and in glycogen content and in tyrosine transaminase at 5 hours.

In contrast with the liver perfusion results above, Sayre, Jensen and Greenberg (1956) found that threonine dehydrase could be induced in 5 hours in perfused liver in the absence of glucocorticoids by threonine in the presence of a complete mixture of amino acids in the perfusion fluid, and Price and Dietrich (1957) found that tryptophan pyrrolase was induced by tryptophan plus casein hydrolysate in 2.5 to 4 hours in perfused liver in the absence of glucocorticoids. Furthermore, tryptophan administration induced both tryptophan pyrrolase and tyrosine transaminase in adrenalectomized rats in which completeness of adrenalectomy was established (Rosen and Nichol 1963).

Thus although excess cortisol acts by a special mechanism to induce tryptophan pyrrolase, it appears to act in another way to induce many of the variety of enzymes which increase over 48 hours after cortisol administration to the intact rat.

Disregarding mechanism, the overall picture of hepatic enzyme induction by glucocorticoids as summarized by ROSEN and NICHOL includes increases in activity of 6 transaminases, 5 urea cycle enzymes, fructose diphosphatase, and glucose 6-phosphatase; all of these are associated with gluconeogenesis. It is interesting that glutamic dehydrogenase, the enzyme catalyzing the pathway through which most of the nitrogen handled in transamination reactions is thought to pass to ammonia, is so little influenced by glucocorticoid administration. FAZEKAS and DOMJÁN (1962) have reported recently that there is only a 29 % increase in this enzyme in livers of rats treated with 5 mg hydrocortisone acetate per day for 14 days; the activity was reported per unit weight of liver without regard to total liver weight.

Changes in glucose 6-phosphatase were reviewed by ASHMORE and WEBER (1959) with respect to increases in the diabetic state and in response to glucocorticoids, growth hormone, thyroxine, and diets high in protein, galactose or fructose. These reviewers noted the fact that increased hepatic glucose production precedes any measurable change in the amount of hepatic glucose 6-phosphatase after hydrocortisone is administered to adrenalectomized alloxan-diabetic rats.

HENNING, SEIFFERT and SEUBERT (1963) have reported that oral administration of cortisol to rats leads to increased hepatic pyruvate carboxylase activity. This increase is visible in 2 hours and is at a maximum 6 hours after cortisol is given. Thus it occurs subsequent to the increase in pyruvate carboxylase activation reported later from the same laboratory.

A quite separate question is whether or not the intact adrenal is necessary for an increase in the conversion of amino acids to glucose by the liver. This question was answered in the negative long ago, 1940, by WELLS and KENDALL, who showed that the phlorizinized adrenalectomized rat fed a casein diet excretes as much urinary glucose and nitrogen as the phlorizinized intact animal. Thus the metabolism by the liver of exogenously supplied amino acids is not dependent upon the presence of the glucocorticoids, whereas the mobilization of endogenous protein is so dependent.

4. Gluconeogenesis from non-nitrogenous precursors. LONG and SMITH (1962) have recently summarized certain discrepancies which are evident between the rate of protein catabolism, as judged by increased urinary nitrogen excretion, and glucose balance, as judged by body carbohydrate stores and urinary glucose excretion. The earlier work in this regard includes that of LEWIS, KUHLMAN, DELBUE, KOEPF and THORN, published in 1940, who studied the urinary glucose excretion of phlorizinized adrenalectomized rats after the

administration of glucose, lactate, or pyruvate, and the effect of glucocorticoids on this process. The untreated animals excreted 69% of injected glucose
in the urine but only 26% of lactate and 34% of pyruvate as glucose; phlorizinized normal rats excreted 81% of injected glucose and also 71% of injected
lactate and 100% of injected pyruvate as glucose. Treatment of the adrenalectomized animals with glucocorticoids restored the picture seen in normal
rats. From this and other evidence the authors concluded that in addition to
promoting gluconeogenesis from protein, the glucocorticoids a) decrease the
utilization of glucose, b) increase the conversion of 3-carbon intermediates to
glucose, and c) decrease the utilization of 3-carbon intermediates.

Ingle and Thorn, in 1941, presented evidence, derived from experiments
with insulin-deficient normal and adrenalectomized rats, that the glucose
excretion following glucocorticoid administration could not all be accounted
for on the basis of increased glucose formation from protein. Engel, in 1951,
reported that urea formation in rats was first seen to be increased in the
4 to 6-hour period after glucocorticoid administration, whereas Long and
Smith, in 1962, found that cortisol in both adrenalectomized and adrenal-
ectomized-alloxan-diabetic rats increased blood glucose concentration during
the first 3 hours after its injection.

From this and other evidence, Long and Smith concluded that cortisol
either augments gluconeogenesis or decreases glucose utilization under circumstances in which increased protein catabolism does not occur. Since a large
body of evidence was available (see a summary in Winternitz, Dintzis and
Long 1957), indicating a lack of effect of glucocorticoids on the glucose uptake
of eviscerated animals, the evidence suggested to them that cortisol either
augments glucose production or decreases glucose utilization *by the liver* under
conditions in which increased protein catabolism does not occur.

Glenn, Miller and Schlagel (1963) have summarized recently experiments of their own along similar lines.

"The first measurable metabolic alteration detected in plasma of fasted
adrenalectomized rats injected with hydrocortisone was a rise in plasma glucose
concentration within 30—60 minutes. Liver glycogen did not begin to accumulate for at least 2 to 4 hours. Plasma free fatty acids, urea nitrogen, lactate,
protein and α-amino nitrogen concentrations were not significantly altered
within the same time intervals." A notable omission from this list is glycerol,
which recently has been shown to issue from adipose tissue unaccompanied
by free fatty acids, under certain circumstances (see previous section).

A separate line of evidence has been developed which indicates an effect of
the adrenal glucocorticoids on glucose synthesis by the liver from 3 carbon
non-nitrogenous precursors. Koepf, Horn, Gemmill and Thorn, in 1941,
reported that the increase in total carbohydrate content of liver slices incubated
for 2 hours with pyruvate or lactate as substrate was enhanced markedly by

prior treatment of either normal or adrenalectomized rats with adrenal cortical extract; however, liver slices from untreated adrenalectomized rats formed carbohydrate from lactate at nearly the same rate as liver slices from normal rats.

METZ and SALTER, in 1962, reported that rat liver slices, when incubated for 90 minutes with various hexose substrates (including the phosphorylated glucose derived from hepatic glycogen when fed rats were used), released less pyruvate and lactate to the medium when glucagon (50 µg/ml), epinephrine (60 µg/ml) or cyclic 3,5-AMP (50 µg/ml) was added. The possible relationship of this finding to the increased hexose synthesis from pyruvate or lactate caused by the glucocorticoids, deserves investigation.

Recent experiments reported by HAFT (1964) also may have pertinence to the above findings. Liver perfusions carried out using, as donors, alloxan-diabetic rats deprived of insulin maintenance for one day prior to the experiment showed a normal rate of $C^{14}O_2$ production from L-lactate-C^{14}, but a higher than normal rate of production of hexose-C^{14} from lactate-C^{14}. The formation of C^{14} fatty acids from lactate-C^{14} and acetate-C^{14} were measured also, and the impairment of lipogenesis was found to be much greater from lactate than from acetate. The conversion of pyruvate to acetyl coenzyme A appeared to be lessened, and this is consonant with the passage of more pyruvate than usual toward gluconeogenesis.

The fixation of CO_2 is obligatory in gluconeogenesis from 3 carbon compounds, so it is expected that the amount of gluconeogenesis going on in liver could be measured by measuring the amount of $C^{14}O_2$ fixed in glucose and glycogen under standard conditions.

In 1962, LANDAU, MAHLER, ASHMORE, ELWYN, HASTINGS and ZOTTU reported that normal rats pretreated with cortisone for 5 days yielded liver slices which incorporated twice as much of the C^{14} of pyruvate-2-C^{14} into glycogen plus glucose as did slices from untreated normal rats, when pyruvate was present in the medium at 0.6 mM concentration. Much larger stimulating effects of cortisone pretreatment were obtained when rats were injected intraperitoneally with pyruvate-2-C^{14} or bicarbonate C^{14}, and the C^{14} incorporation into blood glucose and liver glycogen was subsequently measured. About 4 times as much C^{14} was estimated to be present at 30 and 60 minutes in the total body free glucose of the cortisone-treated rat, and about 14 times as much C^{14} in the total liver glycogen in both the bicarbonate-C^{14} and pyruvate-2-C^{14} experiments.

Results in line with the above have been reported recently by SEGAL and LOPEZ (1963). DE MEUTTER and SHREEVE (1963) found that diabetic patients converted more of an injected dose of pyruvate-C^{14} or lactate-C^{14} to plasma glucose than did normal subjects.

Wagle and Ashmore, in 1963, extended these studies on the control of gluconeogenesis. In *in vivo* experiments, bicarbonate-C^{14} was injected intraperitoneally in rats and blood samples were collected each 30 minutes for the next 2 hours. The amount of C^{14} present as glucose per unit volume of blood was reported; this was higher than in normal rats at all points in cortisol-treated intact rats, diabetic rats, and intact rats treated with anti-insulin serum $1/_2$ hour before bicarbonate-C^{14} injection; in adrenalectomized diabetic rats it was near normal.

Wagle and Ashmore, in 1964, followed up the observations with anti-insulin serum. Using the incorporation of the C^{14} of bicarbonate-C^{14} into medium glucose in the absence of added substrate as the criterion, they found that 12 hours after anti-insulin serum was given to rats, liver slices prepared from these rats incorporated in 90 minutes about $2^1/_2$ times as much C^{14} as slices from untreated normal rats; the rate of incorporation in slices from alloxan-diabetic rats was about 3 times the normal rate. When alanine-C^{14}, succinate-C^{14} or fumarate-C^{14} was used as substrate at a concentration of 1 mg/ml of medium, the conversion of these substances to CO_2 was unchanged in alloxan-diabetic rats whereas the incorporation of the C^{14} from these compounds into medium glucose was increased 2- to 3-fold. Phosphopyruvate carboxykinase activity was measured in $105\,000 \times g$ supernates of KCl homogenates of liver; the activity of this enzyme, expressed per gram of liver tissue, was 5-fold normal in rats treated 12 hours previously with anti-insulin serum and 7-fold normal in alloxan-diabetic rats.

Inasmuch as the earlier *in vivo* findings of these authors suggested that anti-insulin serum increases gluconeogenesis in $1/_2$ hour, the increase in phosphopyruvate carboxykinase activity which is visible only after 12 hours would appear to be an adaptive enzyme change of the kind reflecting increased traffic through an enzymatic pathway, but not necessary for the initial increase in traffic. These findings direct attention once again toward the possibility that a metabolite or metabolites present in the liver cell in increased amounts following the administration of anti-insulin serum is responsible for the increased rate of gluconeogenesis in the *in vivo* situation. Other relationships discussed earlier suggest that non-esterified fatty acids, glycerol and pyruvate may be involved.

5. **Rapid changes in hepatic gluconeogenesis caused by insulin and other hormones.** Short-term as well as long-term changes in the rates of gluconeogenesis are provoked when an imbalance is created between the glucocorticoids and insulin. For example, as described more fully in section II, B 4 above, after injection of anti-insulin serum in the rat, increased gluconeogenesis is evident *in vivo* in $1/_2$ hour (Wagle and Ashmore 1963). Likewise, the administration of glucocorticoids to normal rats is followed in $1/_2$ hour to one hour by a rise in blood glucose concentration (Glenn, Miller and Schlagel 1963).

Conversely, the removal of an adrenal transplant, situated for removal with minimal trauma, from adrenalectomized, alloxan-diabetic rats is followed in 4 hours by the beginning of a rapid decline in blood glucose concentration toward the normal level (LONG and SMITH 1962).

These findings suggest that effects of insulin and contra-insulin agents on gluconeogenesis might be found using perfused livers or liver slices incubated *in vitro*. In fact such an effect was first reported in 1937 by BACH and HOLMES, who found that liver slices, low in glycogen, from 10- to 24-hour-fasted rats, when incubated 2 hours at 37^0 in Ringer-bicarbonate medium, increased in total carbohydrate content; insulin, present in the incubation medium at 0.5 to 3.0 U/ml inhibited this increase by about 43% and simultaneously inhibited urea production by the slices by about 56%. Insulin also inhibited the increased rate of carbohydrate synthesis and urea production by the slices which was seen when DL-alanine was present at 0.4% concentration in the incubation medium.

MORTIMORE (1963) has contributed a careful study of the effects of insulin added to the perfusion fluid on the release of glucose and urea by the perfused rat liver. In his experiments it was found that insulin gave near maximal effects on glucose release when added to the perfusate at only 2 milliunits per hour. In livers from fasted donors, where liver glycogen was low, and at initial perfusate glucose concentrations of about 45 mg/100 ml, insulin in the perfusion fluid reduced urea production (in the 30 to 150 minute period) by 42 μmole and simultaneously reduced glucose release by 20 μmole. In livers from fed rats, where total glycogen was still 700 to 800 μmole at the end of the experiment, and at initial perfusate glucose concentrations of about 45 mg/100 ml, insulin in the perfusion fluid reduced urea production by 83 μmole and simultaneously decreased glucose release by 124 μmole. From these results MORTIMORE concluded a) that insulin inhibits the catabolism of amino acids to urea, whether by increasing protein synthesis or decreasing protein breakdown, and b) that insulin in livers from fed rats has an additional effect on glucose release, whether by decreasing the net loss of glycogen (which, however, could not be supported in a statistically significant way by the results of glycogen analyses) or by decreasing glucose production from precursors other than glycogen.

MILLER (1961) has summarized his work, and that of his collaborators, using the isolated perfused rat liver. GREEN and MILLER (1960) found that livers of 16- to 18-hour-fasted rats which had been given DL-leucine-1-C^{14} 18 hours prior to sacrifice contained a large fraction of the residual C^{14} in proteins. When such livers were used in perfusion experiments, those from diabetic rats lost more C^{14} as CO_2 than those from normal rats. Labeled plasma proteins were obtained also from rats injected previously with leucine-C^{14}; when unlabeled livers were perfused with fluid containing such plasma proteins more $C^{14}O_2$ was produced by the livers from diabetic donors. Similarly when unlabeled livers were perfused and L-leucine-U-C^{14} was included in the perfusion fluid, essentially in trace amount, the diabetic livers converted the same amount of labeled leucine to CO_2 but converted less labeled leucine into plasma proteins than did normal livers.

The free α-amino nitrogen concentration of perfused livers, whether from normal or diabetic rats, remained unchanged over a 6-hour period of perfusion at a level about 10 times as high as the free α-amino nitrogen concentration of the perfusing fluid; for both normal and diabetic rats, the α-amino nitrogen concentration of the perfusion fluid rose about 100% during the 6-hour perfusion period.

Increased urea production was observed in perfused diabetic livers, and it was concluded that this was due to an increased rate of catabolism of both plasma and liver proteins. Although not discussed by the authors, the fact that the same amount of labeled leucine was converted to CO_2 by both normal and diabetic livers probably means that more total free intracellular leucine was converted to CO_2 in the livers from diabetic donors, since there was more protein catabolism, yielding more unlabeled leucine to dilute the leucine-C^{14}. Likewise their additional observation of decreased incorporation of labeled leucine into proteins in the diabetic livers is capable of being explained on the basis of a reduced intracellular leucine specific activity in these livers rather than as a decreased rate of protein synthesis.

In his summary article, Miller (1961) presented evidence that insulin added to the perfusion fluid (in the presence of glucose) prevents, in the case of livers from normal rats, the accumulation of α-amino nitrogen in the perfusion fluid during the 6-hour perfusion period. No evidence was provided as to whether this represented increased protein synthesis or decreased protein catabolism.

Miller (1960) found that the rate of urea production of perfused livers from fed donor rats was increased by addition of glucagon (1 µg/hr) to the perfusion fluid. The glucagon produced this effect even though it also caused glycogenolysis and raised the concentration of glucose in the perfusion fluid to a level known to inhibit urea production. In his summary article Miller (1961) presented evidence that, in livers from 18-hour-fasted normal rat donors, hydrocortisone added to the perfusion fluid did not increase urea production over the 6-hour perfusion period; however, when glucagon was also added to the perfusion fluid, urea production was then increased. Also, in livers from adrenalectomized, 18-hour-fasted, rat donors, hydrocortisone added to the perfusion fluid failed to increase urea production. However, when hydrocortisone was administered for several hours prior to glucagon, the small response of the livers from adrenalectomized donors with regard to the urea-production-increasing effect of glucagon was restored to that seen in liver from normal donors. From these findings MILLER concluded that glucagon is a protein catabolic hormone, and that hydrocortisone, which by itself is incapable of promoting protein breakdown in the liver, enhances the protein catabolic action of glucagon. The free amino acid concentration of the perfusion fluid at the end of the experiment, in the case of the livers from adrenalectomized rat donors, was lower than normal; it was not increased when either glucagon alone or hydrocortisone alone was added to the perfusion fluid but was increased to near normal when both hormones were present in the perfusion fluid.

Also in his summary article, MILLER (1961) presented evidence for the antagonistic actions of glucagon and insulin, both being added *in vitro*, on urea production by perfused livers of normal fed rats. Insulin, added initially (4 units to the 70 or so ml of perfusion fluid) and infused continuously (at 1.5 to 1.7 units per hour) prevented the increased urea production caused by glucagon added (beginning after $1^1/_2$ hours) at 1 µg per hour. However when the rate of addition of glucagon to the perfusion fluid was increased to 10 µg per hour, its effect to increase urea production was again seen. It is clear that these provocative findings would be much more useful if they were extended, and were presented in detailed form.

6. Hepatic protein synthesis and gluconeogenesis. Use in protein synthesis is a fate alternative to deamination for free amino acids present in liver cells in the intact animal. However, it is only over a limited time period, while total liver protein content is shifting to a new steady state, that hepatic protein synthesis is thought to affect the net amount of free amino acids available to the liver cell for gluconeogenesis. When a new steady state has been reached, hepatic protein synthesis again is thought to equal hepatic protein breakdown, and the amounts of amino acids utilized for protein synthesis in the liver are thought to equal the amounts released in the liver by protein hydrolysis.

The concept of "storage protein" and of the synthesis in the liver of protein for use by other tissues is in disfavor. The liver, at 3 to 5 % of the body weight, is no longer regarded as a more important locus for protein synthesis than is indicated by a) the weight of its proteins (and the weight of certain plasma proteins produced by the liver), and by b) the turnover rate of these proteins. Nevertheless, tissue culture experiments (FRANCIS and WINNICK 1953, SIMMS and PARSHLEY 1959) have demonstrated that extrahepatic cells can be nourished by medium protein, indicating, perhaps, that a modification of the current view may become necessary in the future.

7. Summary. Recently developed evidence indicates that enzyme activities at several steps in gluconeogenesis are determined by the concentrations of certain cellular intermediates. In the light of this knowledge it appears that lactate, pyruvate and glycerol, when supplied to the liver in increased amounts from the extrahepatic tissues, may promote gluconeogenesis by instantly increasing the activities of gluconeogenic enzymes in addition to furnishing carbon for the hexose units which are formed. Compounds such as non-esterified fatty acids, which are capable of furnishing acetyl coenzyme A when degraded in the liver, also may instantly increase the activity of a critical enzyme of gluconeogenesis although they do not furnish carbon in net amount for hexose synthesis. Amino acids arriving in the liver in excess stimulate gluconeogenesis in the same ways after their degradation to α-keto acids; in this case a part, only, of the amino acid carbon (i.e. that of the "glucogenic" amino acids) is capable of conversion in net amount to hexose. However, the "ketogenic" amino acids may also increase the activity of a critical enzyme for gluconeogenesis by furnishing acetyl coenzyme A.

Hormones which increase the flux of metabolites to the liver change the rate of hepatic gluconeogenesis so rapidly that the induction of gluconeogenic enzymes lags behind. Thus induction of these enzymes is probably a consequence rather than a cause of the increased rate of gluconeogenesis. Insulin acts in this regard by opposing the flux of metabolites to the liver.

In those instances in which gluconeogenesis in isolated liver tissue responds to insulin, cortisol, or glucagon added *in vitro*, mechanisms in which intra-cellularly-produced intermediates are the immediate agents which influence enzyme activities may be operating. Insulin, by increasing protein synthesis may decrease the stimulus to gluconeogenesis furnished by deaminated amino acids. Insulin may further act in this way by decreasing lipolysis or increasing triglyceride synthesis in the liver cell itself. Cortisol and glucagon (and very likely epinephrine) oppose these effects of insulin and may do so by promoting protein breakdown and lipolysis or inhibiting protein synthesis and triglyceride formation in the liver cell itself.

The early increase in hepatic gluconeogenesis after cortisol administration *in vivo* (and possibly also the early increase after the injection of anti-insulin serum *in vivo*) probably results from an increased arrival in the liver cell of metabolites other than amino acids, since in the early period after cortisol injection increased urea formation is not seen in the *in vivo* situation. The metabolites in question may be increased amounts of lactate, pyruvate, glycerol and non-esterified fatty acids.

After the first few hours of increased cortisol action, amino acids also arrive in the hepatic cell in increased amounts from extrahepatic tissues; this is the case also in insulin deprivation. These amino acids, after oxidative deamination, then contribute their share to the activation of gluconeogenesis and some of the amino acids contribute carbon for the hexose which is formed. Protein synthesis in the extrahepatic tissues is increased by insulin and inhibited by the presence of the glucocorticoids; protein breakdown in the extrahepatic tissues is stimulated by glucagon and epinephrine and these actions are potentiated by the presence of the glucocorticoids and thyroid hormones. These factors contribute to the control of the transfer of amino acids to the liver. Growth hormone accentuates the effect of insulin to increase the net amount of amino acids retained as protein in the extrahepatic tissues. Whether this is entirely a potentiation of protein synthesis as stimulated by insulin is not clear. Since the mechanisms for protein breakdown are unknown, the influence of controls on protein breakdown (including the possible actions of insulin or growth hormone in this regard) on net protein balance tend to be overshadowed by the better knowledge of the control of protein synthesis.

There is some indication that the capture of circulating free amino acids by the liver is increased by glucagon and by the glucocorticoids; there is also some indication that important amounts of amino acids may arrive in the hepatic cell as a result of the hepatic intracellular hydrolysis of proteins which have traveled as such from extrahepatic tissues to the liver.

The extensively studied increases in amounts of hepatic enzymes of gluconeogenesis which are brought about by cortisol administration in the rat may be secondary consequences of changes in concentrations of hepatic cell substrates; in many instances these substrates increase in liver as a consequence of events in distant tissues. Cortisol does not organize the response of the liver of the rat to its injection by bringing about the induction of many related enzymes by way of a single "operon". Rather, the rat appears to respond to cortisol excess in an organized way, by a variety of mechanisms.

The major impact of the evidence just summarized is to offer alternatives to the thought that insulin exercises its restraining influence on gluconeogenesis solely as a secondary consequence of its effect to increase glucose uptake and utilization by the extrahepatic tissues or by the liver.

C. Hepatic protein synthesis

Between 1952 and 1963, KRAHL (1952, 1953, 1956) and PENHOS and KRAHL (1962, 1963) contributed evidence that the incorporation of the C^{14} of glycine-C^{14} or leucine-1-C^{14} into liver proteins by liver slices or perfused livers from alloxan-diabetic rats suffers a net impairment which increases in severity with the duration of the diabetic state. Liver from rats partially pancreatectomized 3 months earlier (fasting blood glucose: 120 to 160 mg/100 ml) suffers an impairment in leucine incorporation *in vitro* which is partly overcome by high (200 mg/100 ml) glucose concentration in the nourishing medium and completely overcome by high glucose concentration plus the presence of insulin (0.01 U/ml) in the nourishing medium. Insulin in the absence of glucose is ineffective. Insulin plus glucose cannot, however, correct the more severely impaired leucine-C^{14} incorporating capacity of liver tissue from rats which have been partially pancreatectomized 6 months previously (fasting blood glucose: 250 to 300 mg/100 ml).

PRYOR and BERTHET (1960a and b) found that the incorporation of tagged amino acids into liver slice proteins in phosphate buffer was inhibited by glucagon (33 µg/ml) or by cyclic 3,5-AMP (3.3×10^{-4} M) or by epinephrine present in the medium.

ROBINSON (1961), in KRAHL's laboratory, found that alloxan-diabetic rats, controlled with insulin until 72 hours prior to sacrifice, yielded liver microsomes with deficient leucine-1-C^{14} incorporating activity; no deficiency was found with respect to the ability of the soluble fractions from such livers to support incorporation of leucine-1-C^{14} by microsomes from normal rat liver. These animals were severely diabetic at sacrifice, with ketonuria, and blood glucose levels over 300 mg/100 ml. Prior treatment of normal rats with insulin for 20 hours prior to sacrifice yielded liver microsomes which incorporated leucine-1-C^{14} at a higher than normal rate. The latter effect was described earlier by DOELL (1959) in a preliminary report; KORNER, in 1960, in a publication concerned with effects of insulin administered to hypophysectomized rats, also mentions unpublished work of his own indicating this effect of injected insulin in normal rats.

D. Hepatic protein synthesis as related to enzyme induction

Liver tissue has been much used in investigations into the mechanism of protein synthesis because subcellular fractions for investigation *in vitro* are readily obtained from liver (ZAMECNIK and KELLER 1954). In such studies it has been found that DNA-dependent RNA polymerase, amino acid activating enzymes, amino acyl transfer RNA synthetase and an enzyme for utilizing amino acyl transfer RNA in protein synthesis in ribosomes, are involved in protein synthesis from amino acids (cf. WOOL 1964).

Much consideration has been given to the concept that an increase in the rate of production of specific messenger RNA molecules, in the process of enzyme induction, is the instrument for increasing the synthesis of specific proteins, whereas considerably less thought has been given to the concept that increased protein synthesis, brought about by changes in enzyme activity or substrate concentration, might result in an increase in the rate of synthesis of all of the messenger RNA of the cell. In analogy with lipogenesis and gluconeogenesis, where increased utilization of a metabolic pathway often precedes an increase in the enzymes involved in the pathway, such a mechanism might be expected to occur, leading to an increase in the amount of one or another of the enzymes concerned with protein synthesis. Nevertheless, it is often tacitly assumed that an agent under study has not affected the amount or activity of one or more protein synthesizing enzymes but has, instead, uncovered the power of an "operon" to make use of an already existing capacity for protein synthesis; this control is assumed to be mediated by the furnishing of a specific messenger RNA template for use by the ribosomes.

TATA (1964), who studied the effect of triiodothyronine (given *in vivo*) to increase (after 36 hours) the incorporation of amino acids by rat liver microsomes and mitochondria *in vitro,* has described an early increase (after 10 hours) in the activity of DNA-dependent RNA polymerase in the cell nuclei. The activity of this protein enzyme was increased before a general increase in amino acid incorporating capacity was seen.

Similar doubts as to whether increased protein synthesis in general or increased synthesis of specific proteins induced *via* specific messenger RNA molecules occurs first is evident in the studies of insulin-stimulated protein synthesis. In diaphragm muscle both protein synthesis and RNA synthesis are increased by insulin added *in vitro* in 5 minutes (WOOL 1964). In liver tissue, under certain circumstances, insulin added *in vitro* increases protein synthesis, the maximum effect being evident in 2 to 3 hours (KRAHL 1961). However, increased protein synthesis by microsomes isolated from the livers of rats (hypophysectomized) previously treated with insulin *in vivo* is barely apparent when the rats are killed 3 hours after insulin treatment and is much greater when 6 hours have been allowed to elapse (KORNER 1960). KORNER (1964) in a recent publication dealing with the enhancing effect of growth hormone injected *in vivo* on protein synthesis by liver ribosomes of hypophysectomized rats *in vitro* mentions preliminary results indicating that insulin injected *in vivo,* like growth hormone injected *in vivo,* acts on this system by stimulating the synthesis of RNA. The important features of the growth hormone effect (seen 24 hours after growth hormone *in vivo*) were a) polyuridylic acid (synthetic "messenger RNA") stimulation of phenylalanine incorporation by isolated ribosomes was as great with ribosomes from hypophysectomized rats as with ribosomes from normal rats, whereas when ribosomes were incubated in cell sap alone the incorporation by ribosomes from the hypophysectomized rats was less and this was increased by growth hormone treatment, b) polysomes (a fraction of the ribosomes separable by differential centrifugation which is supposed to differ from other ribosomes in that polysomes already have messenger RNA attached) from hypophysectomized rats are equally as effective as polysomes from normal rats and polysomes from growth hormone treated hypophysectomized rats in incorporating amino acids; the difference is that the number of polysomes is less in the hypophysectomized rats and this number is increased by growth hormone treatment. Growth hormone injected 12 hours previously increases orotic acid incorporation *in vivo* into liver nuclear and cytoplasmic RNA, including a fraction thought to be messenger RNA. KORNER postulated from these findings that growth hormone might act by increasing the activity of RNA polymerase (as shown by TATA for the triiodothyronine effect, see above) by one of several mechanisms: a) by the combination of a small piece of the growth hormone molecule with RNA polymerase so as to increase its activity, or b) by increasing the concentration of a small molecule metabolite which then would combine with RNA polymerase to increase its activity. The postulated mechanism differs in an important way from an effect at the genetic level; genetic induction is thought selectively to increase the synthesis of specific RNA molecules and specific proteins, whereas the mechanism just described implies increased protein synthesis in general, through an increase in the rate of synthesis of all the kinds of messenger RNA for which unrepressed "operons" exist.

It is apparent that this problem has not been resolved, particularly in the case of insulin where the effects on protein synthesis occur much earlier than after growth hormone or thyroid hormone administration. It is necessary to establish also whether or not the effects of thyroid hormone and growth hormone on hepatic protein synthesis can be elicited in the absence of a pancreas which is capable of increasing its rate of insulin secretion.

III. Rapid effects of insulin on carbohydrate metabolism in the liver

Because facilitation of glucose transport across the cell membrane is an important early effect of insulin action in muscle, attempts have been made to discover a similar effect of insulin in liver. These attempts will be reviewed first. Other consequences of insulin action in liver will then be treated. The effects seen in the whole animal, which are easier to observe but which are more difficult to establish as direct effects of insulin on the liver cells, will be described first. The effects which have been seen in isolated perfused liver and in liver slices will then be reviewed, together with material which offers some explanation of the difficulties encountered in demonstrating insulin effects in these preparations. Previous reviews and summary articles in this general area include the following: DE BODO et al. (1963 b), CAMERON (1962), DUNCAN and BAIRD (1960), LUKENS (1959), DE BODO and ALTSZULER (1958), DE DUVE (1956), and LEVINE and FRITZ (1956).

A. Penetration of free glucose into liver

The translocation of glucose molecules between liver water and plasma water is commonly referred to as "diffusion" or "free diffusion". However, it should be understood that the facilitated transport of glucose by a stereospecific carrier has not been ruled out by experiment and, in fact, is probable. CAHILL et al. (1958) found that sucrose, raffinose, and maltose were confined in their distribution in liver water to the extracellular fluid; this makes it unlikely that channels for diffusion are adequate to account for the rapid inward and outward movement of glucose which takes place. The term "free diffusion" seems to have been used at times to distinguish the supposedly insulin-insensitive glucose transport in liver from the insulin-sensitive glucose transport of muscle and a few other tissues. On the other hand, the red blood cell offers an excellent example of a stereospecific facilitated glucose transport which is not insulin sensitive.

In discussing the penetration of glucose into liver, a useful introduction is a picture of glucose penetration into muscle, as this topic is currently understood. This picture is one in which glucose not accounted for in the extracellular fluid space is assumed to be distributed at uniform concentration in the intracellular water. Under usual circumstances there is no intracellular glucose found at all, the explanation being that the free glucose concentration is kept negligibly small by the avidity with which intracellular glucose is phosphorylated. Under special circumstances (epinephrine action, anoxia, insulin action when external glucose concentration is high or when glucose phosphorylation is inhibited as by cold *in vitro*, or in the diabetic state) free intracellular glucose can be measured. In contrast with the liver, muscle tissue contains no enzyme capable of liberating free glucose from glucose 6-phosphate so the intracellular free glucose in muscle is thought to come entirely from plasma glucose. The plasma glucose is thought to enter the intracellular water by crossing the cell membrane on a carrier system which transports glucose faster in the presence of insulin. This carrier system is supposed to facilitate glucose transport in both directions across the cell membrane so that the net flow glucose is in the direction toward the smaller glucose concentration.

It should be noted that the uniform distribution of free glucose in the intracellular water is an assumption, and that it has not been proven that glucose is not concentrated to above the extracellular concentration in some portion of the muscle intracellular water, with the remainder of the intracellular water being less concentrated in glucose than the extracellular water. The experiences upon which the assumption is based are 1. that the overall intracellular free glucose concentration has never been found greater than the extracellular glucose concentration in muscle tissue, and 2. that the kinetics of transport of glucose and of other, nonutilizable, sugars which compete for the glucose carrier system fit satisfactorily with the assumption that both glucose and the nonutilizable sugars are uniformly distributed in the intracellular water.

On the other hand, glucose transport in certain cells (intestinal mucosal epithelial cells) occurs against a glucose concentration gradient by a process associated with Na$^+$ transport in the opposite direction (CRANE, MILLER and BIHLER 1961). In view of the capacity of some subcellular structures to "extrude" Na$^+$ as reviewed recently by LEHNINGER (1962) for mitochondria it might be wondered if some structural elements of muscle cells are not capable of concentrating glucose above the concentration of glucose in the external medium and above the average glucose concentration measured in the intracellular water. Such a concentrating mechanism, if galactose as well as glucose were concentrated under the action of insulin, would eliminate a puzzling aspect referred to by STADIE (1954) of the experiments of LEVINE, GOLDSTEIN, KLEIN and HUDDLESTUN (1949 and 1950). In these experiments, galactose in eviscerated-nephrectomized dogs (in which galactose does not disappear) attained *at equilibrium* a calculated volume of distribution equal to about 40% of the body weight. In the presence of administered insulin the calculated volume of distribution of galactose rapidly became about 70% of the body weight. The interpretation given, which was that insulin allowed the penetration of galactose into cells *from which it was completely excluded at equilibrium* in the absence of administered insulin, is not an entirely satisfying one.

Thus a hypothesis that insulin may increase the intracellular free glucose concentration of liver relative to blood plasma cannot reasonably be faulted simply because this is logically inconsistent with a widely-accepted concept of the effect of insulin on glucose transport in muscle. First, liver cells are not muscle cells and second, the picture of facilitated glucose transport in muscle may not be understood in all details.

GEY (1956) found, in 4 rats fasted 20 to 28 hours, a mean of 75 mg glucose/100 g liver at a time when mean plasma glucose concentration was 79 mg/100 ml. HETENYI and ARBUS (1962), in 21 rats fasted 20 hours prior to sacrifice, found a mean of 84 mg glucose/100 g liver when mean plasma glucose concentration was 84 mg/100 ml. CAHILL, ASHMORE, EARLE and ZOTTU (1958) found 91 mg glucose/100 g liver when plasma glucose concentration was 84 mg/100 ml. Inasmuch as the respective ratios found were 0.95, 1.00 and 1.08, it can be considered well established that the fasting rat has roughly equal amounts of free glucose per 100 ml plasma and per 100 g liver tissue. Since liver contains about 72 g water per 100 g tissue, and plasma about 93 g water per 100 ml, the glucose concentration in liver water (assumed uniform) is about 1.3 times the glucose concentration in plasma water.

If it be assumed that the concentration gradient in the liver is all between intracellular and extracellular water, i.e. that extracellular water has the same glucose concentration as plasma water, and if 22% of liver weight is assigned to extracellular water (MANERY and HASTINGS 1939), then intracellular water has about 1.4 times the glucose concentration of plasma water. Illustrating with a specific example, when plasma glucose concentration is 84 mg/100 ml, the concentration of glucose in plasma water is 90 mg/100 ml. If liver glucose concentration is 84 mg/100 g fresh tissue, 22 g of this tissue consists of extracellular water with a glucose concentration of 90 mg/100 ml; this accounts for 20 mg of the total 84 mg liver glucose. The remaining 64 mg is in the 50 g of water remaining which is not extracellular. Thus the glucose concentration in intracellular water is 128 mg/100 ml, which is 1.4 times the 90 mg/100 ml present in plasma water.

In 21 control fasted rats studied by HETENYI and ARBUS (1962), the individual extracellular fluid spaces of the livers were determined by measuring tissue sodium and the individual total water contents of the liver were measured by loss of weight on drying samples for 20 hours at 120° C. Using these values, which were not given, the mean ratio of glucose in intracellular water to glucose in plasma water was calculated to be 1.21 ± 0.06 for the 21 rats. This result must signify either that the total liver water values which were found varied a great deal in these rats from the values found by MANERY and HASTINGS (1939) or that some obscurity in calculation is involved. The final result even falls below the ratio (1.3) of glucose concentration in total liver water to glucose concentration in plasma water which is obtained if total liver water is simply taken as 72% of liver weight, with no correction for the hepatic extracellular fluid of lower glucose concentration. An additional indication that something was amiss in these studies is furnished by the finding that, in the alloxan-diabetic rat, glucose concentration in cell water was calculated (HETENYI et al. 1963) to be lower than glucose concentration in plasma water even though net flow of glucose from liver was known to be going on.

HETENYI et al. (1963) have advanced the view that insulin increases hepatic intracellular glucose concentration by causing the uptake of glucose from the plasma against a concentration gradient. Some of their experiments involved intraperitoneal (and in some instances intravenous) injection of insulin in rats 30 minutes prior to simultaneous determination of free glucose concentration in plasma water and hepatic cell water. With smaller insulin doses (less than 1 U/kg) mean plasma glucose concentration at sacrifice was about 30 mg/100ml; the ratio of glucose concentration in hepatic intracellular water to that in plasma water was more than twice the ratio in control fasted rats; a larger similar effect was seen at higher (more than 2 U/kg) insulin doses when plasma glucose concentration was about 18 mg/100 ml.

A large increase in the rate of glucose release by the liver in response to insulin-induced hypoglycemia is seen, at least in the dog (DE BODO et al. 1963 b); thus the effect described above might well have been due not to insulin itself, but to an increased amount of glucose being released inside the liver cells in response to hypoglycemia. In this connection, HETENYI et al. (1963) made similar measurements in rats given either glucagon (2.5 mg/kg) or epinephrine (0.5 to 1.0 mg/kg) subcutaneously 30 minutes prior to sacrifice. The glucagon treated rats had a mean plasma glucose concentration at sacrifice at 190 mg/100 ml and showed a very small increase in ratio of glucose concentration in hepatic cell water to glucose concentration in plasma water. The epinephrine treated rats had a mean plasma glucose concentration at sacrifice of 377 mg/100 ml and showed no increase at all in the ratio in question. From these results the conclusion was drawn that the extra glucose released from liver glycogen by these agents "does not accumulate in liver cells but diffuses out rapidly". This conclusion was intended to discount the possibility that the effect observed with insulin might be due to increased glucose release inside the hepatic cells in response to hypoglycemia. However, it was not demonstrated that increased release of glucose from the liver was, in fact, still going on, in the glucagon- or epinephrine-treated animals, at the time the rats were sacrificed and the liver samples taken for analysis.

Other experiments by Hetenyi, Kopstick and Retelstorf (1963) reinforce the thought that increased glucose production in response to hypoglycemia, rather than a primary action of insulin, is responsible for the increased (relative to plasma) glucose concentration in the livers of insulinized animals. Alloxan-diabetic rats given 6 U/kg insulin had plasma glucose values 30 minutes later of 206 mg/100 ml; similar rats given 22 U/kg insulin had plasma glucose values 30 minutes later of 143 mg/100 ml. In neither case was hypoglycemia produced, and in neither case was the liver glucose (relative to plasma) increased, as it was in normal rats treated the same way. Also, in adrenalectomized animals, in which increased glucose production in response to severe hypoglycemia is deficient (de Bodo et al. 1963 b), Hetenyi et al. (1963) found less than the usual increase in liver glucose concentration after insulin injection.

Other observations of Hetenyi et al. (1963) involved injection of glucose-C^{14} intravenously in rats. At intervals between 7.5 and 30 minutes after glucose-C^{14} injection, the glucose of liver tissue and of plasma, respectively, was isolated and its C^{14} content was determined. In fasted rats the specific activity of liver glucose was 0.8 to 0.9 that of plasma glucose; in insulinized fasted rats the ratio was 0.5 to 0.7. This is in keeping, as stated by the authors, with the idea that insulin-induced hypoglycemia was causing more rapid glucose-C^{14} output by the livers of the insulinized animals. On the other hand, when the ratios of absolute glucose-C^{14} concentration (cpm/ml H_2O) in hepatic cell water and plasma water were compared, the fasted rats had a ratio of 1.0 during the whole period from 7.5 to 30 minutes, whereas the insulinized rats had a ratio of 1.7 to 1.9 in the 15 to 30 minute period. Since plasma glucose specific activity was declining to one-fourth its initial value during this period, the facts allow at least two interpretations. The one preferred by the authors is that, under the influence of insulin, glucose-C^{14} was being concentrated in liver water against the glucose-C^{14} concentration gradient. Another possibility is that during insulin-induced hypoglycemia the absolute glucose-C^{14} concentration of liver water, which had been brought, early in the experiment, to a level near to that of plasma, lagged in its subsequent fall behind the falling absolute glucose-C^{14} concentration of the plasma. In the case of the absolute glucose-C^{14} concentration of the plasma water, increased glucose uptake by the peripheral tissues results in its rapid decline; this situation is to be distinguished carefully from any consideration of changes in the specific activity of the glucose of the plasma water. The latter is not affected by an increased rate of glucose outflow from the plasma.

Cahill et al. (1958) injected various amounts of glucose intravenously in rats and analyzed plasma and liver for glucose content 5 minutes later. During this interval, plasma glucose concentration falls rapidly as a result of the distribution of the injected glucose in the glucose space of the body. For this reason the difference in glucose concentration between plasma and liver was being obliterated rapidly by processes having no relationship to the equilibration of glucose concentration between plasma water and liver water. Nevertheless it was possible to see that this equilibration was proceeding rapidly. For example, when enough glucose was injected so the glucose concentration in plasma water 5 minutes later was 468 mg/100 ml, the glucose concentration in liver water had become 414 mg/100 ml. The initial glucose concentration in liver water had been about 126 mg/100 ml. When the amount of glucose injected was such that the glucose concentration in plasma water after 5 minutes was about 144 mg/100 ml, the glucose concentration in liver water after 5 minutes was also about 144 mg/100 ml. Injection of less glucose than this resulted, after 5 minutes, in the glucose concentration in plasma water being less than that in liver water, as had been the case prior to glucose injection.

In their discussion of these findings the authors ignored the probability that a transient state was being observed at the 5 minute interval. Proceeding as if a new steady-state equilibrium had been shown to be established, they reasoned that above a glucose concentration of 144 mg/100 ml, net glucose uptake by liver into the phosphorylated intermediates is rapid enough to reverse the concentration gradient between liver water and plasma water. This conclusion cannot be accepted since the evidence lies against, rather than in favor of, the existence of a new steady state at the time the observations were made.

Similar findings of rapid equilibration between liver and plasma glucose were furnished by the same authors by intravenous injection of glucose-C^{14} in dogs. Here samples were collected at intervals from 10 to 60 minutes after injection. No consistent difference in absolute glucose-C^{14} concentration (cpm/ml H_2O) between liver water and plasma water was seen over this time period, during which the absolute glucose-C^{14} concentration of both plasma water and liver water fell to one-third the value which existed at the beginning of the period. The same authors also furnished evidence that alloxan-diabetic rats are nearly normal in the rapidity with which glucose-C^{14} becomes equilibrated between liver and plasma water. In 3 such animals, 5 minutes after injection of glucose-C^{14}, the absolute glucose-C^{14} concentration of plasma water was only 8 % higher than the glucose-C^{14} concentration of liver water.

BERTHET, JACQUES, HERS and DE DUVE (1956) estimated the glucose space of rabbit liver slices both by measuring total glucose concentrations at equilibrium in slice and medium and by measuring absolute glucose-C^{14} concentrations at equilibrium in slice and medium. These slices were important because it was with these that a stimulatory effect of insulin added *in vitro* on glucose-C^{14} incorporation into glycogen was seen. By both criteria the glucose space was found equivalent to 50 % of tissue weight, that is, about five-sevenths of the total amount of liver slice water. These findings are in direct disagreement with those of CAHILL et al. (1958), who found that liver slice water rapidly arrived at the same glucose concentration as the water of the bathing medium (Ringer-bicarbonate). However, CAHILL et al. discounted their own findings because inulin and raffinose were found to distribute in more than the extracellular fluid water of their slices, whereas in liver removed from the living animal this was not the case. BERTHET et al. found no influence of insulin on the calculated space occupied by glucose in their liver slices.

It can be concluded that free glucose molecules move in and out of the liver cell rapidly; however, that the outward movement is not facile enough to prevent the existence of a considerable concentration gradient between liver water and plasma water *in vivo* in the fasting state when net glucose flow out of the liver is going on. It is not known how free glucose is distributed in liver cell water *in vivo*, or if the distribution is affected by insulin, in part because transient states have been studied rather than steady states. The evidence from liver slices, after equilibration with medium, indicates that free glucose is distributed in less than the total intracellular liver water. This distribution was not altered by insulin under conditions in which insulin was seen to enhance the incorporation of glucose-C^{14} into glycogen.

B. Whole animal experiments

1. General comments on methods for measuring hepatic glucose production. Two general experimental approaches have recently yielded new knowledge regarding the effects of insulin on glucose production and uptake by the liver *in situ*.

Transhepatic catheterization, advanced notably by SHOEMAKER et al. (1959) and by MADISON [MADISON et al. (1960)], is an extension of the surgical approach which has traditionally contributed in this field of investigation. SHOEMAKER's contribution was the preparation of dogs with indwelling plastic catheters in portal and hepatic veins. These catheters can be kept functional for days so that experiments can be done on unanesthetized animals recovered from the effects of surgery. MADISON's contribution was the adoption of the procedure of permanent, end-to-side, portal vein to inferior vena cava anastomosis, in conjunction with hepatic vein catheterization, *via* an external jugular vein and the right heart, which can be done without major surgery, but which is usually done, in the dog, under general anesthesia. Both MADISON's and SHOEMAKER's approaches make use of the dye (usually bromsulphonphthalein) extraction procedure introduced by BRADLEY et al. (1945) for measurement of hepatic blood flow.

The second approach, used first by FELLER, STRISOWER and CHAIKOFF (1950) involves the use of isotopically labeled glucose and measures the dilution of the circulating blood glucose (utilizing peripheral vein blood) by new, unlabeled, glucose produced by the liver.

An ancillary technique, collection by percutaneous biopsy of samples of liver under local, nerve-block, anesthesia in conjunction with labeled circulating blood glucose measurements, is new to this field but has produced encouraging results (BISHOP, STEELE, ALTSZULER, DUNN, BJERKNES and DE BODO 1965).

It is useful to make some general comments about what is measured by the procedures. Hepatic and portal vein blood glucose concentrations, taken together with arterial blood glucose concentration and total hepatic blood flow (LANDAU, LEONARDS and BARRY 1961, SHOEMAKER, MAHLER, ASHMORE, PUGH and HASTINGS 1959) allow the estimation of net hepatic glucose production. Net hepatic glucose production is lowered by glucose uptake by liver and is elevated by the new glucose produced in the liver. Under certain circumstances, and in contrast with new, unlabeled, glucose production, it can have a negative value (i.e. when uptake exceeds new glucose production). It is superior, as a metabolic measurement, to net splanchnic glucose production (BEARN, BILLING and SHERLOCK 1953, BONDY, JAMES and FARRAR 1949) which can be measured without the portal vein catheter. The superiority becomes crucial when the action of administered insulin is measured, because glucose uptake from the blood in the area drained by the portal vein becomes a significant factor at that time, obscuring what the liver is doing.

Net hepatic glucose production is also measured in the portocaval anastomosis (MADISON, COMBES, ADAMS and STRICKLAND 1960) procedure; here the portal vein catheter is not needed because only arterial blood flows into the liver. A major practical advantage of this technique is the restricted blood flow through the liver, which elevates the difference in blood glucose concentration across the liver and thus facilitates its measurement. This feature of altered blood supply to the liver invites attacks on the physiological validity of the results obtained. In this connection a recent review by BOLLMAN (1961) summarizes the changes which have been seen in the Eck fistula animal.

The isotope dilution technique measures the amount of unlabeled glucose which enters the circulating blood. Besides the liver, the kidney has long been known to contribute glucose in net amount to the blood in the eviscerated or hepatectomized rat and dog (COHN and KOLINSKY 1949, MACKLER, AMMENTORP, GRAUBARTH and GUEST 1951, REINECKE 1943—1944). It was found not to do so appreciably unless evisceration or hepatectomy was carried out (ROBERTS and SAMUELS 1944). The recent very careful studies of McCANN and JUDE (1958) in the unanesthetized dog with indwelling polyethylene catheters in the renal vein and the abdominal aorta have shown that renal glucose is produced at about 5 to 12 % of the standard rate (150 to 200 mg glucose/kg body weight per hr) of hepatic glucose production in the postabsorptive state. It might be supposed that unlabeled glucose release by the kidney would exceed the *net* glucose output of that organ. However, in 3 of the 4 eviscerated rabbits which were maintained by DRURY, WICK and MACKAY (1950) at normal plasma glucose levels by the continuous infusion of labeled glucose at about 100 mg/kg body weight per hr, the specific activity of the circulating plasma glucose at the end of 8 hours of infusion was in each case lowered by an amount indicating that 11 mg/kg body weight per hr of unlabeled glucose was being produced by the kidney. This quantity of unlabeled glucose is about the same fraction of the normal hepatic glucose output as the *net* renal glucose output estimated by McCANN and JUDE by transrenal catheterization. However, when severe hypoglycemia was allowed to develop, in the rabbits of DRURY et al., renal unlabeled glucose production rose to much higher levels, in accord with previous observations of considerable *net* renal glucose output under these conditions by ROBERTS and SAMUELS (1944).

It may be concluded from these findings that in the normal animal in the postabsorptive state the liver accounts for 88 to 95 % of both *net* glucose production and unlabeled glucose release to the blood.

In using the isotope dilution technique, the relationship between abrupt changes in the rate of release of unlabeled glucose and the corresponding changes in net hepatic glucose output is not obvious. This comes about, in part, because the exact arrangement of the compartments of phosphorylated glucose intermediates in the liver cell and the flow of glucose through them is not known. If it is assumed that there is a single pool of liver glucose-6-phosphate through which glucose must pass, both in entering and leaving the liver, the following considerations arise. Increased inward flow of labeled glucose from the blood into glucose-6-phosphate might be brought about by some mechanism. When the blood contains glucose-C^{14}, the resulting increase in glucose-6-phosphate specific activity would reduce the amount of unlabeled glucose leaving the liver even though the total number of glucose molecules set free from glucose-6-phosphate were to remain unchanged. However, a very large change in inward flow would be required to produce a near-cessation of unlabeled glucose release by this means. Here unlabeled glucose release would be a relatively insensitive and also a nonproportional indicator of increased glucose uptake.

On the other hand, a decrease in unlabeled glucose release may also occur in another way. The flow of unlabeled hexose into the phosphorylated intermediates from endogenous sources (glycogenolysis and gluconeogenesis) may be caused to decrease by some mechanism, and as a result there may be a real decrease in the number of glucose molecules set free from glucose 6-phosphate. Such a real decrease would be measured by the isotope dilution technique.

The foregoing assumes the traditional view of the glucose intermediates; additional considerations arise when this view is challenged. BELOFF-CHAIN and her co-workers (1953, 1964) suggest that glucose-6-phosphate is not included in the pathway of glycogen synthesis from glucose in rat diaphragm muscle. FIGUEROA, PFEIFER and NIEMEYER (1962) presented preliminary evidence, derived by use of rabbit liver homogenates, that conversion of glucose to glucose-6-phosphate is not a necessary step in glycogen synthesis. SHAW and STADIE (1957), using rat diaphragm muscle, obtained evidence that there are two discrete Embden-Meyerhof systems for handling glucose taken up from the medium, one leading to lactate only, and the other both to glycogen and to lactate; only the latter pathway was considered to be stimulated by insulin. On quite different grounds, SEGAL and LOPEZ (1963) have suggested recently that there may be two separate pools of glucose-6-phosphate in liver, one involved in the pathway of glycogen synthesis and the other in the pathway of glycogen breakdown. It is clear that if it is finally established that glucose taken up by the liver is shunted into a pool of intermediates which is not in direct communication with the pool of glucose-6-phosphate upon which glucose-6-phosphatase is acting to release free glucose, the above considerations of the relationship of net hepatic glucose output to the release of unlabeled glucose will have to be re-evaluated. Under such conditions, changes in the rate of unlabeled glucose release will be less likely to be influenced by changes in the rate of uptake of labeled glucose than is now assumed to be the case.

2. **Interpretation of the results of isotope dilution experiments.** SEARLE, STRISOWER and CHAIKOFF (1954), after experience with the single intravenous injection of glucose-C^{14}, adopted a procedure in which a continuous infusion of glucose-C^{14} was carried out subsequent to an initial priming injection. Sampling began about an hour after the priming dose was injected; a successful experiment demanded a correct balance between priming and infusion doses so that a nearly constant plasma glucose specific activity was obtained; glucose pool size was not calculated from such an experiment, alone, but from the combined results of the continuous infusion experiment and an earlier single injection experiment on the same dog. The glucose pool size so obtained was consistent with available physiological knowledge of the distribution of glucose in the body, i.e. the glucose pool was about the right size to be contained in the extracellular fluid at the glucose concentration of the blood plasma.

STEELE, WALL, DE BODO and ALTSZULER (1956) utilized the priming dose and continuous infusion procedure and showed that the pool size could be approximated by the backward extrapolation of the curve subsequent to 60 minutes after injection of the priming dose. In practice they chose a ratio of priming dose to continuous infusion which resulted in a nearly constant plasma glucose specific activity subsequent to 60 minutes. Figure 3 shows the plasma glucose specific activities measured in one such experiment; the curving portion of the plot prior to 60 minutes is considered to represent the continued mixing of the injected glucose-C¹⁴ with the body pool of free glucose, a time-consuming process. A similar early, more rapid, fall in plasma glucose specific activity is evident in experiments in which a single injection of glucose-C¹⁴ is given to the rabbit (Berson and Yalow 1957), the rat (Baker, Shipley, Clark and Incefy 1959, Feller, Strisower and Chaikoff 1950), and the cow (Baxter, Kleiber and Black 1955). Subsequent to this initial period, in the single injection experiment, glucose specific activity in the dog falls as

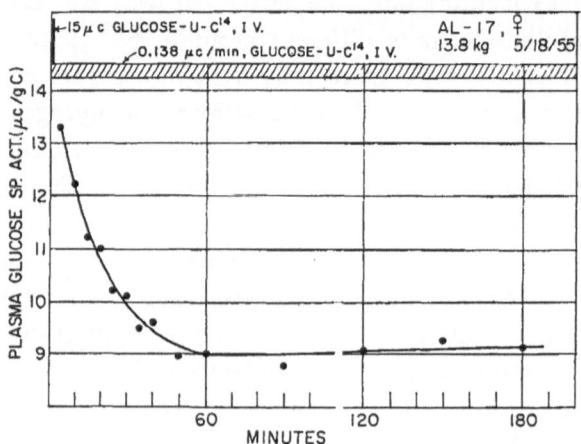

Fig. 3. The time course of the specific activity of plasma glucose in a dog injected intravenously, at zero time with a priming dose (about 0.75 mg) of tagged glucose, and infused continuously, intravenously, with tagged glucose, starting at zero time. The amount of tagged glucose infused per minute was equal to 1/109th of the priming dose. This is the experiment from which Fig. 2 of Steele, Wall, de Bodo and Altszuler (1956) was derived

a single exponential function for several hours (Searle, Strisower and Chaikoff 1954). It is of interest to consider why a long-continued constancy in unlabeled glucose release is attainable at all, in the living animal.

The uptake of glucose from the plasma into the phosphorylated intermediates of the liver of the dog in the resting postabsorptive state is small, as judged a) by the identity of absolute glucose-C¹⁴ content ($\mu c/ml$) of portal and hepatic vein plasma (Leonards et al. 1961), and b) by the low rate of accumulation of C¹⁴ in the liver constituents (Bishop et al. 1965) which prevails when the circulating glucose is labeled. This is one necessary condition lying behind the fact that the rate of unlabeled glucose output by the liver remains essentially constant for 5 hours or more while glucose-C¹⁴ is present in the circulating blood. Another necessary condition is that the extent of recycling of carbon atoms of glucose taken up by other tissues (blood glucose → extrahepatic cell intermediates → intermediates in blood → intermediates in hepatic cell → liver glucose → blood glucose) back into the glucose output of the liver be relatively small. Recent measurements indicating the extent of this recycling process are available (see Reichard, Moury, Hochella, Patterson and Weinhouse 1963).

Many investigations have used the single injection procedure, discarding the plasma glucose specific activities observed prior to 45 to 60 minutes (in the case of the dog). In such procedures glucose pool size is measured by backward extrapolation of the exponential curve (observed between 45 to 60 minutes and 180 to 300 minutes) to zero time; this determined pool size, together with the decay constant of the exponential curve, is used to calculate glucose-C^{14} release by the liver in weight of glucose per unit time. SEARLE, STRISOWER and CHAIKOFF (1954) showed that this procedure, in the dog, gives pool sizes about half again as large as the priming dose and continuous infusion method.

In the priming dose and continuous infusion procedure, the measurement of glucose-C^{12} release in the steady state is relatively independent of pool size and becomes absolutely independent of pool size when an absolutely constant plasma glucose specific activity is maintained, since this means that the body glucose pool is labeled to an unchanging extent with glucose-C^{14} and that the mixture of endogenous glucose-C^{12} and exogenous glucose-C^{14} (in "weightless" amount) entering the systemic circulation has the same specific activity as the glucose of the uniformly labeled pool. When this situation prevails, the rate of glucose-C^{12} release, in mg/min, is simply the C^{14} infusion rate, in μc/min, divided by the plasma glucose specific activity, in μc/mg glucose.

This independence of pool size disappears when there are rapid changes in either plasma glucose specific activity or concentration, as after insulin injection. It is then necessary to assume an arbitrary value for the effective pool size through which mixing occurs over the 10 to 20 minute period between observations. WALL, STEELE, DE BODO and ALTSZULER (1957) decided to use one-half the pool size measured by the priming dose and continuous infusion procedure. Others (DUNN, FRIEDMAN, MAASS, REICHARD and WEINHOUSE 1957), who have utilized the single injection procedure and who have measured the altered slope of the plasma glucose specific activity curve after insulin injection, have used the entire glucose pool size as measured by the single injection procedure in the pre-insulin period, i.e. about 3 times as large a value. In other respects, the calculations are similar by the two methods, as shown by STEELE (1959), and give qualitatively similar results.

In either case, as well as in the calculation of glucose uptake by extrahepatic tissues in the trans-hepatic catheterization procedure and in many glucose tolerance procedures, the total amount of free glucose in the animal is assumed to increase and decrease in proportion to the plasma glucose concentration.

When large amounts of glucose are infused, either to stimulate endogenous insulin secretion or to counteract insulin-induced hypoglycemia, the resulting disturbance in plasma glucose specific activity due to the slow mixing of the new glucose with a part of the body pool can be avoided, in the continuous infusion experiment, by using glucose for infusion which has the same specific activity as that of the glucose already present in the body glucose pool. If glucose-C^{12} is used, there is an immediate overdilution of the glucose-C^{14} of the circulating plasma glucose, followed by a period in which plasma glucose specific activity tends to increase again at the expense of the glucose of higher specific activity in the less rapidly mixing portion of the pool. This tendency, counteracting the diluting action of a continuing influx of glucose-C^{12} from endogenous sources, is capable of giving the impression that glucose-C^{12} inflow has stopped when it has not done so, as shown by STEELE, BISHOP and LEVINE (1959).

The nature of the body glucose pool is such that it is not yet possible to measure rapid changes in unlabeled glucose production and total glucose uptake in a way which can proven to be unequivocally correct.

Disturbed by the usual assumption that the glucose pool size changes in direct proportion to plasma glucose concentration after insulin injection, WRENSHALL, HETENYI and BEST (1961) introduced a new concept and a new technique. The concept was to measure the changes in a central pool of rapidly equilibrating glucose; the technique was the repeated single injection of labeled glucose, and the fitting of the early (10 to 40 minute) plasma glucose specific activity curve to a single exponential function. The concept fails to take into account the physiological information which one wishes to obtain; if it were indeed possible to measure the rate of appearance of unlabeled glucose in a central, rapidly equilibrating, compartment, this rate of appearance would include not only the unlabeled glucose produced by the liver, but also the un-labeled glucose entering the central compartment from the peripheral compart-ment of unlabeled glucose with which the central compartment is intermixing.

The technique, moreover, actually is not capable of measuring the changes in the central pool of rapidly equilibrating glucose, since the shape of the plasma glucose specific activity curve during the time period selected is influenced not only by the influx of glucose-C^{12} into the central pool, but also by the recycling of some labeled glucose back into the central compartment from the peripheral compartment of the body glucose pool, as pointed out by STEELE (1964). An empirical pool size and an empirical exponential decay constant apparently were found by WRENSHALL et al. which interacted to give a correct inflow rate of unlabeled glucose when unlabeled glucose was infused at a measured rate in the anesthetized liverless dog. However, under different conditions of glucose uptake, or when changes in peripheral circulation resulting from responses to hypoglycemia are affecting the inter-mixing of glucose between the central and peripheral pools, the procedures which were found suitable for the anesthetized liverless dog might or might not continue to give valid results.

The application of this technique to insulin-treated dogs was held to show (HETENYI, WRENSHALL and BEST 1961) that the rapidly-equilibrating glucose pool size an hour or so after insulin administration is not in proportion to the plasma glucose concentration (which has returned to normal) but is very much greater than is indicated by the final plasma glucose concentration. This implies either that free glucose comes to occupy nearly all the body water at the concentration found in plasma or that glucose is concentrated in certain water spaces at higher than the plasma glucose concentration. A major difficulty in accepting these findings lies in the fact that insulin has not been found to produce a correspondingly large effect on glucose distribution in isolated normal muscle tissue *in vitro*. Measurements of hepatic intracellular glucose concentration after insulin injection in rats by HETENYI and his co-workers (1962, 1963) have been interpreted to indicate that a part of the increased glucose space can be accounted for by an effect of insulin to cause the transport of glucose from plasma into the hepatic cells against a concentration gradient. However, the amount of glucose involved (liver weight is 3 to 5 % of whole body weight) is not enough to account for the total increase in glucose space claimed for the whole dog. The hepatic glucose measurements involved have been described in the preceding section.

An alternative explanation of the apparent change in glucose space is that the intermixing process in the body glucose pool affects the shape of the plasma glucose specific activity curve differently when conditions are different, as during the long-continued increased glucose uptake after insulin administration. Under these changed conditions the extrapolation of the early plasma glucose specific activity curve back to zero time, as done by WRENSHALL et al. (1961), may give a calculated glucose pool size which is no longer in the same proportion to plasma glucose concentration as under the original conditions, even though the actual amount of free glucose in the body is in fact in proper proportion to the plasma glucose concentration. It is important that future investigation should establish the truth about the possible effect of insulin on glucose pool size; however, the information will have to be gained by a direct method of measuring pool size rather than by an isotope dilution procedure carried out on the intact animal (see STEELE 1964).

BAKER and HUEBOTTER (1964), in an attempt to measure glucose pool size directly, extracted liver and extrahepatic tissues of mice separately with 70 % ethanol. Enzymatic assay of the boiled extract of extrahepatic tissues for glucose gave a value of 3.9 mg/25 g mouse, which together with the free glucose of blood and liver, accounted for the amount of total free glucose expected on the basis of an extracellular distribution of free glucose at the concentration prevailing in plasma. However, deproteinization of the extrahepatic tissue extract was stated to release additional free glucose amounting to 5.3 mg/25 g mouse. Paper chromatography of such a deproteinized ethanolic extract of extrahepatic tissues 5 minutes after intravenous injection of a trace amount of glucose-U-C[14] indicated that half of the glucose-C[14] had been converted to unidentified nonglucose, nonpolysaccharide, nonlipidic compounds. This evidence in mice for a large reservoir of nonglucose material in the extrahepatic tissues which is able to exchange with molecules of free extracellular glucose and so influence the shape of a glucose-C^{14} disappearance curve in a way not connected with glucose utilization in the ordinary sense is not easily reconciled with other existing knowledge derived from studies in many species of animals.

Even in the undisturbed steady state, the glucose pool size cannot be determined exactly by any isotope dilution procedure (see STEELE 1964). Instead, a small range of possible values is obtained, with the exact value depending upon the relative rates of uptake of glucose from the rapidly and slowing mixing compartments. This ratio is presently unknown. The same conclusion was stated, without giving the mathematical proof, by BAKER et al. (1959) for a similar model of the rat glucose pool.

3. Glucose production and uptake by liver. The *net* glucose output of the liver of the normal dog was shown long ago to be decreased by glucose administration (CHERRY and CRANDALL 1937, SOSKIN, ESSEX, HERRICK and MANN 1938). It is readily comprehended that hepatic glucose uptake, in the absence of any change in the uptake machinery, is very likely to be increased by increased plasma glucose concentration. For this reason, and because *net* glucose output is decreased when glucose uptake by the liver is increased, it was not a logical necessity to postulate a decrease in new glucose production by the liver, or an action of extra insulin secreted in response to hyperglycemia, in order to understand these observations.

In more recent years this effect of increased plasma glucose concentration to decrease *net* hepatic glucose output has been demonstrated again by transhepatic catheterization by COMBES, ADAMS, STRICKLAND and MADISON (1961), and by LANDAU, LEONARDS and BARRY (1961). In this more recent work, *net* hepatic glucose output was also found to be inhibited by insulin infusion at plasma glucose concentrations which were kept normal by simultaneous glucose infusion (LEONARDS et al. 1961, MADISON et al. 1960). To explain these observations it sufficed to consider that glucose uptake by liver was stimulated by insulin at normal glucose concentration, and hence that increased insulin secretion probably also played at least some part in the decreased *net* hepatic glucose output seen at elevated plasma glucose concentrations when glucose was infused. It was still not a logical necessity to postulate a decrease in new glucose production by the liver as an effect of insulin.

The concept of decreased new glucose production by the liver came more naturally to those using the isotope dilution procedures. However, as discussed in the preceding section (II B 1), a decrease in the rate of release of unlabeled glucose can come about because of increased labeling of phosphorylated intermediates as a result of increased glucose uptake, as well as because of a real decrease in the flow of glycogenolytic or gluconeogenic glucose into the phosphorylated intermediates and thence to free glucose. Because of this complication, a decrease in the rate of glycogenolysis and/or gluconeogenesis in the liver cannot finally be proven by establishing the existence of a decrease in glucose-C^{12} release as a result of insulin action. Additional evidence is required, such as has been furnished recently by the percutaneous biopsy procedure for the collection of a series of liver samples during an isotope dilution experiment, as reported by BISHOP, STEELE, ALTSZULER, DUNN, BJERKNES and DE BODO (1965).

The first suggestion that injected glucose decreases unlabeled glucose release was furnished by SEARLE and CHAIKOFF (1952), who reported what appeared to be an immediate complete cessation of unlabeled glucose release following the intravenous injection of a glucose-C^{12} load in the dog previously given a trace injection of glucose-C^{14}. These results were confirmed at a later date by REICHARD, FRIEDMAN, MAASS and WEINHOUSE (1958). However, it was observed subsequently by STEELE, BISHOP and LEVINE (1959) that the intravenous injection of a glucose-C^{12} load in the eviscerated dog previously given a trace injection of glucose-C^{14} gave a similar apparent cessation of unlabeled glucose inflow even though, in this case, an infusion pump was delivering unlabeled glucose to the blood at a constant rate. The delayed mixing of injected glucose with a part of the body glucose pool is responsible for this phenomenon (see II B 2). Also, the intravenous injection of a glucose load labeled to the same extent as the glucose pool did not cause, during the short (30 minutes) time period studied, a cessation of unlabeled glucose release, as shown by STEELE and MARKS (1958). This appeared to clarify the situation, but in further experiments, using suitably labeled glucose loads, DE BODO and his coworkers (DE BODO, ALTSZULER et al. 1959, DE BODO, STEELE et al. 1959, 1963a) reported that two of three dogs were found to respond to glucose administration by a decrease in endogenous unlabeled glucose release. Furthermore, in sheep, continuous infusion of suitably labeled glucose was shown to decrease endogenous unlabeled glucose release very markedly (ANNISON and WHITE 1961). In a recent series of experiments by STEELE, BISHOP, DUNN, ALTSZULER, RATHGEB and DE BODO (1965) involving 9 experiments in unanesthetized dogs, using the priming dose and continuous infusion (of glucose-C^{14}) procedure, administration of suitably labeled glucose was found to decrease endogenous unlabeled glucose release in all cases. Four of the 9 experiments were done after the animals had been kept on a high carbohydrate diet for a considerable period of time (see below for significance). Glucose was infused intravenously in these experiments and decreased the rate of endogenous unlabeled glucose release by half when glucose was infused at $1/_2$ to $1^1/_2$ times the resting endogenous rate of glucose production. In the second and third hour, with the rate of glucose infusion increased to 2 to 5 times the resting endogenous glucose production rate, endogenous unlabeled glucose release was reduced by $^3/_4$ or more. These findings are considered to be primarily the results of increased endogenous insulin secretion rather than the results of increased plasma glucose concentration acting directly on the liver to increase hepatic glucose uptake. The reasoning for this is deferred until after the description of the results with injected insulin.

An effect, feeble and poorly reproducible, of injected insulin to decrease endogenous unlabeled glucose production in normal dogs was reported, first in 1956, by WALL, STEELE, DE BODO and ALTSZULER (1956, 1957) for animals given a single intravenous injection of insulin (0.025 to 0.10 U/kg) during experiments utilizing the priming dose and continuous infusion (of glucose-C^{14}) technique. The effect was seen 5 to 10 minutes after insulin injection and was rapidly superseded by increased unlabeled glucose release called forth by insulin-induced hyperglycemia. The necessity for avoiding hypoglycemia in order better to demonstrate the true effect of insulin was not appreciated immediately. Further experiments in the dog by DUNN, FRIEDMAN, MAASS, REICHARD and WEINHOUSE (1957), by REICHARD, FRIEDMAN, MAASS and WEINHOUSE (1958), and by HETENYI, WRENSHALL and BEST (1961) and in the rabbit by BERSON and YALOW (1957), and by BERSON, WEISENFELD and PASCULLO (1959), were interpreted as yielding evidence for a greater or lesser degree of inhibition of unlabeled glucose release when insulin was given in the absence of exogenous glucose to prevent hypoglycemia. On the other hand, other experiments in the dog by SHOEMAKER, MAHLER and ASHMORE (1959), by TARDING and SCHAMBYE (1958), and by DE BODO, ALTSZULER, DUNN, STEELE, ARMSTRONG and BISHOP (1959) and in the rat by ASHMORE, CAHILL, EARLE and ZOTTU (1958), were interpreted in such a way as to discount the physiological importance of any small initial inhibition of unlabeled glucose release which was seen. In the human subject the decrease in unlabeled glucose release was prolonged, and continued in the face of considerable hypoglycemia, as shown by JACOBS, REICHARD, GOODMAN, FRIEDMAN and WEINHOUSE (1958), by REICHARD, JACOBS, KIMBEL, HOCHELLA and WEINHOUSE (1960), by SEARLE, MORTIMORE, BUCKLEY and REILLY (1959), and by KALANT, CSORBA and HELLER (1963). Meanwhile, the slow infusion of insulin (0.1 U/kg over 90 to 120 minutes) in the human was shown by MADISON, UNGER and RENCZ (1960) to result in severe hypoglycemia while having no significant effect on the femoral A-V glucose difference. This difference in the response of the human is discussed further below.

During about the same period of time covered by the above investigations, the decrease in *net* hepatic glucose output in response to injected insulin was found, by trans-hepatic catheterization, not to occur in the dog when hypoglycemia was allowed full development (FINE and WILLIAMS 1960, SHOEMAKER, MAHLER and ASHMORE 1959, TARDING and SCHAMBYE 1958). However, when hypoglycemia was limited to about 20 mg/100 ml or less below the preinsulin level a large and reproducible decrease in *net* hepatic glucose output was observed in porta-caval shunt dogs by MADISON, COMBES, ADAMS and STRICKLAND (1960), and in normal dogs previously maintained on a high carbohydrate diet (LEONARDS, LANDAU, CRAIG, MARTIN, MILLER and BARRY 1961). MADISON et al., but not LEONARDS et al., were able to accomplish this by infusing insulin intravenously at a low rate without an accompanying glucose infusion. It is possible that MADISON's portacaval shunt dogs may have been different in this respect because factors (e.g. glucagon) secreted in response to hypoglycemia did not enter the liver directly *via* the portal vein in these animals. GENES, CHARNAYA and YURCHENKO (1962) observed decreased *net* hepatic glucose output in dogs, by trans-hepatic catheterization, during severe insulin hypoglycemia. They attributed their success in demonstrating this effect to the deep amytal anesthesia utilized in their experiments, which they felt inhibited the increased secretion of the hormones normally called forth by hypoglycemia.

The above findings served to support the emphasis which BOUCKAERT and DE DUVE (1947) had placed on the avoidance of hypoglycemia in examining the effects of insulin on the liver. The earlier findings of BEARN, BILLING and SHERLOCK (1953) and of BONDY and his co-workers (1949), obtained by hepatic vein catheterization in the human, also indicated an effect of insulin on glucose balance across the liver but were less secure because glucose uptake by splanchnic adipose tissue under insulin action was capable of contributing to the positive findings.

In view of the above and also in view of the observation of the "restraining action" of continuing insulin infusion on the increased unlabeled glucose release brought about by hypoglycemia as noted by DUNN, ALTSZULER, DE BODO, STEELE, ARMSTRONG and BISHOP (1959, 1960), experiments were undertaken by STEELE, BISHOP, DUNN, ALTSZULER, RATHGEB and DE BODO (1965) to study the effect of infused insulin on unlabeled glucose release while controlling hypoglycemia by the simultaneous infusion of glucose. These experiments were done using the priming dose and continuous infusion (of glucose-C[14]) technique. The bulk glucose infusions, which were required to maintain plasma glucose concentration at a mean value of only 13 ± 3 mg/100 ml less than the pre-insulin level, were labeled with glucose-C[14] to match the prevailing specific activity in the body glucose pool. In 9 experiments, of which 2 were with dogs kept prior to the experiment on a high carbohydrate diet, large decreases in endogenous unlabeled glucose release were seen in every case. Unlabeled glucose release was decreased by half during the first hour of insulin infusion at 0.1 U/kg per hr or more; during the second hour of insulin infusion, unlabeled glucose release was decreased by $^3/_4$ or more.

In conjunction with the experiments just reviewed, serial samples of liver were obtained by percutaneous needle biopsy without general anesthesia by

Bishop, Steele, Altszuler, Dunn, Bjerknes and de Bodo (1965). The liver samples were analyzed for glycogen content, for total C^{14} content, and for C^{14} present in the isolated glycogen. By this means it was found that the decrease in unlabeled glucose release during the first hour of insulin infusion was accompanied by a decrease in the net loss of unlabeled glycogen (137 mg/kg body wt per hr) which was more than sufficient to account for the observed decrease (97 mg/kg body wt per hr) in unlabeled glucose release. The accumulation of plasma glucose-C^{14} units in liver glycogen during this time was increased from zero (in the preinsulin period) to 20 mg/kg body wt per hr. In the next $1^1/_2$ hours of insulin infusion there was no further decrease in the net loss of unlabeled glycogen, yet unlabeled glucose release was further decreased by 55 mg/kg body wt per hr. The accumulation of plasma glucose-C^{14} units into glycogen was sharply increased in this period to a total of 89 mg/kg body wt/hr. From these results it was concluded that an increased uptake of labeled plasma glucose for glycogen synthesis in the first hour of insulin infusion could not have been responsible for the decreased rate of net glycogen loss, since in that event a larger increase in C^{14} accumulation in glycogen would have been seen. Net glycogen breakdown must then have been decreased either by an inhibition of glycogenolysis or by an extensive resynthesis of unlabeled glycogen from unlabeled glucosyl units in the liver metabolic intermediates. Either process would leave less unlabeled glucose precursor available for release from the liver intermediates as free glucose. Hence it was deduced that a real decrease in new glucose production and release by the liver was brought about during the first hour of insulin infusion. In contrast with this, in the next $1^1/_2$ hours of insulin infusion, there was no further decrease in the net loss of unlabeled glycogen to account for the observed further decrease in unlabeled glucose release. At this time there was a large increase in glucose uptake by the liver, the rate of accumulation in glycogen alone being enough to account for 12% of the elevated (3.7-fold, to 670 mg/kg body wt per hr) rate of whole body glucose uptake. Thus it was concluded that an increased specific activity of the phosphorylated intermediates, due to increased labeled glucose uptake, could account for the additional decrease in unlabeled glucose release brought about in the later period of insulin infusion. However, this did not confirm or deny the possibility that decreased unlabeled glucose production *via* gluconeogenic routes, from unlabeled amino acids for example, did not also play a part in the further decrease in unlabeled glucose release.

The increase in uptake of plasma glucose by the liver, most of this accumulating in liver glycogen, fits with the idea that glycogen synthetase becomes activated at this time, but does not prove this point.

This finding, in the dog, of an early real decrease in new glucose production and release when insulin is infused (along with glucose to prevent hypoglycemia) fits in with the observations of Searle and his coworkers (1960) in human subjects. Insulin was given subcutaneously (0.15 U/kg), and simultaneously an intravenous infusion of unlabeled glucose was begun at a rate equal to the rate of endogenous unlabeled glucose production prior to insulin. After $^1/_2$ hour the unlabeled glucose infusion rate was lowered by 40%. These manipulations were done during the continued infusion of "weightless" glucose-C^{14} at a rate which had sustained plasma glucose specific activity at a constant level for several hours prior to insulin; plasma glucose concentration and specific activity remained constant under these circumstances for several hours after insulin also (G. L. SEARLE, personal communication). The findings indicate that endogenous unlabeled glucose production was zero for the first half hour after subcutaneous insulin, and was 40% of the pre-insulin level for the next 2 hours, and that at the same time no change in glucose uptake occurred, either by the peripheral tissues or by the liver itself.

The obvious differences between the findings in these experiments using human subjects and in the experiments of STEELE et al. using dogs were a) in dogs increased glucose uptake by the peripheral tissues accompanied the reduction in new glucose production, b) in dogs the suppression of new glucose production was seldom complete, and c) in dogs an increased uptake of glucose by the liver itself was evident after the first hour of insulin infusion. The last of the three differences above could have been due in part to the larger amount of insulin infused in the dog experiments. Otherwise the differences are to be ascribed to a greater responsiveness of the human liver to the inhibitory effect of insulin on new glucose production and release. In the normal human in the post-absorptive state a small insulin dosage rate can be arrived at, apparently, which inhibits new glucose production without increasing peripheral glucose uptake, whereas in the normal dog this has been seen extremely rarely in the course of scores of experiments using small insulin infusion rates.

It is suitable to return at this point to the experiments of STEELE et al. (1965) reviewed earlier, in which glucose infusion was shown to decrease endogenous unlabeled glucose release. Hepatic biopsy samples (BISHOP et al. 1965) were taken in conjunction with these experiments also. The similarity of the results with the results of experiments in which insulin was infused was striking. The especially important similarity was the early large decrease in net loss of unlabeled glycogen without much increase in the accumulation of glucose-C^{14} units in glycogen, and the later large increase in the accumulation of glucose-C^{14} units in glycogen with no further decrease in the net loss of unlabeled glycogen. As stated previously, it is easy to believe that increased glucose uptake by the liver is likely to be caused by increased plasma glucose concentration. Nevertheless it is probable that the early decrease in net glycogen loss in these glucose infusion experiments was not brought about by an increased uptake of labeled glucose for glycogen synthesis, for if this had been so, the accumulation of glucose-C^{14} units in glycogen would have been very much greater.

Hence it was concluded that endogenous insulin, secreted in response to the hyperglycemia caused by glucose infusion, brought about the same early actual decrease in new glucose production in the liver that was seen in the experiments in which insulin was infused directly.

The observation that a high carbohydrate diet, fed to the animal previously, accentuates the inhibition of *net* hepatic glucose output by infused glucose was given two possible interpretations by LANDAU, LEONARDS and BARRY (1961). One was a greater secretion of insulin in response to hyperglycemia in the dog maintained on a high carbohydrate diet and the other was an inherently greater glucose uptake by the liver of such a dog at elevated glucose concentration, due to alterations in the level of hepatic enzymes concerned with glucose utilization. The isotope dilution experiments just reviewed included too few animals kept on the high carbohydrate diet to add any substantial weight in favor of the concept of a greater responsiveness of this animal to glucose infusion. However, they do contribute some evidence, as just reviewed, that, of the two possibilities advanced by LANDAU et al., increased endogenous insulin secretion in response to hyperglycemia may be the more important factor, at least in the early period of glucose infusion.

BERTHET, JACQUES, HERS and DE DUVE presented evidence in 1956 for an *in vivo* effect of insulin on the incorporation of plasma glucose-C^{14} into liver glycogen. In these experiments 3 rabbits which had been fasted 24 hours prior to the experiment were given 30 U of insulin, intravenously, 30 minutes after a glucose-C^{14} injection, then a continuous infusion of glucose at 0.9 g/kg per hr together with glucose-C^{14} at 2 μc/kg per hr. Control animals were given the priming injection of glucose-C^{14}, then 30 minutes later a continuous infusion of glucose-C^{14} at 0.65 μc/kg per hr. These treatments resulted in a roughly constant plasma glucose concentration and plasma glucose-C^{14} specific activity in the two groups. One hour after the insulin injection the rabbits were killed and liver glycogen was isolated. In the control animals the mean accumulation of plasma glucose-C^{14} in glycogen was 0.43 mg/g liver per hr; in the insulin-injected group it was 2.6 mg/g liver per hr. These experiments were carried out in conjunction with liver slice experiments, which will be reviewed below.

WILLIAMS, HILL and CHAIKOFF, in 1960, infused glucose in rats (control group) at about 3 g/kg per hr alone, or together with insulin given as a single injection of 100 U/kg followed by a continuous infusion at 48 U/kg per hr (experimental group). The rats were alloxan-diabetic animals maintained on insulin until 14 hours prior to the experiment. The infusion of glucose or glucose and insulin was given into the portal vein under sodium pentobarbital anesthesia. Rats from the control group, sacrificed initially and at 10, 30, and 60 minutes, had livers containing the same amount of glycogen (about 0.3 g per 100 g liver) whereas at 30 and 60 minutes the insulin-treated rats had livers containing $2^{1}/_{2}$ to 3 times this amount of glycogen. These experiments were done in conjunction with *in vitro* experiments on slices of the excised livers (as reviewed below) in which it was shown that the insulin infusion, in 30 minutes, increased by 4-fold the ability of the slices to incorporate glucose-C^{14} from the medium into glycogen.

The evidence reviewed so far in this subsection offers only one relatively weak reason for believing that insulin acts directly on the hepatic cells to produce these effects, and that is the rapidity with which some of them are seen after insulin injection. Since about 40 % of a small dose of insulin injected into the portal vein is removed from the blood by the liver in the first pass, as shown by MORTIMORE, TIETZE and STETTEN (1959) (see also ELGEE et al. 1954 and HAUGAARD, VAUGHAN et al. 1954), using I^{131} tagged insulin, the possibility exists of establishing the directness of the insulin effect on liver by observing the differential effects of intraportally *vs.* peripherally injected insulin. This course has been explored with what appears to be as positive a finding as could be expected. MADISON and UNGER (1958), using sodium pentobarbital anesthetized dogs, injected glucagon-free insulin (0.07 U/kg body wt) over a 2-minute period into either a peripheral (foreleg) vein or into a tributary of the portal vein. Arterial and venous blood samples were drawn from a femoral artery and the contralateral femoral vein at intervals after insulin injection. When insulin was given peripherally the mean increase in A-V difference across the hind limb was 6.9 mg/100 ml, when insulin was given intraportally the increase was 4.6 mg/100 ml; the difference between these two results was statistically significant. The overall lowering of arterial blood glucose concentration caused by the insulin was greater at 20 minutes to 40 minutes when the peripheral route of insulin injection was used, but not significantly so. The conclusion drawn from this fact was that the lesser peripheral uptake when insulin was given intraportally was nearly balanced by a greater depressing effect of the insulin on *net* hepatic glucose output. This indicated that there was a direct effect of insulin on the glucose balance across the liver, since the influence of the liver could be accentuated, and the influence of the peripheral uptake diminished, by giving the insulin intraportally. Previous investigations by WEISBERG, FRIEDMAN and LEVINE (1949), in which dogs were infused continuously with an insulin (Novo) low in glucagon content at 0.1 U/kg per hr for 2 hours indicated that peripherally administered insulin was more effective in the overall lowering of the blood sugar level than intraportally administered insulin. However, in view of the small number of dogs used (intraportal and peripheral infusions were compared in 2 dogs) this work cannot be said to invalidate the conclusions reached by MADISON and UNGER. The findings of GALANSINO and his coworkers (1958), using anesthetized dogs, were also at variance with the conclusion that intraportally administered insulin has about as much overall blood sugar lowering effect as peripherally administered insulin. However, MARTIN, LEONARDS and MILLER (1959) using unanesthetized dogs, found that in 24-hour-fasted animals there was only a barely significant difference in blood sugar lowering in favor of the peripheral route, although in 8-hour-fasted animals the difference was greater. Thus it seems reasonable that MADISON and UNGER (1958) should have observed equal blood sugar lowering whether insulin was given peripherally or intraportally in their particular dogs. Since their peripheral A-V glucose differences showed there was less peripheral uptake with intraportally administered insulin, their interpretation of a greater hepatic effect with intraportally administered insulin seems to stand. Peripheral blood flow was not measured, but the consequences of this seem unable to vitiate the interpretation (cf. discussion by MADISON and SHOEMAKER 1959). More conclusive evidence indicating a direct action of insulin on liver cells is furnished by investigations using isolated livers or liver slices as reviewed in later subsections.

4. Hepatic glycogenolysis. The mechanisms whereby glucagon and epinephrine stimulate glycogenolysis by way of cyclic 3,5-AMP and the activation of phosphorylase have been reviewed by SUTHERLAND and RALL (1960). An effect of insulin to counter phosphorylase activation in an *in vitro* system has been sought for by CRAIG and LARNER (1964), but was not found.

SOKAL and co-workers (1964) on the basis of experiments utilizing perfused rat liver, have concluded recently that glucagon plays a physiological role as a hepatic glycogenolytic agent, whereas endogenous epinephrine under physiological conditions affects only muscle glycogen. They found that the highest value of systemic epinephrine concentration reported to occur naturally is below the threshold for stimulation of hepatic glycogenolysis as measured in the liver perfusion experiments.

An effect which was interpreted to represent an anti-glycogenolytic action of insulin was observed by ISSEKUTZ in 1924. These experiments were elaborated by ISSEKUTZ and SZENDE, in 1934, and it was demonstrated that the livers of frogs (kept at 30—33° C) which had been given 1 U of insulin 4 hours prior to removal of the livers for perfusion released less glucose to the Ringer perfusate than livers from untreated frogs. It was not found possible to produce this effect by perfusion of livers *in vitro* for 4 hours with insulin-containing perfusion fluid. Later experiments having to do with this effect in frogs have been reported by SMITH (1953). The bearing of these studies on the effects seen in mammals, where events move so much more rapidly, is not at present apparent.

It is clear that no specific biochemical or physiological mechanism is known whereby insulin can counter the breakdown of hepatic glycogen to glucose-1-phosphate. On the other hand, it is possible that such a mechanism, though presently unknown, will be discovered. Several lines of physiological evidence in addition to those given above suggest this possibility.

The "restraining action" of insulin (on the increase in unlabeled glucose output which occurs during hypoglycemia) was observed during prolonged continuous infusion of insulin by DUNN, ALTSZULER et al. (1959) and DUNN, STEELE, ALTSZULER, DE BODO, ARMSTRONG and BISHOP (1960). This phenomenon was also observed later, with regard to *net* hepatic glucose output (by use of a transhepatic catheterization technique) by FINE and WILLIAMS (1960). The restraint on the increased glucose output of the dog in insulin-induced hypoglycemia is terminated abruptly (within 5 to 10 minutes) when the continuous infusion of insulin is stopped.

A similar effect of insulin infusion, that is a slowing to normal of the accelerated rate of unlabeled glucose release brought about in the dog by acute phlorizin poisoning, was observed by DUNN, STEELE, ALTSZULER, DE BODO, BISHOP and ARMSTRONG (1960).

In the normal dog in the postabsorptive state there is a continuous net loss of liver glycogen. It is this loss which is stopped by insulin during the early period of infusion as reviewed in the preceding subsection. In the postabsorptive state, hypoglycemia, as this term is generally understood, is not evident yet something appears to be bringing about hepatic glycogenolysis. In this connection the experiments of KOLODNY, KLINE and ALTSZULER (1962) are illuminating. In these experiments *net* hepatic glucose output in dogs was found elevated during phlorizin glucosuria in the absence of an observable lowering of arterial or portal vein plasma glucose concentration. In such animals, *net* hepatic glucose output could be kept down to the prephlorizin level by the infusion into the inferior vena cava of the same quantity of glucose being lost to the urine, again without any significant change in arterial or portal vein plasma glucose concentration being seen. These experiments suggest the existence of a site which contains specialized cells sensitive to very small changes in blood glucose concentration, together with an associated neurohumoral apparatus for increasing hepatic glucose production. The same mechanism may account for glycogenolysis in the normal animal in the postabsorptive state in the absence of any evident hypoglycemia. Glucagon secretion has been reported by UNGER, EISENTRAUT, MCCALL and MADISON (1962) to be increased in the dog during phlorizin glucosuria as evidenced by increased (immunologically measured) glucagon present in the plasma at that time.

The rapid increase in blood glucose concentration which follows the injection of anti-insulin serum, as shown by ARMIN, GRANT and WRIGHT (1960), suggests that endogenous insulin present in the postabsorptive state may be exerting a restraining action on hepatic glucose production in opposition to a factor which is exerting an influence to increase glucose output. Experiments using the priming dose and continuous infusion (of glucose-C^{14}) technique have indeed shown a prompt increase in unlabeled glucose release after injection of anti-insulin serum in the normal dog, as reported by ALTSZULER, STEELE, TOBIN, RATHGEB and DE BODO (1964). The adrenalectomized dog maintained on DOCA and cortisol responded similarly in these studies, indicating that epinephrine secretion was not involved in the response to anti-insulin serum.

Attempts have been made to establish the existence of a restraining action of insulin on the increased unlabeled glucose release brought about by administered glucagon. These, if sucessful, would have offered strong evidence for a direct intervention by insulin in the hepatic cell against enhanced glycogenolysis. At first (DUNN, STEELE, ALTSZULER, DE BODO, BISHOP and ARMSTRONG 1960) such an effect seemed to be obtainable, but this effect has now been found not to be reproducible (STEELE, BISHOP, DUNN, ALTSZULER, RATHGEB and DE BODO 1965).

5. Hepatic glycogen synthesis. The knowledge of the enzymes which convert glucose-1-phosphate to uridine diphosphate glucose and thence to glycogen have been reviewed by LELOIR (1964) and by MANNERS (1962).

LARNER (1964) has reviewed recently his work and that of his collaborators, VILLAR-PALASI, ROSELL-PEREZ, RICHMAN and FRIEDMAN with regard to the activation of glycogen synthetase. In contrast with phosphorylase, where phosphorylation of a form of the enzyme which can be inhibited by ATP and glucose-6-phosphate converts it to a form which cannot be inhibited by these compounds, phosphorylation of a glucose-6-phosphate dependent form of muscle glycogen synthetase converts it to a glucose-6-phosphate independent form. Diaphragm muscle after incubation with insulin has a greater fraction of its glycogen synthetase activity in the glucose-6-phosphate independent form. No inhibiting effect of insulin on phosphorylase activation is as yet known. Cyclic 3,5-AMP, which promotes the phosphorylation ("activation") of phosphorylase by ATP, also promotes the phosphorylation ("inactivation") of muscle glycogen synthetase by ATP. This accounts for the action of epinephrine, which affects both systems in the same way as does cyclic 3,5-AMP. In liver there is evidence for a similar conversion of glycogen synthetase from a glucose-6-phosphate dependent form to a glucose-6-phosphate independent form, accompanied by a release of inorganic phosphate from the enzyme. A third form of the enzyme, inactive in the presence of glucose-6-phosphate, was also found in liver.

Inasmuch as it has not proven possible to show that insulin interferes with the activity or the activation of phosphorylase, some might think it advisable to try to explain the early inhibition, by infused insulin, of hepatic glycogen loss, as reported by BISHOP et al. (1965) as being due to glycogen synthetase activation, and hence an increased rate of glycogen synthesis.

In doing this is necessary to assume that, although rapid glycogenolysis continues unchecked, net glycogen loss is stopped because the glycogenolysis becomes nearly balanced by an increase in glycogen synthesis. Furthermore, to explain the lack of accumulation of glucosyl-C^{14} residues in glycogen during the early period of insulin action, it is necessary to assume that the postulated increased glycogen synthesis is mostly at the expense of unlabeled glucosyl residues of endogenous origin, especially, perhaps, those just released from glycogen by the continuing action of phosphorylase.

In the later period of insulin action, reviewed above, increased uptake of glucose from the plasma is evidenced by the greatly increased accumulation of glucosyl-C^{14} residues in glycogen. Hence at this time, in addition to the postulated activation of glycogen synthetase, an increased entry of plasma glucose into the phosphorylated intermediates is encountered.

There are several difficulties in the way of ascribing this increase in glucose uptake to a decrease in glucose-6-phosphate concentration created by increased glycogen synthetase action. One is that a reasonably high glucose-6-phosphate concentration is commonly expected to accompany glycogen synthetase action, because it is known that the activity, at least of one of the forms of the enzyme, is thereby increased. Future findings relative to the activation of liver glycogen synthetase may remove this difficulty. Another is that the increase in glucose phosphorylation would seem to have to be a *net* increase, due to the decreased splitting of newly formed glucose-6-phosphate, with glucose phosphorylation remaining unchanged. The argument here is somewhat abstract. In liver, in contrast with the situation in muscle, the hexokinase which is regarded as important for glucose uptake is low affinity glucokinase, an enzyme which is not inhibited, as is muscle hexokinase, by high glucose-6-phosphate concentration. Thus a decrease in glucose-6-phosphate concentration in the liver cell would not be expected to lead to an increase in the rate of glucose phosphorylation, whereas it might lead to a decrease in the rate of splitting of newly formed glucose-6-phosphate.

The above would leave unexplained the eventual induction by insulin of an increased amount of hepatic low affinity glucokinase. It would seem that the induction of this enzyme should follow the familiar pattern by which an increased use of the enzyme precedes an increase in its amount. Two other explanations, not subject to this difficulty, for the increased uptake of blood glucose by the liver suggest themselves. One is that there may be an increase in low affinity glucokinase activity, brought about by an unknown mechanism secondary to insulin action, the other is that free glucose under the influence of a larger amount of insulin may occupy a space in the hepatic cell (a space in which glucokinase is located) from which it is excluded in the presence of a smaller amount of insulin. The second possibility has been looked for, but has not been demonstrated.

VESTER and REINO (1963), and VESTER (1964) reported that a particle-free 0.3 M KCl extract of normal rat liver contains a glucokinase whose activity is increased by incubation for 4 minutes with an amorphous insulin preparation (not by zinc insulin). The data are said to suggest the blocking, by insulin, of a noncompetitive inhibitor of glucokinase activity. These preliminary reports require amplification. The much earlier findings of an effect of insulin added *in vitro* on muscle hexokinase activity have not proven to be reproducible (see review by STADIE 1954).

6. Summary. The earliest effects of insulin on liver carbohydrate metabolism in the intact animal can be summarized in a way which takes into account most of the observations reviewed above.

At the resting blood sugar level in the postabsorptive state both insulin and insulin-opposing factors such as glucagon are continually being supplied at a low rate. A rise in blood glucose concentration brings out more insulin; a fall brings out more of the opposing factors.

The release by the liver to the blood of newly-formed glucose is controlled by the balance of these opposing influences. Insulin restrains this process and the opposing influences further it. In the resting postabsorptive state the balance results in the continuous release of new glucose and a steady depletion of liver glycogen. New glucose release can be increased or decreased in 5 minutes by artificially decreasing or increasing the supply of insulin.

In the normal human, but not in the normal dog, a very small increase in insulin supply (an infusion of insulin) above the endogenous secretion rate in the postabsorptive state increases the restraint on new glucose release from the liver without increasing the uptake of glucose from the blood by the peripheral tissues. In the normal human, new glucose release continues to be restrained by a small increase in insulin supply even when a considerable degree of insulin-induced hypoglycemia is allowed to develop slowly. In the dog, the amount of extra insulin required to further restrain glucose release also increases peripheral glucose uptake. Furthermore, in the dog, the opposing influences brought out by hypoglycemia are stronger; new glucose release, though transiently decreased by insulin infusion, is raised above the resting value in the postabsorptive state as soon as mild hypoglycemia develops. Here the restraining influence of the insulin becomes evident only when the infusion of insulin is stopped. When this is done there is an abrupt further increase in new glucose release by the liver.

Increased uptake of glucose from the blood by the liver, predominantly for glycogen synthesis, is a more slowly developing effect of increasing the insulin supply; it may be that a larger amount of extra insulin is required for this than for the decrease in glucose release. In the human, when a small increase in insulin supply is arranged together with an infusion of the amount of glucose released by the liver in the postabsorptive state, there appears to be little or no increase in glucose uptake by any tissue, including liver, for a period of several hours during which the release of new glucose by the liver is stopped. In the dog, during the first hour of increased insulin supply (whether brought about by a moderate hyperglycemia caused by glucose infusion or by the infusion of insulin together with glucose to prevent hypoglycemia) increased glucose uptake by the liver is small whereas the restraint on new glucose release from the liver (and on net glycogen loss) is strongly developed.

In the perfused liver the rate of gluconeogenesis from amino acids is decreased quickly by insulin. In the intact animal there may be circumstances under which this consequence of insulin action, rather than a decreased rate of hepatic glycogenolysis, may prove to be the dominant factor in decreasing the rate of hepatic glucose production.

Indications that insulin acts directly on the liver to produce effects on carbohydrate metabolism are available from whole animal experiments in which insulin was administered by way of the portal vein. Additional evidence for the direct action of insulin in liver depends upon experiments utilizing isolated liver tissue. The restraint of new (unlabeled) glucose release, as distinct from decreased net hepatic glucose output, has not been demonstrated in isolated liver; however, the rapidity with which this effect develops *in vivo* suggests that this also is the result of the direct action of insulin on the liver.

C. Experiments using isolated liver

Loss of K^+ from hepatic cells is a characteristic occurrence when a whole liver is perfused or when liver slices are incubated. It has long been suspected that this K^+ loss is related in some way to carbohydrate metabolism in the hepatic cell. Formerly the effect of insulin to bring about hepatic cell K^+ uptake was regarded to be a secondary result of the influence of the hormone on carbohydrate metabolism, and K^+ loss by isolated liver tissue was thought of as an occurrence which should be prevented primarily because the altered intracellular cationic environment was expected to change, in a secondary way, the activities of hepatic enzymes of carbohydrate metabolism. FENN, in 1940, summarized the earlier findings in this area.

More recent findings are that the earliest known manifestation of insulin action in the liver is altered cation transport, that this occurs in opposition to a very rapid effect produced by epinephrine and glucagon, and that these effects of insulin and insulin antagonists appear to precede the effects of these agents on carbohydrate metabolism. If the usual situation in isolated liver is that injury to the tissue has resulted in a continuing K^+ efflux which cannot be reversed by insulin added *in vitro*, then it might be anticipated that insulin added *in vitro* usually would be found ineffective in causing changes in carbohydrate metabolism in isolated liver tissue. For this reason a short review of the current knowledge of the effects of insulin, glucagon and epinephrine on net K^+ loss by the liver is pertinent at this point.

1. Significance of K^+ release. D'SILVA first reported in 1934 that epinephrine, injected intravenously, very rapidly increases the rate of release of K^+ from tissues to the blood; in subsequent experiments (D'SILVA 1936) he established that the liver is the source of most of the K^+ released. In his original publication D'Silva reported that the increased serum K⁺ values fall back to normal in less than 5 minutes and continue falling to a maximum below-normal value at 10 minutes. He attributed the fall in K⁺ to increased insulin secretion induced by the high blood sugar values produced by epinephrine. He attempted, unsuccessfully, to demonstrate such an action of administered insulin; however, it is likely that his insulin preparation, at that date, was contaminated with glucagon.

Finder, Boyme and Shoemaker (1964), utilized the transhepatic catheterization technique to demonstrate that intravenously injected glucagon initiates a release of K⁺ by the liver of the dog in a matter of seconds. The same laboratory (Shoemaker and Finder) had reported in 1961, using this technique, a similar effect of epinephrine on K⁺ release by the liver.

Mortimore, in 1961, reported that insulin, at perfusate concentrations of 20 to 200 µg/ml inhibited the loss of K⁺ from an isolated perfused rat liver preparation with which he was also later able to show an effect of insulin on net glucose release (see section III, C 3 below). Lambotte and Shoemaker (1964), in a recent abstract, have reported that intravenous injection of insulin results within 2 minutes in a reversal of the continuous loss of K⁺ from the liver which is characteristic for the fasting dog. The effect is greater when the dog is under barbiturate anesthesia, and in the dibenzyline-treated conscious animal. The injection of anti-insulin serum was reported to result in a large increase in K⁺ release by the liver of the fasting dog.

Inasmuch as Finder et al. (1964) demonstrated that the activation of liver phosphorylase, as a result of glucagon injection, follows by several minutes the enhancement of K⁺ release by the liver, it seems reasonable to suggest that an event which leads to K⁺ release might also be the event which leads to the formation of cyclic 3,5-AMP, and that insulin might intervene, in opposition to glucagon and epinephrine, in this very early area. The cyclizing enzyme which produces cyclic 3,5-AMP is associated with a particulate fraction which includes cell nuclei and portions of cell membrane [see Sutherland and Rall (1960)] and not with mitochondria or microsomes. On the other hand, the transport of cations appears to be more closely linked with electron transport in the mitochondria or with the high energy phosphate compounds which result from electron transport in the mitochondria [see Ussing (1961) and Lehninger (1962)]. Increased oxygen uptake and increased K⁺ uptake in the absence of medium glucose are well documented effects of insulin added *in vitro* to frog muscle [Gourley (1957), Gourley and Fisher (1954), Hall, Fisher and Stern (1954), Manery, Gourley and Fisher (1956), Toye and Manery (1955)]. For a review see Krahl (1961). Increased oxygen uptake has not been seen as an effect of insulin in mammalian muscle except in the isolated instance of minced pigeon breast muscle first reported by Krebs and Eggleston in 1938. The future search for the molecular site of insulin action very probably will take place in this area, one of the obvious questions being whether or not insulin penetrates to the mitochondria rapidly enough that an action there might be held able to account for its very early effects on cation and glucose transport.

The glucose tolerance factor of MERTZ, which has been established to be chromium (III), has been found recently [MERTZ and ROGINSKI (1963)] to be effective, when added *in vitro*, in restoring the capacity of adipose tissue deficient in chromium (III) to respond to insulin by an increase in galactose transport. Chromium (III) *in vitro* is also effective in sensitizing liver mitochondria from chromium (III) deficient rats to the induction of swelling by insulin, as shown by CAMPBELL and MERTZ (1963). Polarographic studies have indicated that insulin reacts with mitochondrial sulfhydryl groups and that this interaction is enhanced by chromium (III) [CHRISTIAN, KNOBLOCK, PURDY and MERTZ (1963)].

It is supposed that the very early manifestation of insulin action in the liver, that is decreased K^+ efflux, may be related to the effects of insulin on membrane polarization and K^+ transport in skeletal muscle [ZIERLER (1957, 1959a, 1959b), ANDRES, BALTZAN, CADER and ZIERLER (1962)] which were first reported by ZIERLER in 1957. So far, in skeletal muscle, these effects have been observed to occur more slowly than the earliest reported effect (in heart muscle) of insulin on glucose transport.

2. **Experiments using liver slices.** Loss of potassium from rat liver slices incubated in artificial media was suspected at an early date by HASTINGS and BUCHANAN (1942) and was held to be a negative influence on the glycogen content of slices after incubation. Substitution of a medium of K^+ content higher than the normal extracellular K^+ content was found to result in higher net glycogen gain on incubation in glucose-containing media and the elimination of net glycogen loss in glucose-free media [BUCHANAN, HASTINGS and NESBETT (1949), HASTINGS and BUCHANAN (1942)]. The suspected loss of potassium (a 40% loss) was confirmed (FLINK, HASTINGS and LOWRY 1950) by analysis of slices after 15-minute incubation in a medium containing 5 mM K^+ (Ringer's solution). It was also discovered in the same investigation that not only did sodium replace the lost potassium in the slices, but also increased further in the slice by a large amount which suggested that a 50% increase in the extracellular fluid of the incubated tissue had taken place. A large part of the lost potassium was regained by the slices on longer incubation (45 minutes) but the excessive sodium uptake was not reversed. Furthermore, in media containing K^+ in excess of 80 mM the slice potassium content increased on longer incubation to above that of fresh liver. This high potassium content was attributed to the presence of potassium at 80 mM in the expanded extracellular fluid space in addition to the potassium present in the hepatic cells. DEANE, NESBETT, BUCHANAN and HASTINGS

(1947) reported that histological examination of transverse sections made after 2 hours of incubation of slices in either low potassium or high potassium medium containing 1000 mg glucose/100 ml showed a sheet of glycogen-containing cells about 0.1 mm thick representing the surface layer of the slice, which was 0.5 mm in thickness. Identifiable mitochondria also persisted only at the periphery, and the cytoplasm of the interior cells was vacuolated. In livers from 24-hour-fasted rats glycogen was sparsely dispersed throughout the slices when these were fixed prior to incubation. In incubated slices from such rats there was much more glycogen (than prior to incubation) in the peripheral sheet of cells whether low or high K^+ medium was used, but the sheet was 12 cells deep in the high K^+ medium and only 8 cells deep in the low K^+ medium. Clearly in the presence of glucose (1000 mg/100 ml) glycogen synthesis in the peripheral two-fifths of the slice was counter-balancing the obliteration of glycogen in the central three-fifths of the slide; glycogen accumulation was going on in the peripheral cells at a rate faster than had prevailed in the liver cells *in vivo*, whether the high K^+ or the low K^+ medium was used for incubation. It appears likely that the intracellular space of the degenerating cells in the interior of the slice accounted for the observed increase in sodium space. In the much later (1958) experiments of CAHILL, ASHMORE, EARLE and ZOTTU, inulin and raffinose also were found to distribute themselves in much more than the extracellular fluid space of incubated liver slices. This deterioration of cells in liver slices suspended in artificial media is not generally appreciated [see discussion by KLEINZELLER, BURCK, CHRISTENSEN und MAIZELS (1961)].

In view of the above it might be thought difficult to select a medium with a K^+ concentration which would have the originally desired effect, i.e. the maintenance of the normal intracellular potassium concentration of the hepatic cells. In fact the recovery of a large part of the original potassium content by the whole slice in media 5 mM in K^+, under conditions in which only two-fifths of the cells may have been concentrating potassium suggests that even in low K^+ medium the glycogen synthesizing cells may have contained as much or more potassium than was present in the liver cells *in vivo*. In later studies (1957) by CAHILL, ASHMORE, ZOTTU and HASTINGS, the effect of the high K^+ (110 mM) medium in increasing the total amount of glycogen present at the end of incubation of rat liver slices was found to be associated with a greater decline of the originally high activated phosphorylase content of the slices. A low K^+-high Na^+ medium was found to result in a less rapid loss of activated phosphorylase.

BRESCIANI and SGAMBATI (1960) have reported that X-irradiation of rat liver slices *in vitro* increases the rate of glycogen loss and glucose release to the medium. They have related this loss to the effect of the irradiation to increase K^+ loss and Na^+ uptake by the slices.

KREBS, EGGLESTON and TERNER, in 1951, studied the factors involved in the loss of K+ from incubated tissues, including liver slices, and arrived at the following conclusion: "The rapid initial loss of K from freshly cut tissue slices bears on the mechanism by which potassium is discharged by the tissue... This (the initial loss) is a much higher rate than the turnover of K in the steady state, and it follows that the 'leakage' rate of K is not constant, the discharge being less in the steady state than in freshly prepared slices. It cannot be assumed, therefore, that the migration of K from the tissue to the medium is solely due to potassium leakage. Actively controlling factors must also be operative, a conclusion which has already been drawn by other workers from observations on the increased rate of discharge following the stimulation of muscle or nerve."

In view of the above it might be expected that liver slices would be more useful in revealing the presence of enzymatic capabilities associated with carbohydrate metabolism than in revealing the fine details of the *in situ* control of these capabilities, as for example, by insulin. In particular, any control affecting both carbohydrate metabolism and the intracellular concentration of potassium might well be expected to be distorted in the liver slice in a way not to be corrected by increasing the K+ content of the suspending medium.

Nevertheless success in influencing the incorporation of medium glucose-C^{14} into glycogen by the addition of insulin *in vitro* has been reported in a few instances. In a preliminary note HASTINGS, TENG, NESBETT and RENOLD reported in 1952 that a medium containing K+ at 40 mM, Na+ at 70 mM, Mg++ at 20 mM and Ca++ at 10 mM caused normal rat liver slices to convert less glucose or pyruvate to glycogen than a similar medium in which all other ions were kept the same but K+ was 110 mM and Na+ was absent. Insulin in the lower K+ medium was reported to have increased glycogen synthesis by 50%, increased glucose uptake from the medium by 40% and decreased glucose-C^{12} release to the medium by 10%. Liver slices from alloxan-diabetic rats (in which intracellular potassium was about 30% below normal) were reported to have shown a similar response to insulin added *in vitro* in the high rather than in the moderate K+ medium. In a later publication, in 1955, from the same laboratory [RENOLD, HASTINGS, NESBETT and ASHMORE (1955)] it was noted that these preliminary findings could not be confirmed.

In 1956, BERTHET, JACQUES, HERS and DE DUVE, exercising great care in the preparation of slices from the livers of nembutal anesthetized, then exsanguinated, fed normal rabbits, utilized the lower K+ medium referred to above to obtain, in 1/2 hour of incubation, a 30% increase in glucose-C^{14} incorporation due to addition of 40 µg insulin (1 U) per ml to the incubation medium. The glucose concentration of the medium in these experiments was 230 mg/100 ml; initial glycogen levels ranged from 2 to 12 g/100 g liver. In almost all cases there was net loss of glycogen from the slices; this loss averaged 0.2 g/100. In the presence of insulin the net glycogen loss was about half as much. However, when initial glycogen was low (0.4 g/100 g; substitution of greens for beets in the diet) there was net glycogen synthesis and insulin had no effect on this nor on incorporation of glucose-C^{14} into glycogen. Glucagon (20 µg/ml), added in place of insulin to the incubation medium, resulted in an increased net glycogen loss (to 0.7 g/100 g) and decreased the incorporation of medium glucose-C^{14} into glycogen to about 30% of the amount observed in the absence of both hormones. Epinephrine (40 µg/ml) had a similar effect. In the absence of added glucose the glucose concentration of the medium had a mean value of 30 mg/100 ml. Under these conditions the incorporation of medium glucose-C^{14} into glycogen was about one-eighth as much as when 230 mg/100 ml glucose were present in the medium and here no effect of insulin or glucagon could be established (2 trials). Glucose release by the slices to the medium was greatest when no glucose was added to the medium. No clear effect of insulin was obtained (3 trials) but glucagon brought about a definite increase in glucose release in both the presence and absence of added glucose in the medium. The amount of insulin per gram of liver introduced in the slice experiments (about 20 U/g) was about 600 times as much as the amount found, by BISHOP et al. (1965), to stop glycogen loss, inhibit glucose release and increase plasma glucose uptake into hepatic glycogen *in vivo*.

The effect of insulin added *in vitro* to increase glucose incorporation into glycogen in rabbit liver slices was stated, in 1959 [cf. p. 270, CAHILL, ASHMORE, RENOLD and HASTINGS (1959)] to have been confirmed.

3. **Experiments with perfused livers.** If liver perfusion is to be done under strictly physiological circumstances, both arterial blood and hepatic venous blood have to be supplied at their usual pressures and flow rates. Defibrinated blood contains powerful vasoconstrictive substances which prevent supplying arterial blood under these physiological circumstances unless the lungs are included, along with the liver, in the perfusion circuit. The use of heparinized rather than defibrinated blood has the deficiency that heparin is known to activate lipoprotein lipase; the consequent release of non-esterified fatty acids from the plasma triglycerides may well disturb hepatic metabolism. Mechanical oxygenators cause varying degrees of hemolysis of the perfusing blood. DE BODO and MARKS, in 1928, perfusing a dog liver-and-lung preparation with defibrinated blood containing glucose at 1000 to 2000 mg/100 ml, were able to obtain, consistently, the storage of glycogen in the perfused liver. Insulin, which at that date undoubtedly was contaminated with glucagon, had an adverse effect on glycogen storage when added to the perfusing blood, and did not abolish the glycogen loss caused by the addition of epinephrine to the perfusing blood. LUNDSGAARD (1938, 1954), in many experiences with

perfused livers, was not able to find any influence of insulin on carbohydrate metabolism when the hormone was added to the perfusing fluid.

D'SILVA and NEIL, in 1954, reported an extensive study of the conditions required to prevent edema and glycogen loss and to preserve the normal histological appearance of liver tissue during perfusion of rat liver through the hepatic veins by way of the thoracic portion of the inferior vena cava. In these preparations they were not able to bring about glycogen storage when glucose at 500 to 1000 mg/100 ml was added to the perfusion fluid, nor did insulin at 8 units/100 ml have any discernible effect on the glycogen content of the perfused liver. As a result of their experiences they concluded that one of the most important criteria for a satisfactory perfusion procedure is that potassium loss from the liver should not occur.

In contrast with the generally negative findings exemplified by the above, several successful demonstrations of insulin effects in perfused livers have been reported in the last few years.

HAFT and MILLER (1958a) after 27 experiments utilizing dilute heparinized blood perfused through the portal vein, and using livers from alloxan-diabetic rat donors, divided their experiments into two groups. In one group of 13 perfusions, where "minimal trauma" attended the removal of the liver for perfusion, positive effects of insulin were reported. In a second group of 14 perfusions "excessive trauma" was recorded, and no insulin effects were observed. The authors state: "The susceptibility of the direct action of insulin on the liver to influences . . . which are probably stress-induced and humoral has an obvious bearing on any studies of hormonal action on surviving liver tissue."

In these experiments rat livers weighing around 11 g were perfused with 110 to 140 ml of rat blood from 18-hour-fasted normal donors. Heparin (60 mg), sodium acetate (35 mg), a volume of Ringer's solution equal to $1/_8$ the volume of blood, and enough extra glucose to bring the glucose concentration to 350 mg/100 ml were added to the perfusing medium. Thus when fed normal rats were liver donors (mean liver glycogen: 3.2 g/100 g liver) and glucose was delivered in net amount (1.1 g glucose/100 g liver) by the perfused liver to the perfusion medium during the first hour, the resultant perfusion fluid glucose concentration was about 430 mg/100 ml. When 18-hour-fasted normal rats were liver donors (mean liver glycogen: 0.1 g/100 g liver) there was no net delivery of glucose to the medium during the first hour and the medium glucose concentration remained about 350 mg/100 ml. When fed, nonketotic, alloxan-diabetic rats were liver donors (mean liver glycogen: 1.1 g/100 g liver) the net delivery of glucose to the medium was more than twice as much during the first hour as for the fed normal rat livers and the glucose concentration of the perfusing medium rose to about 550 mg/100 ml. The difference in glycogen content between the livers of fed and starved normal rats suggests glycogenolysis as the source of the extra glucose released by the livers from the fed animals. However, in the livers from fed diabetic rats, the amount of glucose released during the first hour was twice as much as the amount available from their glycogen stores. Furthermore in a separate study by Sokal, Miller and Sarcione (1958), in which intermediate liver glycogen values were observed, net glycogen loss was not seen to accompany the release of glucose during the early period of perfusion.

Insulin (8.5 U, added by constant infusion during the 4-hour period of the perfusion experiment) had, during the first hour of perfusion, no effect on the amount of glucose released by any of the livers. The effects of insulin were seen during the subsequent 3 hours of perfusion in the experiments of Haft and Miller (1958a); glycogen differences were observed only between samples taken prior to perfusion and samples taken at the end of 4 hours of perfusion. The absence of knowledge of intermediate values of glycogen prevented the observation of any inhibition of glycogen breakdown which may have prevailed in the first hour and any stimulation of glycogen resynthesis which may have occurred during the last 3-hour periods of the experiments. Net glycogen synthesis in the absence of added insulin could actually occur in these livers (both from normal and from diabetic rats) as was shown in later experiments [Haft and Miller (1958b)] in which the initial glucose concentration of the perfusing medium was about 2000 mg/100 ml. In separate experiments [Sokal, Miller and Sarcione (1958)], glycogen synthesis in livers from 24-hour-fasted donor rats was observed both in the absence and presence of insulin at medium glucose concentrations of 100 to 200 mg/100 ml. In other experiments by Miller, Sokal and Sarcione (1959), liver glycogen present at 2 to 3 g/100 g in livers from fed normal rats stayed constant for 3 hours of perfusion in the absence of added insulin at medium glucose concentrations of 200—250 mg/100 ml.

Thus no effect of insulin on glycogen breakdown or resynthesis was established by the experiments of Miller and his co-workers.

The positive effect of insulin on carbohydrate metabolism which was reported by Haft and Miller (1958a), was on net glucose balance (glucose uptake by liver minus glucose release by liver) during the last 3 hours of perfusion of livers from fed alloxan-diabetic rats. In the absence of insulin the glucose concentration of the perfusing fluid remained at about 550 mg/100 ml, whereas when insulin was present it declined to about 450 mg/100 ml, indicating the net uptake of about 1 g of glucose/100 g liver for the 3-hour period. In the case of the livers from fed normal donors, net glucose uptake of about the same magnitude was seen in the last 3 hours of perfusion in the absence of added insulin; added insulin had only a doubtful positive influence here. For fasted normal donors, net glucose uptake was less than $1/3$ as great as this in the last 3 hours of perfusion; again added insulin had only a doubtful positive influence on net uptake of glucose.

MORTIMORE (1963) has contributed carefully reported experiments demonstrating the effect of insulin on glucose release by perfused liver. Rat livers were perfused by way of the portal vein with defibrinated diluted blood. Insulin was added to the perfusing fluid continuously; the degradation of insulin by the perfused liver continually depleted the amount of insulin in the recycling perfusion fluid, yet if this loss was ignored completely the final concentration of insulin at the end of experiments in which a near maximal insulin effect was observed was calculated to be only 135 μU/ml. The effect of insulin to lower the rate of glucose release by the liver was seen whether fed or fasted liver donors were employed and at perfusate glucose concentrations of 33 to 300 mg per 100 ml. In the case of the livers from fasted donors in the virtual absence of liver glycogen, the reduction in glucose release was small enough to be accounted for by the decreased amount of amino acids deaminated, as judged by the decrease in urea production. However, in livers from fed donors, the decrease in glucose release was much greater and indicated "the existence of a second, and quantitatively more important, action of insulin on hepatic glucose metabolism." Owing to the variability in glycogen levels, it could not be established, with significance, that the decreased glucose release of the insulin-treated livers was associated with a decrease in net glycogen loss.

D. Summary

Insulin has a direct effect on the hepatic cell *in vivo* to decrease glycogenolysis and the rate of hepatic glucose release. An increased uptake of plasma glucose, most of which is converted to glycogen, occurs subsequent to the establishment of the decreased rate of glycogenolysis. Many of these effects have occasionally been observed, with great difficulty, in isolated liver tissue. They are readily and consistently observed in whole animals provided that the blood glucose concentration is prevented from falling to a level which results in the increased secretion of antagonistic hormones such as glucagon and epinephrine. The ability of insulin to counter, at a very early time, the loss of hepatic K^+ which glucagon and epinephrine promote suggests a molecular action of insulin in the area concerned with mitochondrial electron transport, cyclic 3,5-AMP production and cation transport.

The ability of insulin to decrease hepatic glycogenolysis may be likened with its ability to inhibit lipolysis in isolated adipose tissue in the absence of glucose substrate; its ability to promote hepatic glycogen synthesis may be likened with its ability to promote protein synthesis in isolated muscle in the absence of glucose substrate. These represent precedents for effects of insulin which do not depend on facilitation of glucose transport or glucose phosphorylation. The increased rate of plasma glucose uptake into glycogen seen as a result of insulin action in liver may be coincident with, or caused by, increased liver glycogen synthesis rather than being the cause of increased liver glycogen synthesis; this is not yet known.

Experiments intended to show whether or not insulin increases the concentration of free glucose in the hepatic cell have not achieved their objective.

IV. Special topics

A. The reciprocal inhibition of fatty acid and glucose catabolism in peripheral tissues

WILLIAMSON and KREBS, in 1961, reported important findings using the perfused rat heart. Acetoacetate, when supplied as substrate, had no effect on the slow uptake of glucose (initial perfusate glucose concentration: 90 mg per 100 ml) observed in the absence of insulin, but nearly halved the rapid uptake of glucose observed when insulin was also present in the perfusing fluid. At the same time acetoacetate increased the formation of lactate, an effect which was attributed to the suppression, by acetoacetate, of glucose oxidation; OTTAWAY and SARKAR (1958) had reported similar findings. Insulin, in the absence of glucose, decreased the oxidation of acetoacetate and caused an endogenous substrate to be substituted for the spared acetoacetate as a fuel for respiration.

On the basis of this and other evidence reviewed by RANDLE and MORGAN (1962), a picture has emerged which assigns to the fatty acids an important part in the regulation of peripheral glucose utilization. The effects of insulin on the fatty acid regulation are secondary consequences of its effects to inhibit lipolysis and to promote the resynthesis of triglycerides from free fatty acids by increasing glucose uptake. Adipose tissue plays an important role in the process since in the absence of insulin both increased lipolysis and decreased triglyceride resynthesis in adipose tissue contribute to an elevated plasma non-esterified fatty acid concentration. Muscle responds directly to the elevated plasma non-esterified fatty acid concentration (and also to any elevation in plasma ketone body concentration caused by the increased flow of non-esterified fatty acids to the liver) by decreasing its overall glucose uptake and by decreasing to an even greater extent its utilization of glucose for respiration, so that additional pyruvate and lactate are released to the blood. The ketone bodies and free fatty acid are equivalent in inhibiting glucose uptake and utilization, their common point of action being the elevation of intracellular acetyl coenzyme A and the depression of free intracellular reduced coenzyme A.

RANDLE, GARLAND, HALES and NEWSHOLME, in 1963, extended this concept to include the effect of insulin (through its facilitation of glucose uptake) to promote the resynthesis to triglycerides of fatty acids split from triglycerides in muscle and so to reduce the intracellular free fatty acid concentration induced by lipolytic agents in the muscle cell itself. Glycerol release by muscle *in vitro* was found not to be inhibited by insulin as is the case in adipose tissue, hence direct inhibition by insulin of epinephrine-stimulated lipolysis in muscle tissue was not included in the mechanism. This extension of the concept has important consequences; a growth hormone regimen, for example, elevates plasma non-esterified fatty acid levels for only a few days, yet its effect to cause a relative inhibition of glucose uptake, insulin resistance, and a net increase in fat catabolism persists. This might, under the extended concept, be attributed to increased lipolysis in the muscle cell proper, and hence to an increased intracellular free fatty acid concentration in the muscle cell in the absence of elevated plasma non-esterified fatty acids.

RANDLE and his co-workers have used the general concept, which they term the "glucose fatty acid cycle", to advance the thought that the primary event in maturity onset diabetes, which frequently occurs in the presence of a normal or supernormal rate of secretion of insulin in response to hyperglycemia, is caused by a defect in glyceride metabolism which leads to the release of excessive amounts of fatty acids in adipose tissue and muscle [see HALES and RANDLE (1963)]. They conclude that a higher rate of fatty acid release from adipose tissue and decreased triglyceride resynthesis in muscle is responsible also for the abnormalities in carbohydrate metabolism seen in muscle in starvation, carbohydrate deprivation, excess growth hormone administration and excess glucocorticoid administration. The abnormalities include impaired sensitivity to insulin, impaired pyruvate tolerance, impaired glucose tolerance, and the "glycostatic" effect of growth hormone. They also suggest that epinephrine inhibition of glucose uptake by muscle may be due to its lipolytic effect in the muscle cell as well as to its effect to raise glucose 6-phosphate concentration by causing muscle glycogenolysis.

One of the partial blocks exerted by the fatty acids on glucose utilization by muscle is said by RANDLE (in press) to be at the enzymatic conversion of pyruvate and reduced coenzyme A to CO_2, acetyl coenzyme A, and two protons, that is at the "pyruvate dehydrogenase" reaction. A partial block at the level of pyruvate in the use of glucose for respiration in diabetes or in glucocorticoid excess had been suggested previously by findings by FRY and BUTTERFIELD (1962), by VILLEE and HASTINGS (1949), by HENNEMAN and BUNKER (1957), by HENNEMAN and HENNEMAN (1958), by THORN, RENOLD and CAHILL (1959), by FRAWLEY et al. (1957), by HENNES, WAJCHENBERG, FAJANS and CONN (1957), and by GLENN, MILLER and SCHLAGEL (1963). For a summary see FAJANS (1961).

WEIL, ALTSZULER and KESSLER reported in 1961 that a growth hormone regimen in the dog results in an elevated fasting blood pyruvate concentration and an impaired disposition of injected pyruvate. They postulated that this effect of growth hormone might be related to its fat mobilizing properties.

An excess of pyruvate itself, as well as of fatty acids or ketone bodies, was found by GARLAND, RANDLE and NEWSHOLME (1963) to lead to the accumulation of citrate in heart muscle. The metabolic fate shared by these substances is conversion to citrate by way of acetyl coenzyme A. It is possible that pyruvate and excess free fatty acids or ketone bodies compete for the available reduced coenzyme A as suggested by RANDLE and MORGAN (1962), with pyruvate being the loser.

Another point of view is furnished by recent experiments of WILLIAMSON, JONES and AZZONE (1964), who found that the block in the further catabolism of citrate, which is brought about at the level of aconitase by fluoroacetate poisoning of the perfused rat heart, is overcome by an excess of pyruvate but not of acetate. Pyruvate overcomes the block by increasing the citrate concentration to even higher levels than are present in fluoroacetate poisoning alone, whereas acetate cannot do this. However, glucose cannot furnish the pyruvate for this purpose in the poisoned muscle because phosphofructokinase is inhibited by the high citrate concentration. MILLER, ISSEKUTZ, PAUL and RODAHL (1964), have reported that intravenous lactate administration lowers plasma non-esterified fatty acid levels even in pancreatectomized dogs; pyruvate does also, but cannot be separately considered because pyruvate injection raises blood lactate concentration.

The reason for the accumulation of citrate in the rat heart in diabetes and starvation and in the presence of excess fatty acids and ketone bodies remains unknown; it may be that the mechanism for the preferred derivation of citrate from fatty acids or ketone bodies rather than from pyruvate in these situations is also not well understood.

A secondary consequence of the accumulation of citrate in muscle is the partial inhibition of the enzyme phosphofructokinase, with a resulting accumulation of the intermediates of glucose utilization prior to fructose-1,6-diphosphate. LOWRY and PASSONNEAU (1964) have reviewed the extensive knowledge of the metabolites, including citrate, which affect phosphofructo-kinase activity.

The accumulation of glucose-6-phosphate, one result of phosphofructokinase inhibition, is thought to inhibit hexokinase. This may be responsible for the partial block to glucose utilization which occurs at the initial step in glucose phosphorylation. RANDLE and MORGAN (1962) have suggested that elevation of glucose-6-phosphate concentration by way of phosphofructokinase inhibition also may explain the myoglycostatic effect of a growth hormone regimen.

PARK (1964), who with his co-workers has made major contributions to the knowledge of the hormonal control of glucose transport and glucose phosphory-lation in muscle, has reviewed the current knowledge recently. Glucose trans-port is slow in the heart from an insulin deficient donor animal; the sensitivity of the transport process in the isolated heart to insulin stimulation *in vitro* is decreased by the prior influence in the whole animal of growth hormone and the glucocorticoids. This effect of growth hormone and the glucocorticoids is probably dependent upon their lipid-mobilizing activities, since addition of these hormones *in vitro* does not impair transport whereas addition of fatty acids *in vitro* can interfere with the insulin-induced increase in glucose transport.

This concept of the effect of fatty acids on glucose utilization allows the interpretation of many relationships which were previously unexplained. In several instances the concept requires supporting evidence, as in the case of the explanation of an inhibition of glucose uptake caused by a high rate of free fatty acid release within the muscle cell itself.

B. Special aspects of hormonal regulation of glucose production and utilization

1. **Increased glucose production in the growth hormone-treated dog.** In the growth hormone-treated dog in the postabsorptive state, in the face of insulin-insensitivity and a relative inhibition of glucose uptake (relative, that is, to the rate of uptake which would prevail in the control animal at the same blood glucose concentration) there is an absolute increase in blood glucose production and uptake which DE BODO, STEELE, ALTSZULER, DUNN and BISHOP (1963 b) have postulated to be accompanied by an elevated rate of insulin secretion. This phenomenon is visible after the effects of growth hormone to cause an elevated plasma non-esterified fatty acid concentration and turnover rate in the postabsorptive state have disappeared, as shown by WINKLER, STEELE, ALTSZULER and DE BODO (1964).

The possible relationship, in this situation, of the insulin-insensitivity and relative inhibition of glucose uptake to increased intracellular lipolysis in the extrahepatic tissues has been mentioned above. As for the increased rate of glucose production at an elevated blood glucose concentration, the growth hormone regimen in this animal as well as in the adrenalectomized animal, increases overall metabolism as shown by an increase in the total CO_2 output, as shown by ALTSZULER, STEELE, WALL, DUNN and DE BODO (1959). However, a similar growth hormone regimen in the hypophysectomized dog, which has an atrophic thyroid gland [DE BODO and ALTSZULER (1958)] also elevates to normal the deficient rate of glucose production characteristic of this animal, and in this case there is little or no increase in the total CO_2 output [ALTSZULER, STEELE, WALL, DUNN and DE BODO (1959)]. Thus increased thyroid hormone secretion is not responsible for the whole effect. The source of the increased amount of glucose produced, paradoxically, in these situations as a result of the action of a nitrogen-retaining hormone was thought in general terms to be the spared carbohydrate resulting from enhanced utilization by the liver of the growth hormone-treated dog of fatty acids in place of carbohydrate. To this hypothesis now may be added the thought that increased lipolysis in the peripheral tissues, with increased utilization for respiration in these tissues of the fatty acids resulting, may result in an increased release of glycerol to the blood for transport to the liver as a source of carbon for gluconeogenesis. In addition to this, a decreased use of glucose for respiration in muscle, brought about by the increase in the availability of free fatty acids, may result in the recycling of more pyruvate and lactate back to the liver from the extrahepatic tissues, so that the increase in glucose uptake seen in the extrahepatic tissues may not be accompanied by a corresponding increase in glucose conversion to CO_2, the extra pyruvate and lactate resulting from this being made available to the liver, for gluconeogenesis. Many attempts have been made to estimate completeness of glucose-C^{14} oxidation *in vivo* by measuring respiratory $C^{14}O_2$, but the interpretation of such experiments is difficult. It is clear that the possibilities mentioned above require experimental verification.

2. **Glycogenolysis and gluconeogenesis as glucose sources in the response to hypoglycemia.** Hepatic glycogenolysis is an alternative to gluconeogenesis as a source of the glucose produced by the liver during the period (18 to 24 hours after the preceding meal) in which many dog experiments are carried out. In a series of experiments on normal dogs reported recently by BISHOP, STEELE, ALTSZULER, DUNN, BJERKNES and DE BODO (1965), in which the rate of glycogen loss in this period was estimated by means of a series of percutaneous biopsies, hepatic glycogen disappearance was enough to account on the average for about 80% of the glucose released by the liver. Large individual variations in this value were encountered and it was not certain whether these were real or were due to sampling errors in the measurements of the small decrements in liver glycogen concentration. Corresponding experiments using the hypophysectomized or adrenalectomized dog in place of the normal dog have not been done.

The hypophysectomized dog has adequate liver glycogen stores as reported by DE BODO and SINKOFF (1953), but is insensitive to the blood sugar elevating action of epinephrine injected *in vivo* as shown by DE BODO, BLOCH and GROSS (1942). That this is due to the concomitant lack of adrenal cortical hormone secretion which exists in the hypophysectomized dog was indicated by the results of DE BODO, KURTZ, SINKOFF and KIANG (1952), who showed that regimens of either ACTH or a glucocorticoid, but not of growth hormone, restored to normal the blood sugar response of the hypophysectomized dog to epinephrine injection.

In the case of increased hepatic glucose output in response to insulin-induced hypoglycemia, there is not much assurance that increased glycogenolysis plays a role which is in proportion to the contribution of glycogenolysis to the resting rate of hepatic glucose output and to the increased rate of glucose output after epinephrine. Indeed, since the epinephrine-unresponsive, growth hormone-treated hypophysectomized dog as well as the glucocorticoid-treated, hypophysectomized dog, responds well to hypoglycemia as reported by DE BODO, STEELE, ALTSZULER, DUNN and BISHOP (1963 b), there is indication that increased gluconeogenesis rather than increased glycogenolysis may play the major role in the increased rate of glucose production in this situation. Also a peculiarity noted in the adrenalectomized dog [(DE BODO, STEELE, ALTSZULER, DUNN and BISHOP (1963 b)], namely an adequate increase in hepatic glucose production in response to a small dose of injected insulin but an inadequate response after injection of a large dose of insulin, may have to be explained with reference to a greater amount of increased hepatic gluconeogenesis occurring in response to hypoglycemia in the presence of the lesser amount of extra insulin, rather than in terms of increased hepatic glycogenolysis. The latter response may very well be deficient in the adrenalectomized dog regardless of whether the greater or lesser amount of insulin is injected to elicit the hypoglycemia.

V. Concluding remarks: the hepatic effects of insulin action

When WEIL-MALHERBE prepared his review for the Ergebnisse of 1955, the evidence available could be reconciled with the postulate of a single action of insulin to facilitate glucose transport and hence to increase glucose phosphorylation and utilization.

In contrast, effects of insulin have since been established which are not mediated by facilitated glucose transport nor by increased glucose uptake by the cells acted upon. With regard to the extrahepatic tissues, insulin a) promotes protein synthesis by isolated muscle in the absence of free glucose, b) inhibits lipolysis in isolated adipose tissue in the absence of free glucose, c) inhibits the use of acetoacetate as a respiratory fuel (and promotes the use of an endogenous substrate in place of acetoacetate) by isolated heart muscle in the absence of free glucose, and d) enhances glycogen synthesis by isolated adipose tissue and muscle more than does an equal increase in glucose uptake brought about in another way.

With regard to liver tissue, in which glucose transport is not known to be influenced by insulin, insulin increases protein, glycogen and triglyceride synthesis, and inhibits glycogen breakdown.

A clue to the action of insulin which may precede these diverse effects is furnished by recently acquired knowledge of the very early effect of insulin to inhibit net K^+ loss by the liver of the intact animal, an effect which is antagonized by epinephrine and glucagon. It may be surmised that insulin acts at the primitive level of mitochondrial electron transport, cation transport, and the control of the intracellular concentrations of inorganic phosphate, AMP, cyclic 3,5-AMP, ADP and ATP, and that its action at this primitive level is antagonized by the epinephrine-like agents.

The action of insulin in the extrahepatic tissues has the eventual consequence of limiting the flow of metabolites from these tissues to the liver. Moreover, since insulin acts on isolated liver tissue to increase protein and triglyceride synthesis, it is probable that, to some extent, decreased amounts of metabolites (fatty acids, glycerol and amino acids) exist in the liver as a consequence of the action of insulin on the hepatic cells themselves.

The decreased flow to the hepatic cells of fatty acids, glycerol, pyruvate, lactate and amino acids limits gluconeogenesis. The decreased flow of fatty acids promotes fatty acid synthesis. The important feature of the control exercised by certain of these metabolites is not their function as substrates, but rather their action to bring about allosteric modifications of hepatic enzymes and immediate changes in the activities of these enzymes.

The control of the rate of hepatic gluconeogenesis is probably exerted at a point subsequent to the deamination of amino acids. Increases in hepatic gluconeogenesis from glycerol, pyruvate and lactate rather than from amino acids may predominate under certain circumstances in which urea formation is insufficient to account for the total rate of gluconeogenesis.

The above ultimate consequences, expressed in the liver, of the action of insulin in remote tissues are essential for the maintenance of the normal complement of hepatic enzymes for lipogenesis and gluconeogenesis. Changes in the rates of these processes during insulin lack are followed more slowly by changes in the amounts of hepatic enzymes associated with these processes. These changes reinforce the more rapid effects of insulin lack.

Normal metabolism is not resumed immediately when adequate insulin action is restored in the insulin-deprived animal in which the late changes in enzyme amounts have occurred; time is required for the restoration of the normal pattern of enzyme amounts.

Insulin increases hepatic protein synthesis promptly, and does so when added *in vitro* to isolated liver. There is evidence that glucagon antagonizes the effect of insulin to bring about net hepatic protein synthesis. The concept that the long-term increase in hepatic gluconeogenesis from protein seen during insulin lack is contributed to by a changed balance between protein synthesis and degradation in the liver itself is not tenable at the present time. However, the evidence derived from liver perfusion experiments indicates that insulin may rapidly bring about a short-term decrease in gluconeogenesis from amino acids derived from hepatic and plasma proteins.

The long-term increase in protein synthetic capacity which insulin brings about in the livers of insulin-deficient animals would seem to be analogous with the changes in the amounts of hepatic enzymes for lipogenesis and gluconeogenesis which result from the increased or decreased use of the enzymatic pathways concerned.

The administration of an excess of a glucocorticoid to a normal rat brings about increases in the amounts of a number of the enzymes associated with gluconeogenesis, and these changes have been studied extensively. These increases in enzyme amounts appear to be brought about in much the same way as the similar changes seen during insulin lack, and the findings are considered to throw light on the mechanism of the increases seen during insulin lack. Induction of the enzyme increases by increased amounts of the respective substrates is the picture which emerges. Recent findings indicate that the inductions often may differ in mechanism from the classical dogma of genetic induction. Substrate inductions in mammalian cells may be mediated most commonly by the more prevalent existence of molecules of the enzyme as enzyme-substrate complex. That the increase in enzyme quantity is ordinarily brought about by a resultant decrease in the rate of degradation of the enzyme is possible, but is not yet clear.

The rapid effects of insulin, in the intact animal, to decrease net K^+ release by the liver, to decrease hepatic glycogenolysis, to decrease hepatic glucose release, and to increase hepatic glycogen synthesis, are not known to be mediated by a decrease in the flow of metabolites from the extrahepatic tissues. Because most of these effects have been demonstrated occasionally in isolated liver tissue, under especially favorable circumstances, they are thought to result from the direct action of insulin in hepatic cells. These effects are antagonized by glucagon and epinephrine. In the whole animal, in the postabsorptive state, acute insulin deprivation (e.g. as produced by anti-insulin serum) causes increased net hepatic K^+ release, increased hepatic glycogenolysis and increased hepatic glucose release. For this reason it appears that the small amount of insulin action going on in the liver in the postabsorptive state is sufficient to counterbalance to a considerable extent the effects of the prevailing rates of epinephrine and glucagon secretion. Increasing the amount of insulin action under these circumstances leads to a decrease in hepatic glucose release. This holds true for each of the several species studied, provided that the blood glucose level is prevented from falling by infusing glucose along with the insulin. Decreased hepatic gluconeogenesis as well as decreased glycogenolysis may be involved in this effect.

Increased insulin action, maintained for some time above the extent of insulin action which prevails in the liver in the postabsorptive state, appears to be required to bring about the synthesis of hepatic glycogen at the normal postabsorptive blood glucose concentration. The uptake of blood glucose by the liver, a process which goes on very slowly in the postabsorptive state, is increased simultaneously with the institution of glycogen synthesis. A large fraction of the blood glucose taken up by the liver at this time is found in the glycogen, although the incorporation of carbon from blood glucose into other liver constituents also is increased somewhat.

A low affinity glucokinase increases in amount in rat liver as a result of the combined actions of insulin and glucose in animals severely (alloxan diabetic) or mildly (fasted) deficient in endogenous insulin. The timing of the response suggests that it occurs subsequent to the increased use of the enzyme for hepatic glucose uptake. The mechanism of the induction is unknown; presumably it is a late-appearing consequence of the direct action of insulin on the hepatic cells, in the presence of glucose.

References

Allen, A., G. Medes, and S. Weinhouse: A study of the effects of growth hormone on fatty acid metabolism in vitro. J. biol. Chem. 221, 333—345 (1956).

Altszuler, N., R. Steele, J. Tobin, I. Rathgeb, and R. C. de Bodo: Effect of anti-insulin serum on glucose production and uptake in dogs. Excerpta Medica Intern. Congr. Ser. 74, Abstr. 333. Vth Intern. Congr. Intern. Diabetes Federat., Toronto, 1964.

— — J. S. Wall, A. Dunn, and R. C. de Bodo: Effect of growth hormone on carbohydrate metabolism in normal and hypophysectomized dogs; studies with C^{14} glucose. Amer. J. Physiol. 196, 121—124 (1959).

Andres, R., M. A. Baltzan, G. Cader, and K. L. Zierler: Effects of insulin on carbohydrate metabolism and on potassium in the forearm of man. J. clin. Invest. 41, 108—115 (1962).

Annison, E. F., and R. R. White: Glucose utilization in sheep. Biochem. J. 80, 162—169 (1961).

Armin, J., R. T. Grant, and P. H. Wright: Experimental diabetes in rats produced by parenteral administration of anti-insulin serum. J. Physiol. (Lond.) 153, 146—162 (1960).

Ashmore, J., G. F. Cahill, A. S. Earle, and S. Zottu: Studies on the disposition of blood glucose. A comparison of insulin and Orinase. Diabetes 7, 1—8 (1958).

—, and G. Weber: The role of hepatic glucose 6-phosphatase in the regulation of carbohydrate metabolism. In: Vitamins and hormones, vol. 17, pp. 91—132. Edit. by R. S. Harris, G. F. Marrian and K. V. Thimann. New York: Academic Press 1959.

Bach, J. S., and E. G. Holmes: The effect of insulin on carbohydrate formation in the liver. Biochem. J. 31, 89—100 (1937).

Baker, B. L.: A comparison of the histological changes induced by experimental hyperadrenal corticalism and inanition. Recent Progr. Hormone Res. 7, 331—364 (1952).

Baker, N., I. L. Chaikoff, and A. Schusdek: Effect of fructose on lipogenesis from lactate and acetate in diabetic liver. J. biol. Chem. 194, 435—443 (1952).

—, and R. Huebotter: Glucose metabolism in mice. Amer. J. Physiol. 207, 1155—1160 (1964).

— R. A. Shipley, R. E. Clark, and G. E. Incefy: C^{14} studies in carbohydrate metabolism: glucose pool size and rate of turnover in the normal rat. Amer. J. Physiol. 196, 245—252 (1959).

Ballard, F. J., and I. T. Oliver: The effect of concentration on glucose phosphorylation and incorporation into glycogen in the livers of foetal and adult rats and sheep. Biochem. J. 92, 131—136 (1964).

Barnabei, O., and F. Sereni: Factors influencing activity of tyrosine-α-ketoglutarate transaminase in isolated rat liver. Biochem. biophys. Res. Commun. 9, 188—191 (1962).

Baruch, H., and I. L. Chaikoff: Lipogenesis and CO_2 formation from acetate in the liver of the hypophysectomized rat. Endocrinology 56, 609—611 (1955).

BAUMAN, J. W., R. HILL, and I. L. CHAIKOFF: Hepatic lipogenesis in the diabetic-hypophysectomized (Houssay) rat. Endocrinology 60, 514—518 (1957).

BAXTER, C. F., M. KLEIBER, and A. L. BLACK: Glucose metabolism in the lactating dairy cow. Biochim. biophys. Acta (Amst.) 17, 354—361 (1955).

BEARN, A. G., B. H. BILLING, and S. SHERLOCK: Response of the liver to insulin; hepatic vein catheterization studies in man. In: Ciba Found. Colloquia on Endocrin., vol. VI, pp. 250—260. Edit. by G. E. WOLSTENHOLME. London: J. & A. Churchill 1953.

BELOFF-CHAIN, A., P. BETTO, R. CATANZARO, E. B. CHAIN, L. LONGINOTTI, J. MASI, and F. POCCHIARI: The metabolism of glucose 1-phosphate and glucose 6-phosphate and their influence on the metabolism of glucose in rat-diaphragm muscle. Biochem. J. 91, 620—624 (1964).

— E. B. CHAIN, D. BOVET, F. POCCHIARI, R. CATANZARO, and L. LONGINOTTI: Metabolism of hexose phosphate esters. Biochem. J. 54, 529—539 (1953).

BERSON, S. A., S. WEISENFELD, and M. PASCULLO: Utilization of glucose in normal and diabetic rabbits. Diabetes 8, 116—127 (1959).

—, and R. S. YALOW: Some remarks on the mechanism of action of the sulfonylureas. Diabetes 6, 274—277 (1957).

BERTHET, J., P. JACQUES, H. G. HERS et C. DE DUVE: Influence de l'insuline et du glucagon sur la synthèse du glycogène hépatique. Biochim. biophys. Acta (Amst.) 20, 190—200 (1956).

BIERMAN, E. L., V. P. DOLE, and T. N. ROBERTS: An abnormality of non-esterified fatty acid metabolism in diabetes mellitus. Diabetes 7, 188 (1958).

BISHOP, J. S., R. STEELE, N. ALTSZULER, A. DUNN, C. BJERKNES, and R. C. DE BODO: Effects of insulin on liver glycogen synthesis and breakdown in the dog. Amer. J. Physiol. 208, 307 (1965).

BLECHER, M., and A. WHITE: Effect of steroids on the metabolism of lymphoid tissue. Recent Progr. Hormone Res. 15, 391—418 (1959).

BLOCH, K., and W. KRAMER: The effect of pyruvate and insulin on fatty acid synthesis in vitro. J. biol. Chem. 173, 871—872 (1948).

BODO, R. C. DE, and N. ALTSZULER: The metabolic effects of growth hormone and their physiological significance. In: Vitamins and hormones, vol. XV, pp. 206—251. Edit. by R. S. HARRIS, F. G. MARRIAN and K. V. THIMANN. New York: Academic Press 1957.

— — Insulin hypersensitivity and physiological insulin antagonists. Physiol. Rev. 38, 389—445 (1958).

— — A. DUNN, R. STEELE, D. T. ARMSTRONG, and J. S. BISHOP: Effects of exogenous and endogenous insulin on glucose utilization and production. Ann. N.Y. Acad. Sci. 82, 431—449 (1959).

— H. I. BLOCH, and I. H. GROSS: The role of the anterior pituitary in adrenaline hyperglycemia and liver glycogenolysis. Amer. J. Physiol. 137, 124—135 (1942).

— M. KURTZ, M. W. SINKOFF, and S. P. KIANG: Effects of ACTH, cortisone and adrenal cortical extract on carbohydrate metabolism of hypophysectomized dogs. Proc. Soc. exp. Biol. (N.Y.) 80, 345—350 (1952).

—, and H. P. MARKS: The action of insulin on the perfused mammalian liver. J. Physiol. (Lond.) 65, 48—62 (1928).

—, and M. W. SINKOFF: Anterior pituitary and adrenal hormones in the regulation of carbohydrate metabolism. Recent Progr. Hormone Res. 8, 511—563 (1953).

— R. STEELE, N. ALTSZULER, A. DUNN, D. T. ARMSTRONG, and J. S. BISHOP: Further studies on the mechanism of action of insulin. Metabolism 8, 520—530 (1959).

— — — — and J. S. BISHOP: Effects on insulin on hepatic glucose metabolism and glucose utilization by tissues. Diabetes 12, 16—28 (1963a).

— — — — — On the hormonal regulation of carbohydrate metabolism; studies with C^{14} glucose. Recent Progr. Hormone Res. 19, 445—482 (1963b).

BOLLMAN, J. L.: The animal with an Eck fistula. Physiol. Rev. 41, 607—621 (1961).

BONDY, P. K., and L. R. CARDILLO: The effect of glucagon on carbohydrate metabolism in normal human beings. J. clin. Invest. **35**, 494—501 (1956).

— D. F. JAMES, and B. W. FARRAR: Studies of the role of the liver in human carbohydrate metabolism by the venous catheter technique. II. Patients with diabetic ketosis before and after the administration of insulin. J. clin. Invest. **28**, 1126—1133 (1949).

BORTZ, W. M., S. ABRAHAM, and I. L. CHAIKOFF: Localization of the block in lipogenesis resulting from feeding fat. J. biol. Chem. **238**, 1266—1272 (1962).

—, and F. LYNEN: The inhibition of acetyl-Co A carboxylase by long-chain acyl-Co A derivatives. Biochem. Z. **337**, 505—509 (1963).

BOUCKAERT, J. P., and C. DE DUVE: The action of insulin. Physiol. Rev. **27**, 39—71 (1947).

BRADLEY, S. E., F. J. INGLEFINGER, G. P. BRADLEY, and J. J. CURRY: The estimation of hepatic blood flow in man. J. clin. Invest. **24**, 890—897 (1945).

BRADY, R. O., and S. GURIN: The biosynthesis of radioactive fatty acids and cholesterol. J. biol. Chem. **186**, 461—469 (1950a).

— — Biosynthesis of labelled fatty acids and cholesterol in experimental diabetes. J. biol. Chem. **187**, 589—596 (1950b).

— F. D. W. LUKENS, and S. GURIN: Hormonal influence upon the in vitro synthesis of radioactive fatty acids. Science **113**, 413—415 (1951a).

— — — Synthesis of radioactive fatty acids in vitro and its hormonal control. J. biol. Chem. **193**, 459—464 (1951b).

BRESCIANI, F., e F. SGAMBATI: Variazione del contenuto intracellulare di sodio e potassio e metabolismo del glicogeno in sezione di fegato di ratto X-irradiate. Arch. Sci. biol. (Bologna) **44**, 110—119 (1960).

BUCHANAN, J. M., A. B. HASTINGS, and F. B. NESBETT: The effect of the ionic environment on the synthesis of glycogen from glucose in liver slices. J. biol. Chem. **180**, 435—445 (1949).

BUTCHER, R. W., R. J. HO, H. C. MENG, and E. W. SUTHERLAND: Cyclic 3',5'-AMP levels and free fatty acid (FFA) release. Intern. Union of Biochem. Ser. **32**, 715 (1964).

CAHILL jr., F. G., J. ASHMORE, A. S. EARLE, and S. ZOTTU: Glucose penetration into liver. Amer. J. Physiol. **192**, 491—496 (1958).

— — A. E. RENOLD, and A. B. HASTINGS: Blood glucose and the liver. Amer. J. Med. **26**, 264—282 (1959).

— — S. ZOTTU, and A. B. HASTINGS: Studies on carbohydrate metabolism in rat liver slices. IX. J. biol. Chem. **224**, 237—250 (1957).

CAMERON, J. S.: The effect of insulin on carbohydrate metabolism in the mammalian liver: A review. Guy's Hosp. Rep. **111**, 145—169 (1962).

CAMPBELL, J., and K. S. RASTOGI: Elevation of serum insulin, and production of diabetes by growth hormone in the dog. Excerpta Medica Intern. Congr. Ser. 74, Abstr. 316. Vth Intern. Congr. Intern. Diabetes Federat., Toronto, 1964.

CAMPBELL, W. J., and W. MERTZ: Interaction of insulin and chromium (III) on mitochondrial swelling. Amer. J. Physiol. **204**, 1028—1030 (1963).

CARLSON, L. A., and J. ÖSTMAN: In vitro studies on the glucose uptake and fatty acid metabolism of human adipose tissue in diabetes mellitus. Acta med. scand. **174**, 215—218 (1963).

CHERNICK, S. S., and I. L. CHAIKOFF: Insulin and hepatic utilization of glucose for lipogenesis. J. biol. Chem. **186**, 535—542 (1950).

— — Two blocks in carbohydrate utilization in the liver of the diabetic rat. J. biol. Chem. **188**, 389—396 (1951).

— — E. J. MASORO, and E. ISAEFF: Lipogenesis and glucose oxidation in the liver of the alloxan-diabetic rat. J. biol. Chem. **186**, 527—534 (1950).

—, and R. O. SCOW: Synthesis in vitro of glyceride-glycerol by the liver of normal and pancreatectomized rats. J. biol. Chem. **239**, 2416—2419 (1964).

CHERRY, I. S., and L. A. CRANDALL: The response of the liver to the oral administration of glucose. Amer. J. Physiol. 120, 52—58 (1937).

CHRISTIAN, G. D., E. C. KNOBLOCK, W. C. PURDY, and W. MERTZ: A polarographic study of chromium-insulin-mitochondrial interaction. Biochim. biophys. Acta (Amst.) 66, 420—423 (1963).

COHN, C., and D. JOSEPH: Effect of rate of ingestion of diet on hexose-monophosphate shunt activity. Amer. J. Physiol. 197, 1347—1349 (1959).

—, and M. KOLINSKY: Effect of blood sugar levels and insulin lack on gluconeogenesis by the kidney of the dog. Amer. J. Physiol. 156, 345—348 (1949).

COMBES, B., R. H. ADAMS, W. STRICKLAND, and L. L. MADISON: The physiological significance of the secretion of endogenous insulin into the portal circulation. IV. Hepatic uptake of glucose during glucose infusion in nondiabetic dogs. J. clin. Invest. 40, 1706—1718 (1961).

CONN, J. W., and S. S. FAJANS: Influence of adrenal cortical steroids on carbohydrate metabolism in man. Metabolism 5, 114—127 (1956).

CRAIG, J. W., and J. LARNER: Influence of epinephrine and insulin on uridine diphosphate glucose-α-glucan transferase and phosphorylase in muscle. Nature (Lond.) 202, 971—973 (1964).

CRANE, R. K., D. MILLER, and I. BIHLER: The restrictions on possible mechanisms of intestinal active transport of sugars. In: Membrane transport and metabolism, pp. 439—449. Edit. by A. KLEINZELLER and A. KOTYK. New York: Academic Press 1961.

DAVIDSON, I. W. F., J. M. SALTER, and C. H. BEST: The effect of glucagon on the metabolic rate of rats. Amer. J. clin. Nutr. 8, 540—545 (1960).

DEANE, H. W., F. B. NESBETT, J. M. BUCHANAN, and A. B. HASTINGS: A cytochemical study of glycogen synthesized from glucose or pyruvate by liver slices in vitro. J. cell. comp. Physiol. 30, 255—269 (1947).

DOELL, R. G.: The effect of insulin on the amino acid incorporating system of rat liver. Fed. Proc. 18, 37 (1959).

DOUGHERTY, T. F., and A. WHITE: Functional alterations in lymphoid tissue induced by adrenal secretion. Amer. J. Anat. 77, 81—116 (1945).

DRURY, D. R., A. N. WICK, and E. M. MACKAY: Formation of glucose by the kidney. Amer. J. Physiol. 163, 655—661 (1950).

DUNCAN, L. J. P., and J. D. BAIRD: Compounds administered orally in the treatment of diabetes mellitus. Pharmacol. Rev. 12, 91—158 (1960).

DUNN, A., N. ALTSZULER, R. C. DE BODO, R. STEELE, D. T. ARMSTRONG, and J. S. BISHOP: Mechanism of action of insulin. Nature (Lond.) 183, 1123—1124 (1959).

— R. STEELE, N. ALTSZULER, R. C. DE BODO, D. T. ARMSTRONG, and J. S. BISHOP: Effects of insulin on glucose utilization and production: studies with C^{14} glucose. Proc. Ist Intern. Congr. of Endocrinol. 1960, p. 1251.

— — — — J. S. BISHOP, and D. T. ARMSTRONG: An effect of insulin on production of glucose during hepatic glycogenolysis. Nature (Lond.) 188, 236—237 (1960).

DUNN, D. F., B. FRIEDMAN, A. R. MAASS, G. A. REICHARD, and S. WEINHOUSE: Effects of insulin on blood glucose entry and removal rates in dogs. J. biol. Chem. 225, 225—237 (1957).

DUVE, C. DE: The hepatic action of insulin. In: Ciba Found. Colloquia on Endocrin., vol. XI, pp. 203—222. Edit. by G. E. W. WOLSTENHOLME and C. M. O'CONNOR. London: J. & A. Churchill 1956.

ELGEE, N. J., R. H. WILLIAMS, and N. D. LEE: Distribution and degradation studies with insulin-I^{131}. J. clin. Invest. 33, 1252—1260 (1954).

ELLIS, S.: The metabolic effects of epinephrine and related amines. Pharmacol. Rev. 8, 485—562 (1956).

ENGEL, F. L.: A consideration of the roles of the adrenal cortex and stress in the regulation of protein metabolism. Recent Progr. Hormone Res. 6, 277—308 (1951).

— On the nature of the interdependence of the adrenal cortex, non-specific stress and nutrition in the regulation of nitrogen metabolism. Endocrinology 50, 462—477 (1952).

EWALD, W., H.-J. HÜBENER u. E. WIEDEMANN: Weitere Untersuchungen über die Enzym-Induktion durch Cortisol in der Leber. Hoppe-Seylers Z. physiol. Chem. **333**, 57—70 (1963).

FAIN, J. N., R. O. SCOW, and S. S. CHERNICK: Effects of glucocorticoids on metabolism of adipose tissue in vitro. J. biol. Chem. **238**, 54—58 (1963).

—, and A. E. WILHELMI: Effects of adrenalectomy, hypophysectomy, growth hormone and thyroxine on fatty acid synthesis in vivo. Endocrinology **71**, 541—548 (1962).

FAJANS, S. S.: Some metabolic actions of corticosteroids. Metabolism **10**, 951—965 (1961).

FAZEKAS, A. GY., and GY. DOMJÁN: Biochemical examination of the effect of cortizol on gluconeogenesis. Enzymologia **24**, 267—274 (1962).

FELLER, D. D., E. H. STRISOWER, and I. L. CHAIKOFF: Turnover and oxidation of body glucose in normal and alloxan-diabetic rats. J. biol. Chem. **187**, 571—588 (1950).

FENN, W. O.: The role of potassium in physiological processes. Physiol. Rev. **20**, 377—415 (1940).

FIGUEROA, E., A. PFEIFER, and H. NIEMEYER: Incorporation of ^{14}C-glucose into glycogen by whole homogenate of liver. Nature (Lond.) **193**, 382—383 (1962).

FINDER, A. G., T. BOYME, and W. C. SHOEMAKER: Relationship of hepatic potassium efflux to phosphorylase activation induced by glucagon. Amer. J. Physiol. **206**, 738—742 (1964).

FINE, M. B., and R. H. WILLIAMS: Effect of an insulin infusion on hepatic output of glucose. Amer. J. Physiol. **198**, 645—648 (1960).

FITCH, W. M., and I. L. CHAIKOFF: Extent and patterns of adaptation of enzyme activities in livers of normal rats fed diets high in glucose and fructose. J. biol. Chem. **235**, 554—557 (1960).

FLINK, E. B., A. B. HASTINGS, and J. K. LOWRY: Changes in potassium and sodium concentration in liver slices accompanying incubation in vitro. Amer. J. Physiol. **163**, 598—604 (1950).

FRANCIS, M. D., and T. WINNICK: Studies on the pathway of protein synthesis in tissue culture. J. biol. Chem. **202**, 273—289 (1953).

FRANKLIN, M. J., and E. KNOBIL: The influence of hypophysectomy and of growth hormone administration on the oxidation of palmitate-1-C^{14} by the unanesthetized rat. Endocrinology **68**, 867—872 (1961).

FRAWLEY, T. F. S., S. SEGAL, M. M. CAMUS, and J. FOLEY: A concept of the mechanism of sulfonylurea-induced hypoglycemia based on studies of glucose and pentose distribution in man. Ann. N.Y. Acad. Sci. **71**, 81—96 (1957).

FRITZ, I. B.: Factors influencing the rates of long-chain fatty acid oxidation and synthesis in mammalian systems. Physiolog. Rev. **41**, 52—129 (1961).

FROESCH, E. R., and J. L. GINSBERG: Fructose metabolism of adipose tissue. I. Comparison of fructose and glucose metabolism in epididymal adipose tissue of normal rats. J. biol. Chem. **237**, 3317—3324 (1962).

FRY, I. K., and W. J. H. BUTTERFIELD: Carbohydrate metabolism in diabetics. A possible intracellular block. Lancet 1962 II, 66—68.

GALANSINO, G., G. D'AMICO, D. KANAMEISHI, and P. P. FOA: Mode of action of insulin, carbutamide and tolbutamide. Proc. Soc. exp. Biol. (N.Y.) **99**, 447—451 (1958).

GARLAND, P. B., P. J. RANDLE, and E. A. NEWSHOLME: Citrate as an intermediary in the inhibition of phosphofructokinase in rat heart muscle by fatty acids, ketone bodies, pyruvate, diabetes and starvation. Nature (Lond.) **200**, 169—170 (1963).

GENES, S. G., P. M. CHARNAYA, and M. Z. YURCHENKO: Effect of insulin, chlorpropamide and chlorisopropamide on homeostatic function of the liver and on passage of blood sugar into the portal system. Fiziol. Zh. (Mosk.) **48**, 1113 (1962).

GEY, K. F.: The concentration of glucose in rat tissues. Biochem. J. **64**, 145—150 (1956).

GIBSON, D. M., and D. W. ALLMANN: Adaptive changes in fatty acid biosynthesis. In: Advances in enzyme regulation, vol. I, pp. 183—185. Edit. by G. WEBER. Oxford: Pergamon Press 1963.

GLASSER, S. R., and I. L. IZZO: The influence of adrenalectomy on the metabolic actions of glucagon in the fasted rat. Endocrinology 70, 54—61 (1962).

GLENN, M., W. L. MILLER, and C. A. SCHLAGEL: Metabolic effects of adrenocortical steroids in vivo and in vitro: relationship to anti-inflammatory effects. Recent Progr. Hormone Res. 19, 107—191 (1963).

GOLDSTEIN, L., E. J. STELLA, and W. E. KNOX: The effect of hydrocortisone on tyrosine-α-ketoglutarate transaminase and tryptophan pyrrolase activities in the isolated perfused rat liver. J. biol. Chem. 237, 1723—1726 (1962).

GOODMAN, H. M., and E. KNOBIL: Mobilization of fatty acids by epinephrine in normal and hypophysectomized monkeys. Proc. Soc. exper. Biol. (N.Y.) 100, 195—197 (1959).

GORDON, E. E.: The rate of generation of reduced nicotine adenine dinucleotide and reduced nicotine adenine dinucleotide phosphate in the liver of normal and alloxan-diabetic rats. J. biol. Chem. 238, 2135—2140 (1963).

GORDON jr., R. S.: See discussion in: Symp. on Fat and Diabetes. Diabetes 7, 190—192 (1958).

GOURLEY, D. R. H.: Combination of insulin with frog skeletal muscle. Amer. J. Physiol. 189, 489—494 (1957).

—, and K. C. FISHER: Role of citrate in stimulation of oxygen consumption by insulin in frog muscle. Amer. J. Physiol. 179, 378—385 (1954).

GREEN, M., and L. L. MILLER: Protein catabolism and protein synthesis in perfused livers of normal and alloxan-diabetic rats. J. biol. Chem. 235, 3202—3208 (1960).

GREENBAUM, A. L., and R. F. GLASCOCK: The synthesis of lipids in the livers of rats treated with pituitary growth hormone. Biochem. J. 67, 360—365 (1957).

HAFT, D. E.: Evidence for inhibition of acetyl coenzyme A formation from pyruvate in diabetic rat liver. Biochim. biophys. Acta (Amst.) 90, 173—175 (1964).

—, and L. L. MILLER: Alloxan diabetes and demonstrated direct action of insulin on metabolism of isolated perfused rat liver. Amer. J. Physiol. 192, 33—42 (1958a).

— — Enhanced sugar uptake fails to simulate the insulin effect on lipogenesis in the isolated perfused rat liver. Amer. J. Physiol. 193, 469—475 (1958b).

HALES, C. N., and P. J. RANDLE: Effects of low carbohydrate diet and diabetes mellitus on plasma concentrations of glucose, non-esterified fatty acid, and insulin during oral glucose tolerance tests. Lancet 1963I, 790—794.

HALL, J. C., K. C. FISHER, and J. R. STERN: Stimulation of oxygen consumption by insulin in intact isolated frog muscle. Amer. J. Physiol. 179, 29—35 (1954).

HASTINGS, A. B., and J. M. BUCHANAN: The role of intracellular cations on liver glycogen formation in vitro. Proc. nat. Acad. Sci. (Wash.) 28, 478—482 (1942).

— C. TENG, F. B. NESBETT, and A. E. RENOLD: Further observations on potassium and carbohydrate metabolism. Fed. Proc. 11, 227 (1952).

HAUGAARD, E. S., and N. HAUGAARD: The effect of hyperglycemic-glycogenolytic factor on fat metabolism in liver. J. biol. Chem. 206, 641—645 (1954).

—, and W. C. STADIE: The effect of hyperglycemic-glycogenolytic factor and epinephrine on fatty acid synthesis. J. biol. Chem. 200, 753—757 (1953).

HAUGAARD, N., M. VAUGHAN, E. S. HAUGAARD, and W. C. STADIE: Studies of radioactive injected labeled insulin. J. biol. Chem. 208, 549—563 (1954).

HAUSBERGER, F. X., and A. J. RAMSAY: Islet hypertrophy in mice bearing ACTH-secreting tumors. Endocrinology 65, 165—171 (1959).

HAYNES, R. C.: Relation of L-alanine metabolism to the action of triamcinolone in liver slices. Endocrinology 75, 602—607 (1964).

HENNEMAN, D. H., and J. P. BUNKER: The pattern of intermediary carbohydrate metabolism in Cushing's syndrome. Amer. J. Med. 23, 34—45 (1957).

—, and P. H. HENNEMAN: Depression of serum and urinary citric acid level by 17-hydroxy-corticosteroids. J. clin. Endocr. 18, 1093 (1958).

HENNES, A. R., B. L. WAJCHENBERG, S. S. FAJANS, and J. W. CONN: The effect of adrenal steroids on the blood levels of pyruvic and alpha-ketoglutaric acids in normal subjects. Metabolism 6, 339 (1957).

HENNING, H. V., W. HUTH, and W. SEUBERT: An in vitro effect of cortisol on pyruvate carboxylase and gluconeogenesis. Biochem. biophys. Res. Commun. 17, 496—501 (1964).
— I. SEIFFERT, and W. SEUBERT: Cortisol induzierter Anstieg der Pyruvatcarboxylase-activität in der Rattenleber. Biochim. biophys. Acta (Amst.) 77, 345—348 (1963).
HETENYI jr., G., and G. S. ARBUS: The effect of insulin on the distribution of glucose between the blood plasma and the liver. J. gen. Physiol. 45, 1049—1063 (1962).
— F. K. KOPSTICK, and L. J. RETELSTORF: The effect of insulin on the distribution of glucose between the blood plasma and the liver in alloxan-diabetic and adrenal-ectomized rats. Canad. J. Biochem. 41, 2431—2439 (1963).
— G. A. WRENSHALL, and C. H. BEST: Rates of production, utilization, accumulation and apparent distribution space of glucose. Diabetes 10, 304—311 (1961).
HIATT, H. H., and T. B. BOJARSKI: The effects of thymidine administration on thymidy-late kinase activity and on DNA synthesis in mammalian tissues. Cold Spr. Harbor Symp. quant. Biol. 26, 367—369 (1961).
HILL, R., J. W. BAUMAN, and I. L. CHAIKOFF: Dietary repair of defective lipogenesis and cholesterogenesis (from C^{14}-acetate) in the liver of the hypophysectomized rat. Endocrinology 57, 316—332 (1955).
— W. W. WEBSTER, J. M. LINAZASORO, and I. L. CHAIKOFF: Time of occurrence of changes in the liver's capacity to utilize acetate for fatty acid and cholesterol synthesis after fat feeding. J. Lipid Res. 1, 150—153 (1960).
HILZ, H. W., W. TARNOWSKI, and P. AREND: Glucose polymerisation and cortisol. Biochem. biophys. Res. Commun. 10, 492—497 (1963).
HOFERT, J., J. GORSKI, G. C. MUELLER, and R. K. BOUTWELL: The depletion of liver glycogen in puromycin-treated animals. Arch. Biochem. 97, 134—137 (1962).
INGLE, D. J., and G. W. THORN: A comparison of the effects of 11-desoxycorticosterone acetate and 17-hydroxy-11-dehydrocorticosterone in partially depancreatized rats. Amer. J. Physiol. 132, 670—678 (1941).
— E. O. WARD, and M. H. KUIZENGA: The relationship of the adrenal glands to changes in urinary nonprotein nitrogen following multiple fractures in the force-fed rat. Amer. J. Physiol. 149, 510—515 (1947).
ISSEKUTZ, B. v., u. J. SZENDE: Die Wirkung des Insulins auf die Zuckerproduktion der überlebenden Froschleber. Biochem. Z. 272, 412—416 (1934).
IZZO, J. L., and S. R. GLASSER: Comparative effects of glucagon, hydrocortisone, and epinephrine on the protein metabolism of the fasting rat. Endocrinology 68, 189—198 (1961).
— — Interrelationship of action of glucagon and somatotropin on protein metabolism in the fasted hypophysectomized rat. Endocrinology 72, 701—708 (1963).
JACOB, F., and J. MONOD: Genetic regulatory mechanisms in the synthesis of proteins. J. molec. Biol. 3, 318—355 (1961).
JACOBS, G., G. REICHARD, E. H. GOODMAN jr., B. FRIEDMAN, and S. WEINHOUSE: Action of insulin and tolbutamide on blood glucose entry and removal. Diabetes 7, 358—364 (1958).
JEANRENAUD, B., and A. E. RENOLD: Studies on rat adipose tissue in vitro. VII. Effects of adrenal cortical hormones. J. biol. Chem. 235, 2217—2223 (1960).
JUNGAS, R. L., and E. G. BALL: Studies on the metabolism of adipose tissue. XII. The effects of insulin and epinephrine on free fatty acid and glycerol production in the presence and absence of glucose. Biochemistry 2, 383—388 (1963).
KALANT, N., T. R. CSORBA, and N. HELLER: Effects of insulin on glucose production and utilization in diabetes. Metabolism 12, 1100—1111 (1963).
KLEINZELLER, A., H. C. BURCK, H. CHRISTENSEN, and M. MAIZELS: See discussion in: Membrane transport and metabolism, pp. 594—597. Edit. by A. KLEINZELLER and A. KOTYK. New York: Academic Press 1961.
KNOX, W. E.: Substrate-type induction of tyrosine transaminase, illustrating a general adaptive mechanism in animals. In: Advances in enzyme regulation, vol. II, pp. 311—318. Edit. by G. WEBER. Oxford: Pergamon Press 1964.

KOEPF, G. F., H. W. HORN, C. L. GEMMILL, and G. W. THORN: The effect of adrenal cortical hormone on the synthesis of carbohydrate in liver slices. Amer. J. Physiol. **135**, 175—186 (1941).

KOLODNY, E. H., R. KLINE, and N. ALTSZULER: Effect of phlorizin on hepatic glucose output. Amer. J. Physiol. **202**, 149—154 (1962).

KORNACKER, M. S., and J. M. LOWENSTEIN: Citrate cleavage and acetate activation in livers of normal and diabetic rats. Biochim. biophys. Acta (Amst.) **84**, 490—492 (1964).

KORNER, A.: The effect of the administration of insulin to the hypophysectomized rat on the incorporation of amino acids into liver protein in vivo and in a cell-free system. Biochem. J. **74**, 471—478 (1960).

— Regulation of the rate of synthesis of messenger ribonucleic acid by growth hormone. Biochem. J. **92**, 449—456 (1964).

KRAHL, M. E.: Incorporation of C^{14} amino acids into peptides by normal and diabetic rat tissues. Science **116**, 524 (1952).

— Incorporation of C^{14} amino acids into glutathione and protein fractions of normal and diabetic rat tissues. J. biol. Chem. **200**, 99—109 (1953).

— Functions of insulin and other regulatory factors in peptide formation by animal cells. Recent Progr. Hormone Res. **12**, 199—225 (1956).

— The action of insulin on cells. New York: Academic Press 1961.

KREBS, H. A.: Renal gluconeogenesis. In: Advances in enzyme regulation, vol. I, pp. 385 to 400. Edit. by G. WEBER. Oxford: Pergamon Press 1963.

— C. DIERKS, and T. GASCOYNE: Carbohydrate synthesis from lactate in pigeon-liver homogenate. Biochem. J. **93**, 112—120 (1964).

—, and L. V. EGGLESTON: The effect of insulin on oxidations in isolated muscle tissue. Biochem. J. **32**, 913—925 (1938).

— — and C. TERNER: In vitro measurements of the turnover of potassium in brain and retina. Biochem. J. **48**, 530—537 (1951).

— E. A. NEWSHOLME, R. SPEAKE, T. GASCOYNE and P. LUND: Some factors regulating the rate of gluconeogenesis in animal tissues. In: Advances in enzyme regulation, vol. II, pp. 71—81. Edit. by G. WEBER. Oxford: Pergamon Press 1964.

LAMBOTTE, L. E., and W. C. SHOEMAKER: Effect of insulin on hepatic K movements as influenced by hypothermia, barbiturate and dibenzyline. Physiologist **7**, 184 (1964).

LANDAU, B. R., J. R. LEONARDS, and F. M. BARRY: Regulation of blood glucose concentration; response of liver to glucose administration. Amer. J. Physiol. **201**, 41—46 (1961).

— R. MAHLER, J. ASHMORE, D. ELWYN, A. B. HASTINGS, and S. ZOTTU: Cortisone and the regulation of hepatic gluconeogenesis. Endocrinology **70**, 47—53 (1962).

LANGDON, R. G.: Hormonal regulation of fatty acid metabolism. In: Lipide metabolism, pp. 238—290. Edit. by K. BLOCH. New York: John Wiley & Sons 1960.

LARNER, J.: Control of UDPG α-glucan transferase activity. Intern. Union of Biochem. Ser. **32**, 697—698 (1964).

LEBOEUF, B., A. E. RENOLD, and G. F. CAHILL jr.: Studies on rat adipose tissue in vitro. IX. Further effects of cortisol on glucose metabolism. J. biol. Chem. **237**, 988—991 (1962).

LEHNINGER, A. L.: Water uptake and extrusion by mitochondria in relation to oxidative phosphorylation. Physiol. Rev. **42**, 467—517 (1962).

LELOIR, L. F.: Nucleoside diphosphate sugars and saccharide synthesis. IVth Hopkins Memor. Lect. Biochemic. J. **91**, 1—8 (1964).

LEONARDS, J. R., B. R. LANDAU, J. W. CRAIG, F. I. R. MARTIN, M. MILLER, and F. M. BARRY: Regulation of blood glucose concentration: hepatic action of insulin. Amer. J. Physiol. **201**, 47—54 (1961).

LEVINE, R.: The relation of insulin action to the endocrine balance in diabetes. In: Perspectives in biology, pp. 99—104. Edit. by E. F. CORI, V. G. FOGLIA, L. F. LELOIR and S. OCHOA. Amsterdam: Elsevier Publ. Co. 1963.

LEVINE, R., and L. B. FRITZ: The relation of insulin to liver metabolism. Diabetes 5, 209—219 (1956).

— M. S. GOLDSTEIN, B. HUDDLESTUN, and S. P. KLEIN: Action of insulin on the „permeability" of cells to free hexoses as studied by its effect on the distribution of galactose. Amer. J. Physiol. 163, 70 (1950).

— — S. KLEIN, and B. HUDDLESTUN: The action of insulin on the distribution of galactose in eviscerated nephrectomized dogs. J. biol. Chem. 179, 985—986 (1949).

LEWIS, R. A., D. KUHLMAN, C. DELBUE, G. F. KOEPF, and G. W. THORN: The effect of the adrenal cortex on carbohydrate metabolism. Endocrinology 27, 971—982 (1940).

LOGOTHETOPOULOS, J., B. B. SHARMA, J. M. SALTER, and C. H. BEST: Metaglucagon diabetes in rabbits. NewEngl. J. Med. 261, 423—426 (1959).

LONG, C. N. H., and O. K. SMITH: Some recent studies on the adrenal cortex and carbohydrate metabolism. In: The human adrenal cortex, pp. 268—293. Edit. by A. R. CURRIE, T. SYMINGTON and J. K. GRANT. Edinburgh and London: E. &. S. Livingston 1962.

LOWRY, O. H., and J. V. PASSONNEAU. P-fructokinase and the control of glycolysis. Intern. Union of Biochem. Ser. 32, 705—706 (1964).

LUKENS, F. D. W.: The pancreas: Insulin and glucagon. Ann. Rev. Physiol. 21, 445—474 (1959).

LUNDSGAARD, E.: The metabolism of the isolated liver. Bull. Johns Hopk. Hosp. 63, 90—103 (1938).

— The liver and carbohydrate metababolism. In: Trans. XIIth Conf. on Liver Injury, pp. 11—66. Edit. by F. W. HOFFBAUER. NewYork: Josiah Macy jr. Found. 1954.

McCANN, W. P., and J. R. JUDE: The synthesis of glucose by the kidney. Bull. Johns Hopk. Hosp. 103, 77—93 (1958).

MACKLER, B., P. AMMENTORP, H. GRAUBARTH, and G. M. GUEST: Glucose formation by kidneys in eviscerated dogs. Proc. Soc. exp. Biol. (N.Y.) 78, 479—480 (1951).

MADISON, L. L., B. COMBES, R. ADAMS, and W. STRICKLAND: The physiological significance of the secretion of endogenous insulin into the portal circulation. III. Evidence for a direct immediate effect of insulin on the balance of glucose across the liver. J.clin. Invest. 39, 507—522 (1960).

—, and W. C. SHOEMAKER: See discussion in: Symp. on the Hypoglycemic Agents. Metabolism 8, 481 (1959).

—, and R. H. UNGER: The physiologic significance of the secretion of endogenous insulin into the portal circulation. I. Comparison of the effects of glucagon-free insulin administered via the portal vein and via a peripheral vein on the magnitude of hypoglycemia and peripheral glucose utilization. J. clin. Invest. 37, 631—639 (1958).

— — and K. RENCZ: The physiologic significance of the secretion of insulin into the portal circulation. II. Effect of rate of administration of glucagon-free insulin on magnitude of peripheral and hepatic actions. Metabolism 9, 97—108 (1960).

MANCHESTER, K. L., and F. G. YOUNG: Insulin and protein metabolism. In: Vitamins and hormones, vol. 19, pp. 95—132. Edit. by R. S. HARRIS, D. J. INGLE, G. F. MARRIAN and K. V. THIMANN. NewYork: Academic Press 1961.

MANERY, J. F., D. R. H. GOURLEY, and K. C. FISHER: The potassium uptake and rate of oxygen consumption of isolated frog skeletal muscle in the presence of insulin and lactate. Canad. J. Biochem. 34, 893—902 (1956).

—, and A. B. HASTINGS: The distribution of electrolytes in mammalian tissues. J. biol. Chem. 127, 657—676 (1939).

MANNERS, D. J.: Enzymic synthesis and degradation of starch and glycogen. In: Advances in carbohydrate chemistry, vol. 17, pp. 371—430. Edit. by M. L. WOLFROM. NewYork: Academic Press 1962.

MARTIN, F. I. R., J. R. LEONARDS, and M. MILLER: A comparison of the effect of the intraperitoneal and intravenous administration of I[131] insulin on peripheral blood glucose and serum radioactivity. Metabolism 8, 472—478 (1959).

MASRI, M. S., I. LYON, and I. L. CHAIKOFF: Nature of the stimulating action of insulin on lipogenesis from acetate in fasted rat liver. J. biol. Chem. 197, 621—624 (1952).

MERTZ, W., and E. E. ROGINSKI: The effect of trivalent chromium on galactose entry in rat epididymal fat tissue. J. biol. Chem. 238, 868—872 (1963).

METZ, R., and J. M. SALTER: Effect of glucagon on liver glycolysis. Nature (Lond.) 196, 1094—1095 (1962).

METZGER, R. P., S. S. WILCOX, and A. N. WICK: Studies with rat liver glucose dehydrogenase. J. biol. Chem. 239, 1769—1772 (1964).

MEUTTER, R. C. DE, and W. W. SHREEVE: Conversion of DL-lactate-2-C^{14} or -3-C^{14} or pyruvate-2-C^{14} to blood glucose in humans: effects of diabetes, insulin, tolbutamide and glucose load. J. clin. Invest. 42, 525—533 (1963).

MILLER, H. I., B. ISSEKUTZ jr., P. PAUL, and K. RODAHL: Effect of lactic acid on plasma free fatty acids in pancreatectomized dogs. Amer. J. Physiol. 207, 1226—1230 (1964).

MILLER, L. L.: Glucagon: a protein catabolic hormone in the isolated perfused rat liver. Nature (Lond.) 185, 248 (1960).

— Some direct actions of insulin, glucagon and hydrocortisone on the isolated perfused rat liver. Recent Progr. Hormone Res. 17, 539—564 (1961).

— J. E. SOKAL, and E. J. SARCIONE: Effects of glucagon and tolbutamide on glycogen in isolated perfused rat liver. Amer. J. Physiol. 197, 286—288 (1959).

MONOD, J., J.-P. CHANGEUX, and F. JACOB: Allosteric proteins and cellular control systems. J. molec. Biol. 6, 306—329 (1963).

MORTIMORE, G. E.: Effect of insulin on potassium transfer in isolated rat liver. Amer. J. Physiol. 200, 1315—1319 (1961).

— Effect of insulin on release of glucose and urea by isolated rat liver. Amer. J. Physiol. 204, 699—704 (1963).

— F. TIETZE, and DEW. STETTEN jr.: Metabolism of insulin-I^{131}. Studies in isolated perfused rat liver and hind limb preparations. Diabetes 8, 307—314 (1959).

MUNCK, A.: The effect of cortisol on glucose uptake by rat epididymal fat pads. Endocrinology 68, 178—180 (1961a).

— The effect in vitro of glucocorticoids on net glucose uptake by rat epididymal adipose tissue. Biochim. biophys. Acta (Amst.) 48, 618—620 (1961b).

— Studies on the mode of action of glucocorticoids in rats. II. The effects in vivo and in vitro on net glucose uptake by isolated adipose tissue. Biochim. biophys. Acta (Amst.) 57, 318—326 (1962).

NEJAD, N. S., I. L. CHAIKOFF, and R. HILL: Hormonal repair of defective lipogenesis from glucose in the liver of the hypophysectomized rat. Endocrinology 71, 107—112 (1962).

NIEMEYER, H., L. CLARK-TURRI, E. GARCÉS, and F. E. VERGARA: Selective response of liver enzymes to the administration of different diets after fasting. Arch. Biochem. 98, 77—85 (1962).

— — and E. RABAJILLE: Induction of glucokinase by glucose in rat liver. Nature (Lond.) 198, 1096—1097 (1963).

— N. PÉREZ, E. GARCÉS, and F. E. VERGARA: Enzyme synthesis in mammalian liver as a consequence of refeeding after fasting. Biochim. biophys. Acta (Amst.) 62, 411—413 (1962).

NOALL, M. W., T. R. RIGGS, L. M. WALKER, and H. N. CHRISTENSEN: Endocrine control of amino acid transfer. Science 126, 1002—1005 (1957).

NUMA, S., M. MATSUHASHI u. F. LYNEN: Zur Störung der Fettsäuresynthese bei Hunger und Alloxan-Diabetes. I. Fettsäuresynthese in der Leber normaler und hungernder Ratten. Biochem. Z. 334, 203—217 (1961).

OLIVECRONA, T.: Metabolism of chylomicrons labeled with C^{14}-glycerol-H^3-palmitic acid in the rat. J. Lipid Res. 3, 439—444 (1962).

OTTAWAY, J. H., and A. K. SARKAR: Formation of lactate from acetoacetate in the perfused heart. Nature (Lond.) 181, 1791—1792 (1958).

PARK, C. R.: Some factors regulating the utilization of carbohydrate. Intern. Union of Biochem. Ser. 32, 711—712 (1964).

PENHOS, J. C., and M. E. KRAHL: Insulin stimulus of leucine incorporation into liver protein. Amer. J. Physiol. 202, 349—352 (1962).

— — Stimulus of leucine incorporation into perfused liver protein by insulin. Amer. J. Physiol. 204, 140—142 (1963).

PERRY, W. F., and H. F. BOWEN: The effect of growth hormone on lipogenesis in intact and adrenalectomized rats. Endocrinology 56, 579—583 (1955).

— — Factors affecting the in vitro production of non-esterified fatty acid from adipose tissue. Canad. J. Biochem. 40, 749—755 (1962).

—, and R. J. TJADEN: The effect of various saccharides on the release of non-esterified fatty acids from adipose tissue in vitro. Canad. J. Biochem. 40, 455—458 (1962).

PIETRO, D. L. DI, C. SHARMA, and S. WEINHOUSE: Studies on glucose phosphorylation in rat liver. Biochemistry 1, 455—462 (1962).

—, and S. WEINHOUSE: Hepatic glucokinase in the fed, fasted and alloxan-diabetic rat. J. biol. Chem. 235, 2542—2545 (1960).

POTTER, V. R., and T. ONO: Enzyme patterns in rat liver and Morris hepatoma 5123. Cold Spr. Harb. Symp. quant. Biol. 26, 355—362 (1961).

PRICE jr., J. B., and L. S. DIETRICH: The induction of tryptophane peroxidase in the isolated perfused liver. J. biol. Chem. 227, 633—636 (1957).

PRYOR, J., and J. BERTHET: Action du glucagon sur l'incorporation des acides aminés dans les protéines par le tissu hepatique in vitro. Arch. int. Physiol. Biochem. 68, 227 (1960a).

— — The action of adenosine 3',5'-monophosphate on the incorporation of leucine into liver proteins. Biochim. biophys. Acta (Amst.) 43, 556—557 (1960b).

RAMEY, E. R., and M. S. GOLDSTEIN: The adrenal cortex and the sympathetic nervous system. Physiol. Rev. 37, 155—195 (1957).

RANDLE, P. J.: The glucose fatty acid cycle in obesity and maturity onset diabetes. Ann. N.Y. Acad. Sci. (in press).

— P. B. GARLAND, C. N. HALES, and E. A. NEWSHOLME: The glucose fatty-acid cycle. Its role in insulin-sensitivity and the metabolic disturbances of diabetes mellitus. Lancet 1963I, 785—789.

—, and H. E. MORGAN: Regulation of glucose uptake by muscle. In: Vitamins and hormones, vol. XX, pp. 199—249. Edit. by R. S. HARRIS and I. G. WOOL. NewYork: Academic Press 1962.

REICHARD, G. A., B. FRIEDMAN, A. R. MAASS, and S. WEINHOUSE: Turnover rates of blood glucose in normal dogs during hyperglycemia induced by glucose or glucagon. J. biol. Chem. 230, 387—397 (1958).

— A. G. JACOBS, P. KIMBEL, N. J. HOCHELLA, and S. WEINHOUSE: Effects of insulin on blood glucose entry and removal rates in man. Diabetes 9, 447—453 (1960).

— N. F. MOURY jr., N. J. HOCHELLA, A. L. PATTERSON, and S. WEINHOUSE: Quantitative estimation of the Cori cycle in the human. J. biol. Chem. 238, 495—501 (1963).

REINECKE, R. M.: The kidney as a source of glucose in the eviscerated rat. Amer. J. Physiol. 140, 276—285 (1943—1944).

RENOLD, A. E., A. B. HASTINGS, F. B. NESBETT, and J. ASHMORE: Studies on carbohydrate metabolism in liver slices. J. biol. Chem. 213, 135—146 (1955).

RESHEF, L., and B. SHAPIRO: Effect of epinephrine, cortisone and growth hormone on release of unesterified fatty acids by adipose tissue in vitro. Metabolism 9, 551—555 (1960).

RIZACK, M. A.: Activation of an epinephrine-sensitive lipolytic activity from adipose tissue by adenosine 3',5'-phosphate. J. biol. Chem. 239, 392—395 (1964).

ROBERTS, S.: The influence of the adrenal cortex on the mobilization of tissue protein. J. biol. Chem. 200, 77—88 (1952).

—, and M. B. KELLEY: Metabolism of plasma proteins in vitro. J. biol. Chem. 222, 555—564 (1956).

Roberts, S., and L. T. Samuels: Fasting and gluconeogenesis in the kidney of the eviscerated rat. Amer. J. Physiol. 142, 240—245 (1944).

Robinson, W. S.: Alloxan diabetes and insulin effects on amino acid incorporating activity of rat liver microsomes. Proc. Soc. exp. Biol. (N.Y.) 106, 115—118 (1961).

Rosen, F., and C. A. Nichol: Corticosteroids and enzyme activity. In: Vitamins and hormones, vol. XXI, pp. 135—214. Edit. by R. S. Harris, I. G. Wool and J. A. Loraine. New York: Academic Press 1963.

Russell, J. A.: Hormonal control of amino acid metabolism. Fed. Proc. 14, 696—705 (1955).

Sachs, G., C. de Duve, B. S. Dvorkin, and A. White: Effect of adrenal cortical steroid injection on lysosomal enzymic activities of rat thymus. Exp. Cell Res. 28, 597—600 (1962).

Salas, M., E. Viñuela, and A. Sols: Insulin-dependent synthesis of liver glucokinase in the rat. J. biol. Chem. 238, 3535—3538 (1963).

Salter, J. M.: Metabolic effects of glucagon in the Wistar rat. Amer. J. clin. Nutr. 8, 535—539 (1960).

— J. W. F. Davidson, and C. H. Best: The pathologic effects of large amounts of glucagon. Diabetes 6, 248—252 (1957).

Sayre, F. W., D. Jensen, and D. M. Greenberg: Substrate induction of threonine dehydrase in vivo and in perfused rat livers. J. biol. Chem. 219, 111—117 (1956).

Schimke, R. T.: Studies on factors affecting the levels of urea cycle enzymes in rat liver. J. biol. Chem. 238, 1012—1018 (1963).

Scow, R. O., and S. S. Chernick: Hormonal control of protein and fat metabolism in the pancreatectomized rat. Recent Progr. Hormone Res. 16, 497—541 (1960).

Searle, G. L., L. R. Bristow, and W. A. Reilly: Insulin effects on the turnover and oxidation of glucose in the human as studied with C^{14} glucose. Fed. Proc. 19, 162 (1960).

— and I. L. Chaikoff: Inhibitory action of hyperglycemia on delivery of glucose to the blood stream by liver of the normal dog. Amer. J. Physiol. 170, 456—460 (1952).

— G. E. Mortimore, R. E. Buckley, and W. A. Reilly: Plasma glucose turnover in humans as studied with C^{14} glucose. Diabetes 8, 167—173 (1959).

— E. H. Strisower, and I. L. Chaikoff: Glucose pool and glucose space in the normal and diabetic dog. Amer. J. Physiol. 176, 190—194 (1954).

Segal, H. L., and C. G. López: Early effects of glucocorticoids on precursor incorporation into glycogen. Nature (Lond.) 200, 143—144 (1963).

Shafrir, E., and D. Steinberg: The essential role of the adrenal cortex in the response of plasma free fatty acids, cholesterol and phospholipids to epinephrine injection. J. clin. Invest. 39, 310—319 (1960).

Sharma, C., R. Manjeshwar, and S. Weinhouse: Effects of diet and insulin on glucose-adenosine triphosphate phosphotranferases of rat liver. J. biol. Chem. 238, 3840—3845 (1963).

Shaw, W. N., and W. C. Stadie: Coexistence of insulin-responsive and insulin-non-responsive glycolytic systems in rat liver. J. biol. Chem. 227, 115—134 (1957).

Shimizu, C. S. N., and S. A. Kaplan: Effects of cortisone on in vitro incorporation of glycine into protein of rat diaphragm. Endocrinology 74, 709—713 (1964).

Shoemaker, W. C., and A. G. Finder: Relation of potassium and glucose release from the liver in the unanesthetized dog. Proc. Soc. exp. Biol. (N.Y.) 108, 248—252 (1961).

— R. Mahler, and J. Ashmore: The effect of insulin on hepatic glucose metabolism in the unanesthetized dog. Metabolism 8, 494—511 (1959).

— — — D. E. Pugh, and A. B. Hastings: The hepatic glucose response to insulin in the unanesthetized dog. J. biol. Chem. 234, 1631—1633 (1959).

—, and T. B. van Itallie: The hepatic response to glucagon in the unanesthetized dog. Endocrinology 66, 260—268 (1960).

Shoemaker, W. C., W. F. Walker, T. B. van Itallie, and F. D. Moore: A method for simultaneous catheterization of major hepatic vessels in a chronic canine preparation. Amer. J. Physiol. 196, 311—314 (1959).

Silva, J. L. D'.: The action of adrenaline on serum potassium. J. Physiol. (Lond.) 82, 393—398 (1934).

— The action of adrenaline on serum potassium. J. Physiol. (Lond.) 86, 219—228 (1936).

—, and M. W. Neil: The potassium, water and glycogen contents of the perfused rat liver. J. Physiol. (Lond.) 124, 515—527 (1954).

Simms, H. S., and M. S. Parshley: The effect of proteins and amino acids on the growth of adult tissue in vitro. In: Protein and amino acid nutrition, pp. 143—194. Edit. by A. A. Albanese. New York: Academic Press 1959.

Sinex, F. M., J. MacMullen, and A. B. Hastings: The effect of insulin on the incorporation of C^{14} into the protein of rat diaphragm. J. biol. Chem. 198, 615 (1952).

Smith, C. L.: Action of insulin on the frog (Rana temporaria). Nature (Lond.) 171, 311—312 (1953).

Sokal, J. E., L. L. Miller, and E. J. Sarcione: Glycogen metabolism of the isolated liver. Amer. J. Physiol. 195, 295—300 (1958).

— E. J. Sarcione, and A. M. Henderson: Relative potency of glucagon and epinephrine as hepatic glycogenolytic agents: Studies with the isolated perfused rat liver. Endocrinology 74, 930—938 (1964).

Soskin, S., H. E. Essex, J. L. Herrick, and F. C. Mann: The mechanism of regulation of the blood sugar by the liver. Amer. J. Physiol. 124, 558 (1938).

Spencer, A. F., and J. M. Lowenstein: The supply of precursors for the synthesis of fatty acids. J. biol. Chem. 237, 3640—3648 (1962).

Spiro, R. G.: Studies on carbohydrate metabolism in rat liver slices. XIII. Influence of the pituitary on the insulin-deficient state. J. biol. Chem. 230, 773—779 (1958).

— J. Ashmore, and A. B. Hastings: Studies on carbohydrate metabolism in rat liver slices. XII. Sequence of metabolic events following acute insulin deprivation. J. biol. Chem. 230, 761—771 (1958).

Stadie, W. C.: Current concepts of the action of insulin. Physiol. Rev. 34, 52—100 (1954).

Steele, R.: Influence of glucose loading and of injected insulin on hepatic glucose output. Ann. N.Y. Acad. Sci. 82, 420—430 (1959).

— Reflections on pools. Fed. Proc. 23, 671—679 (1964).

— J. S. Bishop, A. Dunn, N. Altszuler, I. Rathgeb, and R. C. de Bodo: Inhibition by insulin of hepatic glucose production in the normal dog. Amer. J. Physiol. 208, 301 (1965).

— — and R. Levine: Does a glucose load inhibit hepatic sugar output? C^{14} glucose studies in eviscerated dogs. Amer. J. Physiol. 197, 60—62 (1959).

—, and P. Marks: Production of glucose by the liver during hyperglycemia. Nature (Lond.) 182, 1444—1445 (1958).

— J. S. Wall, R. C. de Bodo, and N. Altszuler: Measurement of the size and turnover rate of the body glucose pool by the isotope dilution method. Amer. J. Physiol. 187, 15—24 (1956).

Stein, Y., and B. Shapiro: Uptake and metabolism of triglycerides by the rat liver. J. Lipid Res. 1, 326—331 (1960).

Steiner, D. F., and J. King: Induced synthesis of hepatic uridine diphosphate glucose-glycogen glucosyltransferase after administration of insulin to alloxan-diabetic rats. J. biol. Chem. 239, 1292—1298 (1964).

— V. Rauda, and R. H. Williams: Effects of insulin, glucagon and glucocorticoids upon hepatic glycogen synthesis from uridine diphosphate glucose. J. biol. Chem. 236, 299—304 (1961).

—, and R. H. Williams: Some observations concerning hepatic glucose 6-phosphate content in normal and diabetic rats. J. biol. Chem. 234, 1342—1346 (1959).

SUTHERLAND, E. W., and T. W. RALL: The relation of adenosine-3',5'-phosphate and phosphorylase to the actions of catecholamines and other hormones. Pharmacol. Rev. 12, 265—299 (1960).

TARDING, F., and P. SCHAMBYE: The action of sulfonylureas and insulin on the glucose output from the liver of normal dogs. Endocrinologie 36, 222—228 (1958).

TARRANT, M. E., R. MAHLER, and J. ASHMORE: Studies in experimental diabetes. IV. Free fatty acid mobilization. J. biol. Chem 239, 1714—1719 (1964).

TATA, J. R.: Biological action of thyroid hormones on the cellular and molecular levels. In: Action of hormones on molecular processes, pp. 58—131. Edit. by G. LITWACK and D. KRITCHEVSKY. New York: John Wiley & Sons 1964.

TEPPERMAN, H. M., and J. TEPPERMAN: On the response of hepatic glucose 6-phosphate dehydrogenase activity to changes in diet composition and food-intake pattern. In: Advances in enzyme regulation, vol. I, pp. 121—136. Edit. by G. WEBER. Oxford: Pergamon Press 1963.

TEPPERMAN, J., and H. M. TEPPERMAN: Some effects of hormones on cells and cell constituents. Pharmacol. Rev. 12, 301—353 (1960).

THORN, G. W., A. E. RENOLD, and G. F. CAHILL jr.: The adrenal and diabetes. Some interaction and interrelations. Diabetes 8, 337—351 (1959).

TOYE, K. E., and J. F. MANERY: The effect of bound insulin on the rate of oxygen consumption and potassium concentration of frog muscle. Rev. canad. Biol. 14, 289—290 (1955).

TRAKATELLIS, A. C., A. E. AXELROD, and M. MONTJAR: Actinomycin D and messenger RNA turnover. Science 203, 1134—1136 (1964a).

— — — Studies on liver messenger ribonucleic acid. J. biol. Chem. 239, 4237—4244 (1964b).

UNGER, R. H., A. M. EISENTRAUT, M. S. McCALL, and L. L. MADISON: Meaurements of endogenous glucagon in plasma and the influence of blood glucose concentration upon its secretion. J. clin. Invest. 41, 682—689 (1962).

USSING, H. H.: Transport of ions across cellular and biological membranes. Introduction. In: Membrane transport and metabolism, pp. 115—116. Edit. by A. KLEINZELLER and A. KOTYK. London: Academic Press 1961.

UTTER, M. F., and D. B. KEECH: Formation of oxaloacetate from pyruvate and CO_2. J. biol. Chem. 235, PC17—18 (1960).

— — and M. C. SCRUTTON: A possible role for acetyl CoA in the control of gluconeo-genesis. In: Advances in enzyme regulation, Vol. II, pp. 49—68. Edit. by G. WEBER. Oxford: Pergamon Press 1964.

VESTER, J. W.: Insulin effect on hepatic glucokinase: response to variation in substrate level. Intern. Union of Biochem. Ser. 32, 530 (1964).

—, and M. L. REINO: Hepatic glucokinase: a direct effect of insulin. Science 142, 590—591 (1963).

VILLEE, C. A., and A. B. HASTINGS: The utilization in vitro of C^{14}-labeled acetate and pyru-vate by diaphragm muscle of rat. J. biol. Chem. 181, 131—139 (1949).

VIÑUELA, E., M. SALAS, and A. SOLS: Glucokinase and hexokinase in liver in relation to glycogen synthesis. J. biol. Chem. 238, PC1175—1177 (1963).

WAGLE, S. R., and J. ASHMORE: Studies on experimental diabetes. II. Carbon dioxide fixation. J. biol. Chem. 238, 17—20 (1963).

— — Studies on experimental diabetes. III. Effects of acute insulin insufficiency on ^{14}C-glucose formation from labeled substrates. J. biol. Chem. 239, 1289—1291 (1964).

WAKIL, S. J.: Lipid metabolism. Ann. Rev. Biochem. 31, 369—406 (1962).

WALKER, D. G.: The development of hepatic hexokinases after birth. Biochem. J. 84, 118P—119P (1962).

— On the presence of two soluble glucose-phosphorylating enzymes in adult liver and the development of one of those after birth. Biochim. biophys. Acta (Amst.) 77, 209—226 (1963).

Walker, D. G., and S. Rao: Some factors affecting the hepatic hexokinases. Biochem. J. 88, 17P (1963).

Wall, J. S., R. Steele, R. C. de Bodo, and N. Altszuler: Radioactive glucose and insulin hypersensitivity in the hypophysectomized dog. Fed. Proc. 15, 196 (1956).

— — — — Effect of insulin on utilization and production of circulating glucose. Amer. J. Physiol. 189, 43—50 (1957).

Weber, G., G. Banerjee, and S. B. Bronstein: Selective induction and suppression of liver enzyme synthesis. Amer. J. Physiol. 202, 137—144 (1962).

Weil, R., N. Altszuler, and J. Kessler: Effect of growth hormone on pyruvic acid metabolism. Amer. J. Physiol. 201, 251—254 (1961).

Weil-Malherbe, H.: The mechanism of action of insulin. Ergebn. Physiol. 48, 54—111 (1955).

Weinges, K. F.: Die Wirkung des Glucagons auf die Gesamtaminosäuren im Serum. Naunyn-Schmiedebergs Arch. exp. Path. Pharmak. 237, 17—21 (1959).

Weisberg, H. F., A. Friedman, and R. Levine: Inactivation or removal of insulin by the liver. Amer. J. Physiol. 158, 332—336 (1949).

Wells, B. B., and E. C. Kendall: The influence of the adrenal cortex in phlorhizin diabetes. Proc. Mayo Clin. 15, 565—573 (1940).

Wertheimer, E., and E. Shafrir: Influence of hormones on adipose tissue as a center of fat metabolism. Recent Progr. Hormone Res. 16, 467—486 (1960).

White, A.: See discussion in: Recent Progr. Hormone Res. 20, 242—243 (1964).

Wieland, O., I. Neufeldt, S. Numa u. F. Lynen: Zur Störung der Fettsäuresynthese bei Hunger und Alloxandiabetes. II. Fettsäuresynthese in der Leber alloxandiabetischer Ratten. Biochem. Z. 336, 455—459 (1963).

—, and L. Weiss: Increase in liver acetyl-coenzyme A during ketosis. Biochem. biophys. Res. Commun. 10, 333—339 (1963).

— — I. Eger-Neufeldt, and U. Müller: Ketone formation and inhibition of liver lipid synthesis by chylomicrons. Life Sci. 7, 441—447 (1963).

Williams, W. R., R. Hill, and I. L. Chaikoff: Portal venous injection of insulin in the diabetic rat: time of induction of changes in hepatic lipogenesis, cholesterogenesis, and glycogenesis. J. Lipid Res. 1, 236—240 (1960).

Williamson, J. R., E. A. Jones, and G. F. Azzone: Metabolic control in perfused rat heart during fluoroacetate poisoning. Biochem. biophys. Res. Commun. 17, 696—702 (1964).

—, and H. A. Krebs: Acetoacetate as fuel of respiration in the perfused rat heart. Biochem. J. 80, 540—547 (1961).

Winegrad, A. J.: Endocrine effects on adipose tissue metabolism. In: Vitamins and hormones, vol. XX, pp. 141—197. Edit. by R. S. Harris and I. G. Wool. New York: Academic Press 1962.

— Insulin and lipid metabolism. In: Actions of hormones on molecular processes, pp. 382 to 421. Edit. by G. Litwack and D. Kritchevsky. New York: John Wiley & Sons 1964.

Winkler, B., R. Steele, N. Altszuler, and R. C. de Bodo: Effect of growth hormone on free fatty acid metabolism. Amer. J. Physiol. 206, 174—178 (1964).

Winternitz, W. W., R. Dintzis, and C. N. H. Long: Further studies on the adrenal cortex and carbohydrate metabolism. Endocrinology 61, 724—741 (1957).

Wool, I. G.: Insulin and protein biosynthesis. In: Actions of hormones on molecular processes, pp. 422—469. Edit. by G. Litwack and D. Kritchevsky. New York: John Wiley & Sons 1964.

—, and E. I. Weinshelbaum: Incorporation of C^{14}-amino acids into protein of isolated diaphragms: role of the adrenal steroids. Amer. J. Physiol. 197, 1089—1092 (1959).

— — Adrenal cortical hormone and incorporation of C^{14} from amino acid precursors into muscle protein. Amer. J. Physiol. 198, 360—362 (1960).

WRENSHALL, G. A., G. HETENYI jr., and C. H. BEST: The validity of rates of glucose appearance in the dog calculated by the method of successive tracer injections. II. The influence of intermixing time following tracer injection. Canad. J. Biochem. **39**, 267—278 (1961).

ZAMECNIK, P. C., and E. B. KELLER: Relation between phosphate energy donors and incorporation of labeled amino acids into protein. J. biol. Chem. **209**, 337—354 (1954).

ZIERLER, K. L.: Increase in resting membrane potential of skeletal muscle produced by insulin. Science **126**, 1067—1068 (1957).

— Effects of insulin on membrane potential and potassium content of rat muscle. Amer. J. Physiol. **197**, 515—523 (1959a).

— Hyperpolarization of muscle by insulin in a glucose-free environment. Amer. J. Physiol. **197**, 524—526 (1959b).

Ganglioside

Von

HERBERT WIEGANDT*

Mit 5 Abbildungen

Inhaltsverzeichnis

Einleitung

Hinweis auf die Existenz einer neuen Klasse von Lipiden gab die Farbstoffbildung der in allen Gangliosiden vorkommenden Sialinsäure[1] mit Orcin (Bial's-Reagens) oder p-Dimethylaminobenzaldehyd (Ehrlich-Aldehyd), die zuerst von K. LANDSTEINER und P. A. LEVENE (1925) und von H. THIERFELDER und E. WALZ (1927) beobachtet wurde. Die nähere Erforschung dieser Verbindungen begann erst, als E. KLENK (1935) bei amaurotischer Idiotie des Typs Tay-Sachs und später bei der Niemann-Pickschen Krankheit eine im Hirn gespeicherte Substanz isolieren konnte, die er als saures kohlenhydratreiches Glykolipid charakterisierte. Später wies G. BLIX (1938) eine solche Verbindung in geringer Konzentration auch im normalen Hirn nach. KLENK (1942) vermutete ein Vorkommen dieser Stoffe in den Ganglienzellen des Hirns und gab ihnen daher die Bezeichnung Ganglioside.

A. Beschreibung der Ganglioside
1. Chemische Bestandteile und Struktur

Die Ganglioside bestehen aus

Sphingosin (C_{18} und C_{20}-Base), 1 Mol;

Fettsäure (C_{14} bis C_{24}-Säuren), 1 Mol;

---

* Physiologisch-Chemisches Institut der Universität Marburg.

[1] Als Sialinsäure bezeichnet man die Gruppe der verschiedenen N- bzw. O-acylierten Neuraminsäuren. Der Grundkörper, die Neuraminsäure (vgl. Formel, S. 192) ist eine 5-Amino-3,5-didesoxy-D-glycero-β-D-galakto-nonulosaminsäure.

Kohlenhydrat (Mono- bzw. Oligosaccharid) und
Sialinsäure (NANS bzw. NGNS)[1], 1 oder mehrere Mol.

Folgendes Aufbauprinzip ist bisher bei allen Gangliosiden gefunden worden:

Die Hydroxylgruppe in 3-Stellung des Sphingosins ist unsubstituiert. An der Aminogruppe in 2-Stellung befindet sich, amidartig gebunden, ein Molekül Fettsäure, während die Hydroxylgruppe an C-1 der Base mit Kohlenhydrat glykosidisch verknüpft ist. Der Kohlenhydratteil wiederum trägt die für die Ganglioside charakteristische Sialinsäure in ketosidischer Bindung. Dies zeigt das Formelbild eines Gangliosids aus Menschen- bzw. Rinderhirn:

Gangliosid G_{GNT1}

Die Unterschiedlichkeit der an das Sphingosin gebundenen Fettsäuren und die der Zuckerreste bedingt die Vielfalt der Glykosphingolipide. Die Einheitlichkeit aller bis jetzt dargestellten Ganglioside bezieht sich im wesentlichen nur auf den Kohlenhydratteil, der allerdings auch den stärksten Beitrag zur Charakteristik der einzelnen Verbindungen, sowohl chromatographisch als auch chemisch liefert.

Das *Sphingosin* hat E. KLENK als basischen Bestandteil der Ganglioside erkannt. Nach neueren Untersuchungen findet man daneben geringe Mengen von Dihydrosphingosin[2] (E. G. TRAMS und L. J. LAUTER 1962) und vor allem eine Sphingosinbase (auch Gangliosin, K. SAMBASIVARAO und R. H. McCLUER 1963, bzw. Ikosisphingosin, N. Z. STANACEV und E. CHARGAFF 1962, 1965 genannt) mit einer Kohlenstoffkette von 20 anstatt 18 Atomen. Nach Angaben von K. A. KARLSSON (1964) beträgt das Verhältnis von C_{18}- zu C_{20}-Base bei Gangliosiden aus Menschen- bzw. Rinderhirn 1:2, wobei jeweils $1/10$ in hydrierter Form vorliegt. Milzganglioside dagegen enthalten in überwiegender Menge die C_{18}-Base (J. MENKES 1964). Eine Zusammenfassung der analytischen Bestimmungsmethoden für Sphingosin wurde von N. ZÖLLNER und D. EBERHAGEN (1965) gegeben. (Vgl. auch G. SCHMIDT und A. KIETA-FYDA 1964.)

Die *Fettsäure* der Hirnganglioside ist zu 80—90% die für Lipide der Hirnrinde charakteristische Stearinsäure. Daneben wurden aber, meist durch gaschromatographische Bestimmung, noch geringe Mengen anderer Fettsäuren mit kürzerer und längerer, sowie teilweise ungesättigter Kohlenstoffkette festgestellt. E. G. TRAMS, L. E. GIUFFRIDA und A. KARMEN (1962) untersuchten

---

[1] Vgl. Tabelle der Abkürzungen, S. 194.
[2] synn. Sphingosan.

die Fettsäuren der Hirnganglioside bei verschiedenen Tierarten. Ihre Resultate gibt Tabelle 1.

Tabelle 1. *Fettsäuren der Hirnganglioside* [*]

C-Kettenlänge:C=C Doppelbindungen	14:0	16:0	18:0	20:0	21:1[**]	22:0	24:2[**]	24:1[**]	24:0
Rind		3	82	4		2	2	3	4
Affe		2	89	7		2	1		
Mensch			86	10		3			
Schwein		4	82	7		2	1	2	3
Puter			96	2		2			
Tümmler		1	80	14	1			2	1
Haifisch		5	72	1	4	5		13	
Alligator	1	7	79	4		2		7	

[*] Die Werte geben die Prozent-Zusammensetzung der Fettsäuren.
[**] Vorläufige Identifikation der Fettsäuremethylester auf Grund des Retentionsvolumens an einer Säule.

Es wurden keine Säuren mit ungerader Anzahl von C-Atomen gefunden. Die Fettsäuren der Ganglioside aus Milz und Erythrocyten bestehen nach E. Klenk und G. Padberg (1962) im wesentlichen aus Behen- und Lignocerinsäure. L. Svennerholm (1963) fand dort auch Nervonsäure.

HO—C—COOH

CH$_2$

HC—OH

R—NH—CH   Acetyl = R = CH$_3$. CO—
                 Glykolyl = R = HO . CH$_2$. CO—
O—CH

HC—OH

HC—OH

CH$_2$OH   Sialinsäure, 5-Acyl-
                 amino-3,5-didesoxy-
                 D-glycero-β-D-ga-
lakto-nonulosaminsäure

CH$_2$OH

HC—NH$_2$

HC—OH

C

H   C   H

(CH$_2$)$_{12}$. CH$_3$

Sphingosin, (2S:3R)-2-
Amino-trans-oktadecen-(4)-
diol-(1,3)

Als *Kohlenhydratbausteine* der Ganglioside wurden Glucose, Galaktose, N-Acetylglucosamin und N-Acetylgalaktosamin gefunden. Fucose kommt in sialinsäurehaltigen Glykosphingolipiden noch unbekannter Struktur mit Blutgruppenaktivität aus Menschenerythrocyten vor (Sh. Handa 1963). Eine Übersicht über die bisher in Substanz isolierten und zum Teil strukturell bekannten vollständigen Kohlenhydratteile der Ganglioside gibt Tabelle 4. Die von R. Kuhn und A. Gauhe (1962) aus der Frauenmilch isolierte Lacto-N-neotetraose bildet den Zuckerteil von Gangliosiden aus Rindererythrocyten (R. Kuhn und H. Wiegandt 1964a). Die immunologischen Spezifitäten der Ganglioside beruhen auf ihren Kohlenhydratteilen.

Von den sonstigen Glykosphingolipiden (Cerebrosiden, Aminoglykolipiden) unterscheiden sich die Ganglioside durch ihren Gehalt an Sialinsäure; sie können ein oder mehrere Mol Sialinsäure enthalten. Übersichten orientieren über diese Zuckersäuren (F. ZILLIKEN und M. W. WHITEHOUSE 1958; G. BLIX 1956, 1959; A. GOTTSCHALK 1960 und G. SMITS 1961). Sialinsäuren sind verhältnismäßig stark sauer (pK 2—3) Sie bilden unter Wasserabspaltung Lactone. Sowohl bei der Colominsäure (polymere N-Acetylneuraminsäure, E. J. McGUIRE und S. B. BINKLEY 1964) als auch bei sauren Gangliosidoligosacchariden (R. KUHN und H. WIEGANDT 1964b) wurde die Möglichkeit einer solchen Lactonbildung gezeigt. Sie besteht, wenn ein Molekül N-Acetylneuraminsäure ketosidisch mit der 8-Stellung eines weiteren Moleküls der Säure verknüpft ist. Diese inneren Ester sind sehr labil. Physiologisch könnte die Bildung solcher Lactone von Bedeutung sein, wenn es darum ginge, starke örtlich gebundene Ladungen zu verändern. Titrationsversuche an Gangliosiden (H. C. MELTZER 1964) weisen darauf hin, daß in manchen Präparationen, in denen nur etwa 25 % der vorhandenen Sialinsäure als starke Säure titriert werden kann, diese in zwei funktionellen Formen vorliegt. Auch hier ist vielleicht eine Lactonisierung eines Teils der Sialinsäure anzunehmen. Die Wanderungsgeschwindigkeiten der aus den Gangliosiden gewonnenen sialinsäurehaltigen Oligosaccharide in der Papierelektrophorese bei verschiedenen pH-Werten, ließen erkennen, daß im Fall der Verknüpfung zweier Sialinsäurereste miteinander nur ein Sialinsäuremolekül stark sauer reagiert. Die zweite Molekel besitzt anscheinend einen höheren pK-Wert (H. WIEGANDT, unveröffentlicht). Viele biologische Eigenschaften hängen vom Vorhandensein der Sialinsäure in den Gangliosiden ab.

Die Bindung zwischen Sialinsäuren und Kohlenhydrat wird durch Säuren und oft auch enzymatisch leicht gespalten. Viele der in der Literatur beschriebenen Glykosphingolipide mögen durch accidentelle Abspaltung der Sialinsäure bei der Aufarbeitung aus Gangliosiden entstanden sein. E. KLENK (1957) wies darauf hin, daß auch bei längerer Einwirkung von Formaldehyd auf Ganglioside Sialinsäure abgespalten wird.

2. Nomenklatur der Ganglioside

Für die bisher von den verschiedenen Arbeitskreisen rein dargestellten Ganglioside wurden völlig uneinheitliche Bezeichnungen eingeführt. Es hat sich im Bereiche der Kohlenhydrate sehr bewährt, statt der langen und unübersichtlichen systematischen Benennungen kurze halbtriviale Ausdrücke zu gebrauchen. Beispielsweise wird für die O-β-D-Galaktopyranosyl (1,3)-O-β-D-2-Desoxy-2-Acetaminogalaktopyranosyl (1,4)-3'-N-Acetylneuraminyl-O-β-D-galaktopyranosyl (1,4)Glucopyranose der Ausdruck Monosialo*ganglio-N-tetraose* gewählt.

Meist kommt es nicht auf eine eindeutige Charakterisierung der Fettsäure und der Sphingosinbase bei der Benennung der Sphingolipide an. Wir verbinden in einem solchen Fall den Namen Ceramid, der für eine Sphingosinbase steht, die am Stickstoff ein Molekül Fettsäure trägt, mit dem des Oligosaccharids. Zur Kennzeichnung der Stellung der Sialinsäure am Oligosaccharid markieren wir die Folge der Monosaccharidbausteine vom am Sphingosin gebundenen Ende her mit a, b, c usw. und die Verknüpfungsstellen mit arabischen Ziffern entsprechend der bei Kohlenhydraten üblichen Form. So beispielsweise für VIII[1] ($G_{GNT}1$)

$$\begin{array}{cccc} a & b & c & d \end{array}$$

Sphingosin-O[1]-Glc $(4,1\,\beta)$ Gal $(4,1\,\beta)$ GalNAc $(3,1\,\beta)$ Gal $=$

$\qquad\qquad\qquad\qquad\qquad\qquad\qquad\quad |$

Fettsäure $\qquad\qquad\qquad\qquad\qquad (3,2)$ NANS

Ceramid-(b 3)-sialo-ganglio-N-tetraosid

oder für XI[1] ($G_{GNT}3$)

Sphingosin-O[1]-Glc $(4,1\,\beta)$ Gal $(4,1\,\beta)$ GalNAc $(3,1\,\beta)$ Gal $=$

$\qquad\qquad\qquad\qquad\qquad\qquad\quad | \qquad\qquad\qquad\qquad\quad |$

$\qquad\qquad\qquad\qquad\qquad (3,2)$NANS$(8,2)$NANS $\quad (3,2)$NANS

Fettsäure

Ceramid-(b 3)-disialo[2,8]-(d 3)-sialo-ganglio-N-tetraosid.

Statt sialo- kann zur genauen Angabe der Sialinsäure N-Acetyl- (bzw. N-Glykolyl)-neuraminyl stehen.

Tabelle der Abkürzungen

		Kurzbezeichnung im Text
I	Ceramid-(3)sialo-galaktosid	G_{Gal}
II	Ceramid-(b 3)sialo-lactosid	$G_{Lact}1$
III	Ceramid-(b 3)N-Glykolyl-neuraminyl-lactosid	$G_{Lact}1$ [NGNS]
IV	Ceramid-(b 3)disialo[2,8]-lactosid	$G_{Lact}2$
V	Ceramid-(b 3) sialo-ganglio-N-triosid II	$G_{GNTrII}1$
VI	Ceramid-(b 3)disialo*-ganglio-N-triosid II	$G_{GNTrII}2$
VII	Ceramid-N-Glykolyl-neuraminyl*-lacto-N-neotetraosid	G_{LNnT} [NGNS]
VIII	Ceramid-(b 3)sialo-ganglio-N-tetraosid	$G_{GNT}1$
IX	Ceramid-(b 3,d 3)disialo-ganglio-N-tetraosid	$G_{GNT}2a$
X	Ceramid-(b 3)disialo[2,8]-ganglio-N-tetraosid	$G_{GNT}2b$
XI	Ceramid-(b 3)disialo[2,8]-(d 3)sialo-ganglio-N-tetraosid	$G_{GNT}3$
XII	Ceramid-tetrasialo*-ganglio-N-tetraosid	$G_{GNT}4$

Weitere Abkürzungen: Glc = Glucose, Gal = Galaktose, GlcNAc = N-Acetylglucosamin, GalNAc = N-Acetylgalaktosamin, NANS = N-Acetyl-neuraminsäure (= Lactaminsäure), NGNS = N-Glykolyl-neuraminsäure, Lact = Lactose, GNB = Ganglio-N-biose, GNTr = Ganglio-N-triose, GNT = Ganglio-N-tetraose, LNnT = Lacto-N-neotetraose, Fs = Fettsäure, Sph = Sphingosin.

---

[1] Vgl. Tabelle der Abkürzungen, S. 194.

* Die Stellung eines Moleküls Sialinsäure ist noch unbekannt.

Bei den *Kurzbezeichnungen*[1] steht G für Gangliosid; ein Index, z. B. G_{GNTrII}, zeigt den sialinsäurefreien Kohlenhydratrest. Eine beigefügte arabische Ziffer gibt die Anzahl der Sialinsäurereste. Hinsichtlich der Sialinsäure stellungsisomere Ganglioside werden bei den Kurzbezeichnungen mit kleinen Buchstaben unterschieden, z. B. $G_{GNT}2a$ und $G_{GNT}2b$. Handelt es sich bei der Sialinsäure eines Gangliosids statt NANS um NGNS, so ist der Gangliosidbezeichnung [NGNS] zugefügt. Durch das Präfix Des-NANS bzw. Des-NGNS werden die entsprechenden Glykocerebroside abgeleitet.

Des-Sph- bedeutet, daß es sich um das durch Abspaltung von Fettsäure und Sphingosin aus Gangliosid erhaltene saure Oligosaccharid handelt. Die in der Literatur zu findenden Gangliosidbezeichnungen gibt Tabelle 4.

Die Strukturen der Kohlenhydrate Ganglio-N-triose und Ganglio-N-tetraose zeigen die Formeln auf S. 197. Die Lacto-N-neotetraose ist eine O-β-D-Galaktopyranosyl (1,4)-O-β-D-2-Desoxy-2-Acetamino-Glucopyranosyl (1,3) -O-β-D-Galaktopyranosyl (1,4) Glucopyranose.

3. Physikalische und chemische Eigenschaften der Ganglioside

Die Ganglioside sind kristallisierende Substanzen, die unter Zersetzung schmelzen (z. B. $G_{GNT}1$, $Fp_{Zers.}$ 189—190⁰). Sie bilden untereinander sowie mit anderen Glykosphingolipiden Mischkristalle, so daß Kristallisation allein kein Kriterium für Einheitlichkeit darstellt. Zufolge ihres einerseits hydrophoben (Sphingosin, Fettsäure) andererseits hydrophilen (Zucker, Sialinsäure) Molekülteiles bilden die Ganglioside allein oder im Gemisch mit beispielsweise Lecithin, Lysolecithin oder Cerebrosid in wäßrigen Lösungen Micellen, die in der Ultrazentrifuge einheitlich sedimentieren. E. G. TRAMS und L. J. LAUTER zeigten (1962) durch Sedimentationsbestimmungen, daß solche Micellen nahezu sphärische Gestalt (axiales Verhältnis von 2) und Molgewichte von 200000—250000 besitzen (D. B. GAMMACK 1963). Der Sedimentationskoeffizient ist konzentrationsabhängig.

Die $S_{20,\,w}$-Werte, die für die einzelnen Ganglioside unterschiedlich sind, steigen bei abnehmender Gangliosidkonzentration. Die Sedimentationskonstanten sind in 0,2 m NaCl- bzw. $CaCl_2$-Lösung größer als in Wasser. Es wird angenommen, daß beim Entstehen der Micellen auch H-Brücken zwischen der $C=O$-Gruppe der Fettsäure und der 3-Hydroxygruppe des Sphingosins, wirksam sind. In nichtwäßrigen Lösungen oder solchen mit geringem Wassergehalt nimmt die Größe der Micellen ab. Die Micellen dissoziieren zu Einzelmolekülen. In Dimethylformamid (E. KLENK und W. GIELEN 1960) oder Tetrahydrofuran (R. E. HOWARD und R. M. BURTON 1964a) zeigen Ganglioside eine molekulare Verteilung (Molgewichte zwischen 1000 und 3000).

Die an sich nur geringe Oberflächenaktivität der Ganglioside (kleiner als 10 dyn/cm²) ist bei der kritischen Micellenkonzentration am größten. D. B.

---

[1] Vgl. Tabelle der Abkürzungen, S. 194.

Gammack (1963) bestimmte die kritische Micellenkonzentration aus der Oberflächenaktivität bei verschiedener Gangliosidkonzentration zu 0,015—0,02 %. Dagegen wurde durch Leitfähigkeitsmessung für die Mizellenbildung eine kritische Konzentration von 10^{-5} m gefunden (R. E. Howard und R. M. Burton, 1964a).

Ganglioside sind unlöslich in unpolaren Flüssigkeiten. Ihre Löslichkeit in polaren Lösungsmitteln (MeOH, Wasser) ist um so stärker, je größer der Kohlenhydratrest und der Sialinsäuregehalt sind. Ähnlich wie bei den Phospholipiden (M. B. Abrahamson, R. Katzman und H. P. Gregor 1964, H. St.

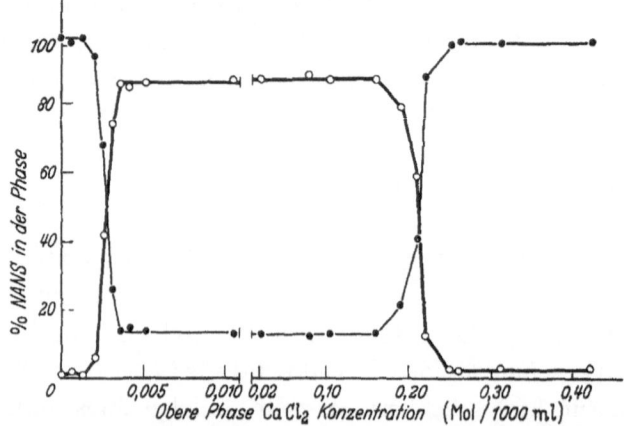

Abb. 1. Entnommen aus R. Quarles und J. Folch-Pi (1965). Die Resultate wurden ausgedrückt in Prozent NANS in der entsprechenden Phase, berechnet als NANS in der Phase/NANS in oberer Phase + NANS in unterer Phase × 100. Die offenen Kreise bezeichnen die Konzentration in der unteren, die geschlossenen Kreise diejenige in der oberen Phase. Die an der Abszisse angegebenen Werte entsprechen den Salzkonzentrationen in der oberen Phase nach der Verteilung, berechnet durch Multiplikation der Konzentrationen der zugegebenen wäßrigen Lösung mit 0,42 (2 ml Salzlösung pro 4,8 ml obere Phase)

Hendrickson und C. E. Ballou 1964, R. M. Dawson 1965, H. St. Hendrickson und J. G. Fullington 1965, E. Rojas und J. M. Tobias 1965, D. O. Shah und J. H. Schulman 1965, M. Wolman und H. Wiener 1965) bzw. Sulfatiden (U. Breyer 1965) hat die Anwesenheit von Ca^{++}-Ionen einen drastischen Einfluß auf die Löslichkeitseigenschaften der Ganglioside: R. Quarles und J. Folch-Pi (1965) verfolgten die Verteilung von Gangliosiden zwischen den beiden Phasen eines „Folch-Pi-Systems" von Chloroform-Methanol-Wasser. Ohne Zusatz von Calcium befinden sich Ganglioside zu etwa 80 % in der oberen, wäßrigen Phase, während bei Calciumkonzentrationen zwischen 0,05—0,18 m gegen 80 % allen Gangliosids in der unteren chloroformreichen, unpolareren Phase sind. Calcium scheint also durch Bildung von Chelaten die Ladung zu verringern. Eine weitere Erhöhung des Calciumgehaltes über 0,2 m hinaus bringt einen neuerlichen Übergang der Ganglioside von der unpolaren in die wäßrige Flüssigkeitsschicht (Abb. 1). Magnesium- und Alkaliionen besitzen auf die Löslichkeitsveränderungen der Ganglioside keinen Einfluß.

Ganglioside lassen sich als freie Säuren mit Alkali titrieren. Liegen Sialin-säurereste als Lacton vor, so erweist sich die Titration in 50% Alkohol als günstig, wobei man erst $^1/_2$ Std unter CO_2-Ausschluß mit einem geringen Über-schuß an n/100 NaOH stehen läßt. Ganglioside sind gegen überschüssiges Alkali und vor allem gegen Säuren empfindlich. Die Einwirkung von Alkali führt zur Abspaltung der Sialinsäure, die α-Pyrrolcarbonsäure bildet (A. GOTTSCHALK 1953, 1955). In der Form der freien Säuren verlieren die Ganglio-side in wäßriger Lösung schon bei Zimmertemperatur unter dem Einfluß der Eigenacidität Sialinsäure. Hierbei entstehen wie auch bei der Spaltung mit Neuraminidase (Neuraminat-glycohydrolase, E.C. 3.2.1.18, RDE = receptor destroying enzyme) die entsprechenden sialinsäurefreien Glyko-sphingolipide. *Partialhydrolyse* mit stärkerer Säure (0,1 n H_2SO_4, 1 Std, 100°) liefert Mono- bzw. Oligosaccharide, deren Strukturen Rückschlüsse auf die des Gangliosids erlauben. So wurde z. B. aus $G_{GNT}1$[1] (0,05 n H_2SO_4, 4 Std, 75°) ein sialinsäurefreies Glykolipid erhalten, das bei stärkerer Hydrolyse in folgende Oligosaccharide zerfiel (R. KUHN und H. WIEGANDT 1963 a) (Abb. 2).

	Abkürzungen
Gal (β,1⟶3) GalNAc (β,1⟶4) Gal (β,1⟶4)Glc 1⟶0¹——Sph——Fs NANS (α,2⟶3) ↓H⁺ (-NANS)	$G_{GNT}1$
Gal (β,1⟶3) GalNAc (β,1⟶4) Gal (β,1⟶4)Glc 1⟶0¹—— Sph——Fs ↓H⁺	Des-NANS-G_{GNT}
Gal (β,1⟶3) GalNAc (β,1⟶4) Gal (β,1⟶4)Glc	GNT
Gal (β,1⟶3) GalNAc (β,1⟶4) Gal	GNTrI
GalNAc (β,1⟶4) Gal (β,1⟶4)Glc	GNTrII
Gal (β,1⟶3) GalNAc	GNBI
GalNAc (β,1⟶4) Gal	GNBII
Gal (β,1⟶4)Glc	Lact

Bei partieller saurer *Methanolyse* wird die glykosidische Bindung der Neuraminsäure weniger leicht angegriffen, so daß es gelingt Zuckerreste abzu-spalten und dadurch größere Ganglioside zu solchen mit kleinerem Kohlen-hydratteil abzubauen. So erhielten R. LEDEEN und K. SALSMAN (1965) aus $G_{GNTrII}1$ das Gangliosid $G_{Lact}1$.

Totalhydrolyse (2n H_2SO_4, 7 Std, 100°) liefert Fettsäure, Sphingosin und Monosaccharide. Zur analytischen Bestimmung der Monosaccharide hat sich die Hydrolyse mit Ameisensäure bewährt (R. KUHN und H. WIEGANDT 1963 a). Neuraminat-glycohydrolase (E.C.3.2.1.18) spaltet aus vielen Ganglio-siden Sialinsäure ab. Die Enzymwirkung läßt sich leicht dünnschichtchromato-graphisch verfolgen (R. KUHN und H. WIEGANDT 1963 a). R. KUHN und H.

---

[1] Vgl. Tabelle der Abkürzungen, S. 194.

Abb. 2. Papierchromatogramm der aus dem Gangliosid G_{GNT1} gewonnenen sialinsäurefreien Oligosaccharide. (Schleicher und Schüll 2043b mgl, Laufmittel: Essigester-Pyridin-Eisessig-Wasser wie 5:5:1:3, Laufzeit 42 Std. Anfärbung: Anilinhydrogenphthalat.) **1** und **9** Glucose, Galaktose, Lactose (zur Markierung); **2** Ganglio-N-tetraose (GNT); **3** Ganglio-N-triose I (GNTrI); **4** Ganglio-N-triose II (GNTrII); **5** Lactose; **6** Ganglio-N-biose II (GNB II); **7** Ganglio-N-biose I (GNB I); **8** Glucose, Galaktose, Lactose (durch Abbau gewonnen)

Wiegandt (1964b) konnten so die folgenden strukturellen Beziehungen zwischen den Gangliosiden aus Menschen- bzw. Rinderhirn feststellen:

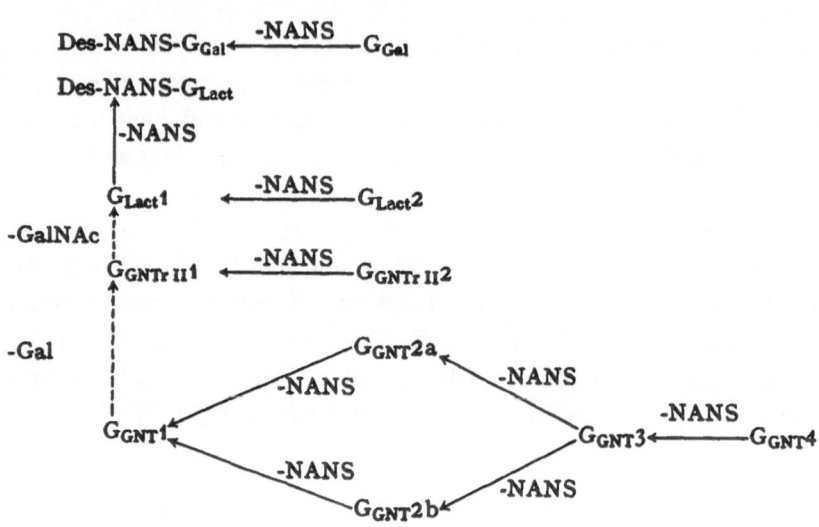

Die Ganglioside $G_{GNTrII}1$ und $G_{GNT}1^1$ erwiesen sich als durch Neuraminidase unspaltbar. Dies ist nach R. KUHN und H. WIEGANDT (1963 a) so
zu deuten, daß ein Zutritt des Enzyms zur glycosidischen Verknüpfung der
N-Acetylneuraminsäure mit der 3-Stellung der Galaktose durch die räumliche
Nähe großer Substituenten in 4-Stellung verhindert wird. Kürzlich wurden
auch niedermolekulare sialinsäurehaltige Glykopeptide erhalten, die ebenfalls
gegen Neuraminidase unempfindlich sind (W. E. MARSHALL und J. PORATH
1965). Am Beispiel der Colominsäure wurde gezeigt (E. J. McGUIRE und S. B.
BINKLEY 1964), daß Sialinsäurereste, deren Carboxylgruppe durch Lactonbildung verestert ist, von Neuraminidase nicht abgespalten werden.

Durch Synthese von reinen α- und β-Ketosiden der N-Acetylneuraminsäure konnten P. MEINDL und H. TUBBY (1965) eindeutig zeigen, daß nur die
optisch weniger stark nach links drehenden, sog. „α"-Ketoside, von Neuraminat-
glykohydrolase aus Vibrio cholerae gespalten werden.

Die enzymatische Hydrolyse des jeweils endständigen Zuckers durch
adaptierte Stämme von Klebsiella aerogenes konnte zur Bestimmung der
Reihenfolge der Kohlenhydratbausteine in Oligosacchariden benutzt werden
(S. A. BARKER und G. J. PARDOE 1964).

Ganglioside reagieren mit *Ozon*. Dabei wird die Doppelbindung des Sphingosins gespalten und es entstehen Produkte, die unter der Einwirkung von
schwachem Alkali bei Zimmertemperatur leicht zerfallen. Es wird hierbei der
gesamte an C-1 des Sphingosins gebundene Zuckerteil frei. Oligosaccharid-
bindungen werden *nicht gespalten und auch die sonst sehr empfindlichen*
Bindungen an Sialinsäuren bleiben unangegriffen (H. WIEGANDT und G. BA-
SCHANG 1965). Durch diese Methode ist eine verschärfte Prüfung der Ganglioside auf Einheitlichkeit möglich. Abb. 3 zeigt dafür einige Beispiele.

Durch partiellen Abbau der Ganglioside mit Essigsäureanhydrid/Essig-
säure unter Zusatz katalytischer Mengen Schwefelsäure wird nur ein Teil der
Sialinsäure abgespalten und es lassen sich nach Ent-O-acetylierung der
Acetolyseprodukte eine Reihe für die Strukturaufklärung wichtiger sialin-
säurehaltiger Oligosaccharide gewinnen (R. KUHN und H. WIEGANDT 1963 a).

Die Verknüpfungsstellen der Monosaccharide miteinander kann man durch
Methylierungen (K. WALLENFELS, G. BECHTLER, R. KUHN, H. TRISCHMANN
und H. EGGE 1963) mit anschließender Hydrolyse und Analyse der methy-
lierten Monosaccharide bestimmen (R. KUHN und H. EGGE 1963, T. YAMA-
KAWA und N. UETA 1964, C. SWEELEY und B. WALKER 1964, E. KLENK und
W. KUNAU 1964). Unter den alkalischen Bedingungen der Methylierung wird
teilweise Sialinsäure abgespalten. Es empfiehlt sich daher das Permethylie-
rungsprodukt einer Reinigung zu unterziehen.

---

[1] Vgl. Tabelle der Abkürzungen, S. 194.

Perjodat, in schwach saurer Lösung zur Verhinderung der „Überoxydation", spaltet alle Glykolgruppen im Zuckerteil der Ganglioside. Anschließende Hydrolyse und Bestimmung der unversehrt gebliebenen Kohlenhydrate läßt Rückschlüsse auf die Gangliosidstruktur zu (R. Kuhn und H.

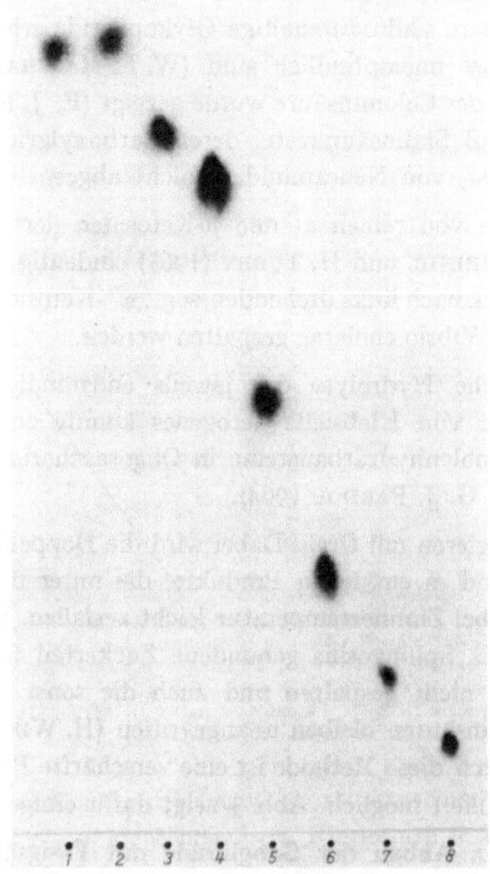

Abb. 3. Papierchromatogramm von aus Gangliosiden gewonnenen vollständigen, sialinsäurehaltigen Kohlenhydratteilen (Schleicher und Schüll 2043b mgl., Laufmittel: Essigester-Pyridin-Eisessig-Wasser wie 5:5:1:4, Laufzeit 209 Std. Anfärbung Anilinhydrogenphthalat). 1 3'-N-Acetylneuraminyl-lactose (zur Markierung); 2 Des-Sph-G_{Lact}1; 3 Des-Sph-G_{Lact}1 [NGNS]; 4 Des-Sph-G_{GNTrII}1; 5 Des-Sph-G_{GNT}1; 6 Des-Sph-G_{GNT}2a; 7 Des-Sph-G_{GNT}2b; 8 Des-Sph-G_{GNT}3

Wiegandt 1963a). Oft ist es günstig nach der Perjodatoxydation mit Natriumborhydrid zu reduzieren, um kleinere Bruchstücke (Glycerinaldehyd, Tetrosen) vor Zerstörung bei der anschließenden Hydrolyse zu bewahren. R. Kuhn und A. Gauhe (1965) bestimmten kürzlich durch Perjodatoxydation die Verknüpfungsstellen von Sialinsäureresten in zahlreichen Oligosacchariden, die diese Säure enthalten.

Zur schnellen Charakterisierung der Ganglioside eignet sich vor allem die Dünnschichtchromatographie z. B. auf Kieselgel. Beim Vergleich der Wanderungsgeschwindigkeiten ist darauf zu achten, daß z. B. G_{Lact}2 und G_{GNTrII}2[1]

---

[1] Vgl. Tabelle der Abkürzungen, S. 194.

gegenüber anderen Gangliosiden unterschiedliche relative Retentionszeiten in verschiedenen Laufmittelsystemen haben können (R. KUHN und H. WIE- GANDT 1964 b) (Abb. 4).

Durch den anionischen Zucker- wie den hydrophoben Lipidteil sind die Ganglioside in hervorragendem Maße zur Komplexbildung befähigt. R. W. ALBERS und G. J. KOVAL (1962) bestimmten anhand der Fluorescenzausbeute bzw. Fluorescenzpolarisation von Gangliosid-Farbstoff-Komplexen die rela- tiven Affinitäten der Bindung von Kationen (Curare, Hyamin, Mytelase) an

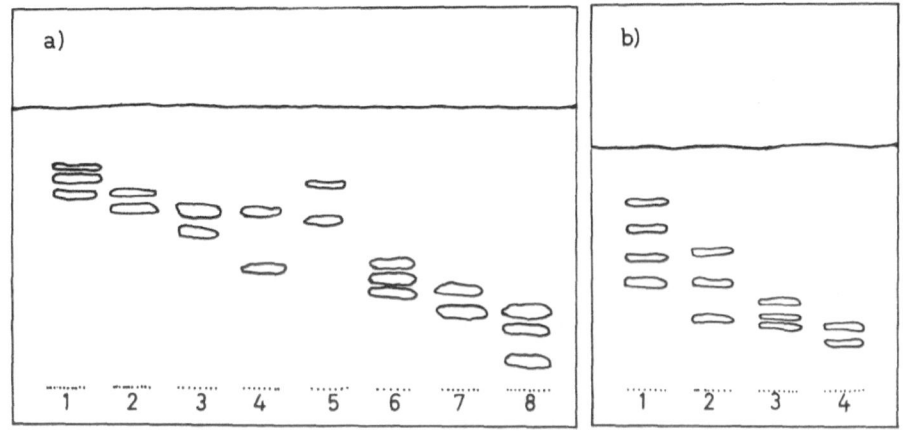

Abb. 4a u. b. Dünnschichtchromatogramm der Ganglioside. (Kieselgel G „Merck" bei 130° aktiviert. Laufmittel: n-Propanol-Wasser wie 7:3. Anfärbung: Ehrlichs Reagens bei 110°.) a 1 G_{Gal}, $G_{Lact}1$, $G_{GNTrII}1$, 2 $G_{GNTrII}1$, $G_{GNT}1$; 3 $G_{GNT}1$, $G_{Lact}2$; 4 $G_{GNT}1$, $G_{GNT}2a$; 5 $G_{Lact}1$ [NGNS], G_{LNnT} [NGNS]; 6 $G_{GNT}2a$, $G_{GNTrII}2$, $G_{GNT}2b$; 7 $G_{GNT}2b$, $G_{GNT}3$; 8 $G_{GNT}3$, $G_{GNT}4$, NANS. b Laufmittel: Chloroform-Methanol- Wasser wie 55:40:8,5; 1G_{Gal}, $G_{Lact}1$, $G_{GNTrII}1$, $G_{GNT}1$; 2$G_{Lact}2$, $G_{GNT}1$, $G_{GNT}2a$; 3$G_{GNTrII}2$, $G_{GNT}2a$, $G_{GNT}2b$; 4 $G_{GNT}2b$, $G_{GNT}3$

Ganglioside. Ausschlaggebend für die Komplexbildung sind Ladung und Lipophilie. Die Autoren konnten weitere Beweise für die Existenz von Ganglio- sid-Cerebrosid-Komplexen erbringen, deren Darstellung W. E. VAN HEYNIN- GEN (1959) angegeben hatte. Chlorpromazin wurde bei dieser Bestimmungs- weise von sialinsäurefreiem Glykolipid, das aus Gangliosid gewonnen war, stärker gebunden. A. F. HARRIS, A. SAIFER und B. W. VOLK (1960) beobach- teten eine Veränderung des U.V.-Spektrums von Chlorpromazin sowie Stelazin bei Komplexierung mit Gangliosiden. Auch Untersuchungen in der Ultra- zentrifuge deuteten auf eine Chlorpromazin-Gangliosid-Zwischenwirkung. Brucin, Thebain und Strychnin bilden ebenfalls mit Gangliosiden Komplexe (W. E. VAN HEYNINGEN 1963). Hinweise auf eine Bindung an d-Tubocurarin erhielten R. L. IRVIN und E. G. TRAMS (1961). Das Komplexierungsvermögen einer Reihe saurer Lipide, unter ihnen auch Ganglioside, für Serotonin unter- suchten R. S. DOMBRO, L. S. BRADHAM, N. K. CAMPBELL und D. W. WOOLEY (1961). Die Lipide besitzen mehr oder weniger starke Vehikelfunktion für Serotonin bei einer Verteilung zwischen wäßriger und organischer Phase.

D. B. Gammack (1963) studierte die Komplexierung des basischen Farbstoffes „Cresyl Fast Violet" durch Ganglioside in der Ultrazentrifuge. Er beobachtete einen Anstieg der Sedimentationskonstante, jedoch keine Steigerung der elektrophoretischen Beweglichkeit. Hieraus wurde geschlossen, daß die Bindung zwischen dem Farbstoff und Gangliosid eher auf Absorption an Teile der Gangliosidmicellen mit geringer Dielektrizitätskonstante als auf ionischer Zwischenwirkung beruht.

Eingehend wurde die Bindung von Gangliosiden an basische Proteine und Polypeptide erforscht (H. McIlwain 1963, Zusammenfassung). Ganglioside lassen sich aus wäßrigen Lösungen mit Chloroform-Methanolgemischen extrahieren. Nach Bindung an Histone, Protamine bzw. an Poly-L-Lysin liegen die Ganglioside in nicht mehr extrahierbarer Form vor. Teilweise entstehen mit basischen Proteinen Niederschläge. Komplexbildung zwischen Gangliosiden und Protaminen verhindert die Präzipitation des Proteins durch Pikrinsäure oder Fluorescein.

B. Vorkommen der Ganglioside

Ganglioside sind in einer Reihe von Organen und im Blut enthalten. In großer Konzentration kommen sie im Hirn von Wirbeltieren vor. Die Gangliosidmenge der weißen Hirnsubstanz beträgt beim Menschen etwa $1/5$ bis $1/4$ des Gehaltes der grauen Substanz. Im Rinderhirn (R. E. Howard und R. M. Burton 1964b) findet man die stärkste Konzentration im Putamen bzw. Nucleus caudatus (bis 0,6% des Frischgewichtes). Sowohl der Gesamtgehalt des Hirns an Gangliosiden wie auch die relative Menge einzelner Ganglioside (K. Suzuki 1964, H. Wiegandt und M. Schöpfner unveröffentlicht, G. G. Honegger und T. A. Freyvogel 1963) ist bei einzelnen Tierarten oft sehr unterschiedlich. Einige Beispiele bringt Tabelle 2.

Bei Kindern, wie auch bei Rindern und Kaninchen, überwiegt $G_{GNT}2a$. Im Alter nimmt, wie Untersuchungen an Menschenhirn gezeigt haben die relative Menge an $G_{GNT}2a$ ab und man findet verstärkt Ganglioside mit höherem Sialinsäuregehalt (K. Suzuki 1964, 1965, H. Wiegandt und M. Schöpfner, unveröffentlicht).

P. B. Dietzel (1957) konnte histochemisch die Ganglioside in Ganglienzellen nachweisen. Mehrere Arbeitskreise haben sich mit der Bestimmung des Gangliosidgehaltes von subcellulären Hirnfraktionen beschäftigt (L. S. Wolfe 1961, J. R. Wherret und H. McIlwain 1962, L. M. Seminario, N. Hren, C. J. Gomez 1964, J. Eichberg, V. P. Whittaker und R. M. C. Dawson 1964, R. M. Burton und J. M. Gibbons 1964, R. M. Burton und R. E. Howard 1964, H. König, D. Gaines, Th. McDonald, R. Gray und J. Scott 1964). Bei vorsichtiger Homogenisierung von Hirngewebe findet sich die größte Gangliosidkonzentration in einer mittels Differential- bzw. Dichtegradientenzentrifugation gewonnenen Fraktion, die im wesentlichen aus Nervenendigungen, sog. Synaptosomen, besteht. Durch Lyse in hypoto-

Tabelle 2. *Gangliosidgehalt* von Menschen- und Tierhirn*

Material: Hirn	Gesamt-G-Gehalt	G_{Lact}[1]	G_{GNTrII}[1]	G_{GNT1}	G_{GNT2}a	G_{GNT2}b	G_{GNT3}	G_{GNT4}	Methode
Maus	640 G-NS								[1]
Huhn	590 G-NS								
Schlange	320 G-NS								
Frosch	220 G-NS								
Fisch	370 G-NS								
Krabbe	1 G-NS								
Rind	940 G-NS								[2]
graue	497 G-NS =								
Substanz	1639 G	53,7	392,6	619,4	240,7	268,0	64,7		[3]
Rind	210 G-NS	—	—	—	—	—	—		[2]
weiße	101,7 G-NS =								
Substanz	318 G	123,1	98,9	44,1	45,24	6,8			[3]
Kaninchen	497 G-NS								
graue	1523 G	66,8	365,6	555,2	234,4	254,4	46,4		[3]
Substanz									
Ratte	1069 G-NS	12,8	144,0	340,0	221,0	282,0	58,8		[4]
Mensch**	283 G-NS =								
graue	946 G	57,6	262,8	216,7	228,6	155,4	24,8		[3]
Substanz									
(erw.)									
Mensch (1 d	229 G-RS =								
alt) Ge-	718 G	62,8	182,4	404,0	69,6	—	—		[3]
samthirn									
Mensch	1070 G-NS =	11,8	117,0	245,0	304,0	316,0	67,4		[4]
graue	3130 G	52,8	585,6	728,7	904,2	725,8	32,9		
Substanz,									
44 Jahre									
Mensch***	348 G-NS =	6,6	302,0	5,9	7,7	10,1	9,7	5,2	[4]
Tay-Sachs	1494 G	25,2	1353,6	29,5	22,9	30,0	22,2	10,2	
2 Jahre									

* Angaben in µg G-NS (Gangliosidsialinsäure) bzw. G (Gangliosid) pro g Gewebefrischgewicht.

** Autopsiematerial, entnommen 2d post mortem.

*** Angaben in µg G-NS bzw. G „per sample"

[1] G. G. HONEGGER und T. A. FREYVOGEL (1963).

[2] H. H. HESS und E. ROLDE (1964).

[3] H. WIEGANDT und M. SCHÖPFNER unveröffentlicht (Methode in N. ZÖLLNER und D. EBERHAGEN 1965).

[4] K. SUZUKI (1965).

nischem Milieu und Zentrifugieren lassen sich aus ihnen zwei gangliosidhaltige Fraktionen gewinnen, von denen eine hauptsächlich Membranstrukturen und die andere die synaptischen Vesikel enthält. Eine besonders weitgehende Trennung der Membranen von synaptischen Vesikeln läßt sich durch Vorfraktionierung mittels der Ultrazentrifuge und anschließende trägerfreie Elektrophorese mit Sucrose- bzw. Sucrose-Elektrolytgradienten erreichen

(H. WIEGANDT, unveröffentlicht). Die Membranen sind Träger der struktur-gebundenen Acetylcholinesterase und Na, K, Mg-abhängiger ATP-ase[1].

Synaptische Vesikel aus Hirncortex, welche die neurochemischen Über-trägersubstanzen enthalten (E. DE ROBERTIS, G. RODRIGUEZ DE LORES AR-NAIZ, L. SALGNICOFF, A. PELLEGRINO DE IRALDI und L. M. ZIEHER 1963, V. P. WHITTAKER, J. A. MICHAELSON und R. J. A. KIRKLAND 1964) sind relativ arm an Gangliosiden. Der Gangliosidgehalt von Membranfraktionen aus der Hirnrinde von Rindern wurde zu etwa 2% des Trockengewichtes bestimmt während in synaptischen Vesikeln weniger als 1% Gangliosid (ber. auf Trockengewicht) enthalten war (H. WIEGANDT, unveröffentlicht). R. M. BUR-TON, R. E. HOWARD und Y. M. BALFOUR (1964) geben allerdings einen Gang-liosidgehalt von 10—12% des Trockengewichtes der synaptischen Vesikel aus Rinderhirncortex an[2].

Über das Vorkommen von Gangliosiden in verschiedenen Hirntumoren berichten H. SEIFERT und G. UHLENBRUCK (1965). Sie fanden in allen unter-suchten Tumoren Ganglioside, bei Meningiomen in größter Menge $G_{Lact}1$, das normalerweise nur in geringer Konzentration im Hirn enthalten ist.

Ganglioside befinden sich auch in der Säugermilz und im Stroma der Erythrocyten, sowie in Leukocyten (H. WIEGANDT unveröffentlicht). Die sialinsäurehaltigen Lipide im Serum (Plasma-G-Gehalt ca. 25 µg/ml), den Gefäßwandungen, den Nieren und Nebennieren (in Rinde und Mark) sind noch wenig untersucht. Auch im Darm wurden gangliosidartige Substanzen fest-gestellt (H. KÖNIG 1962, J. M. MCKIBBIN und W. R. VANCE 1964). Neuer-dings wurde über polare Lipide in der Milch berichtet (D: S. GALAMOS und V. M. KAPOULAS 1965). Man vermutet in ihnen Ganglioside.

C. Gangliosidspeicherkrankheiten

Charakteristisch für eine Reihe von genetisch bedingten Idiotien mit autosomal recessivem Erbgang ist die Speicherung von Sphingolipiden. Bei einigen dieser Krankheiten ragen die Ganglioside und Desialoganglioside an Menge hervor (E. KLENK 1935, Sphingolipidoses 1962). Hierher gehören die

---

[1] Diese beiden Enzyme werden löslich durch Behandeln der Membranfraktion mit Ultraschall und Detergentien. Dabei gehen auch Ganglioside in Lösung. Durch frak-tionierte Fällung mit Ammoniumsulfat und Chromatographie an Sephadex lassen sich dann Gangliosid-Proteinkomplexe erhalten, die Acetylcholinesterase bzw. Na, K, Mg-abhängige ATP-asewirksamkeit zeigen [vgl. dazu H. C. LAWLER 1964, Biochim. biophys. Acta (Amst.) 81, 280, R. KUHN und H. MÜLDNER 1964, Naturwissenschaften 5, 635, R. TANAKA und K. P. STRICKLAND 1965, Arch. Biochem. 111, 583].

[2] Der Reichtum und die enge Verbindung von Gangliosiden, Acetylcholin und Acetylcholinesterase in den synaptischen Strukturen veranlassen R. M. BURTON an-zunehmen, daß die Ganglioside durch Erhöhung der Permeabilität der präsynaptischen Membran eine Rolle bei der Freisetzung des Acetylcholins im Moment der Erregung spielen.

verschiedenen bislang im wesentlichen nur nach dem Erkrankungsalter unter-
schiedenen Formen der amaurotischen Idiotie:

a) Infantile Form (Typ Tay-Sachs). Krankheitsbeginn während des ersten
Lebensjahres. Tod nach 2—3 Jahren.

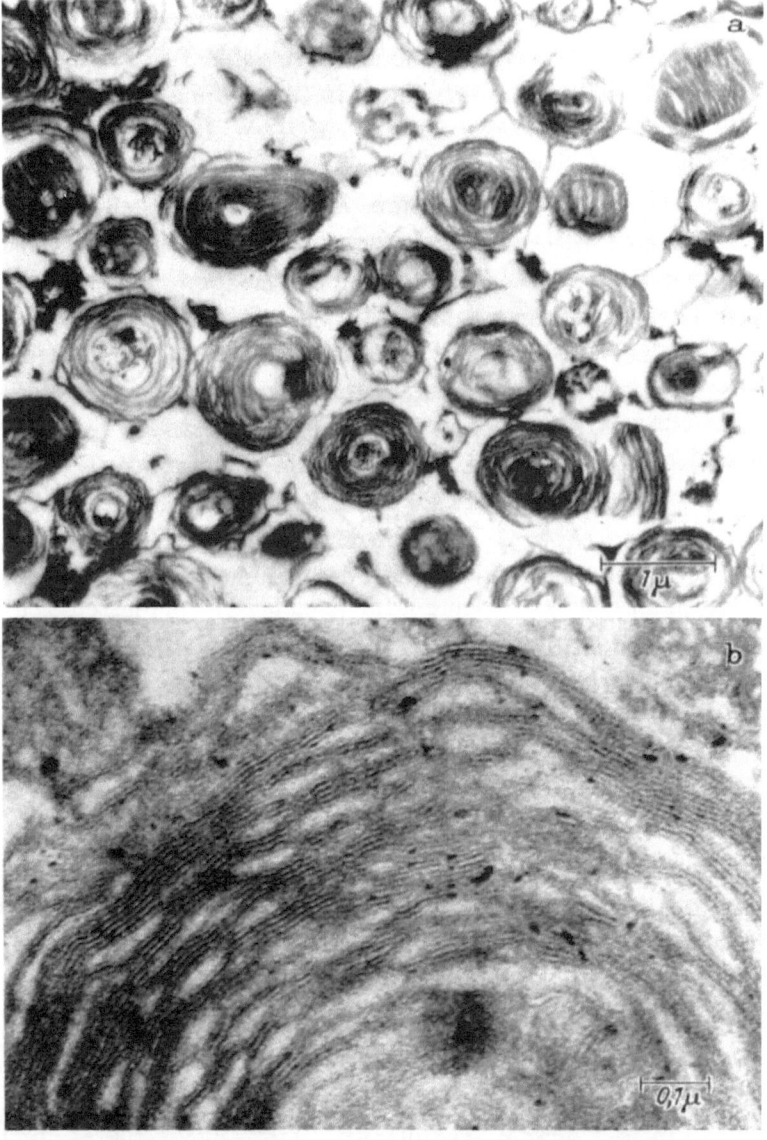

Abb. 5a u. b. Elektronenmikroskopische Aufnahme (Methacrylateinbettung) von Tay-Sachs-Hirn.
a Membrancytoplasmakörper, ×15000. b Detail der Lamellenstruktur der Membrancytoplasmakörper,
×90000. (Aufnahme: Dr. J. Escola, Max Planck-Institut für Hirnforschung, Frankfurt/M.)

b) Spätinfantile Form (Typ Bielschowsky): Krankheitsbeginn nach dem
ersten Lebensjahr. Tod nach 3—4 Jahren (S. R. Korey und R. D. Terry
1963).

c) iuvenile Form (Typ Batten, Spielmeyer-Vogt): Krankheitsbeginn nach dem 5., Tod vor dem 20. Lebensjahr.

d) adulte Form (Typ Kufs): Tod nach dem 20. Lebensjahr.

Zu den „primären" Lipoidosen, die eine direkte Lipidspeicherung in den Zellen und Veränderungen der Ganglioside zeigen, gehören: die Niemann-Picksche Krankheit, Morbus Gaucher (W. D. Suomi und B. W. Agranoff 1965) und der Gargoylismus (Lipochondrodystrophie Typ Pfaundler-Hurler).

Bei diesen Krankheiten kann eine Lipidanhäufung nicht nur im Hirn, sondern im gesamten visceralen System auftreten. Damit verbunden ist oft eine Erhöhung des Gehaltes an lipoidgebundener und z. T. auch an nicht lipoidgebundener Sialinsäure im Serum. Eine Zusammenstellung von Sialin-säurewerten für normales und pathologisches Hirngewebe gab G. Smits (1961).

Besonders gut untersucht ist die amaurotische Idiotie vom infantilen Typ Tay-Sachs. Hier findet man lamellierte Cytoplasmakörperchen des Golgi-apparates und des Ergastoplasmas, Mitochondrien mit Lamellenstruktur und charakteristisch lamellierte Membrancytoplasmakörper. Sie treten haupt-sächlich in Neuronen, aber auch in Gliazellen auf und zeigen keine Verwandt-schaft zu anderen Zellorganellen. Sie kommen in nichtpathologischem Material nicht vor. Ihre Größe beträgt etwa 1 μ, die Periodizität 50—60 Å (J. Escolá-Picó 1961, 1964a—c) (Abb. 5). Die Form der lamellierten Membrancytoplasma-körperchen rührt her von der Eigenschaft der Ganglioside (und anderer Lipide), sich aus physikalischen Gründen zusammenzulagern. Sie lassen sich auch in vitro darstellen und elektronenmikroskopisch sichtbar machen. Die chemische Analyse der lamellierten Membrancytoplasmakörper zeigt, daß sie zu 35—50% ihres Trockengewichtes aus Gangliosiden bestehen und außerdem Phosphatide, Cholesterin, Glucocerebrosid, eine Reihe von Aminosäuren und wenig Protein bzw. Polypeptide enthalten (S. R. Korey und R. D. Terry 1963).

Bei der Tay-Sachs-Lipidose ist die Menge an Gangliosiden in der weißen Hirnsubstanz gegenüber der Norm um das 5—6fache erhöht, während die graue Substanz 2—3fach an Gangliosid angereichert ist (ber. aus Sialinsäure unter Annahme von etwa 23% der Säure im Gangliosid). Besonders vermehrt ist $G_{GNTrII}1^1$ (L. Svennerholm 1962, K. Suzuki 1964), ein Gangliosid, das in normalem Hirn nur in geringer Menge vorkommt und das sich von den Hauptgangliosiden der G_{GNT}-Reihe durch das Fehlen der endständigen Galaktose unterscheidet (R. Kuhn und H. Wiegandt 1963a). Bei der spät-infantilen Lipidose ist statt $G_{GNTrII}1$ im wesentlichen $G_{GNT}1^*$ gespeichert (H. Jatzkewitz und K. Sandhoff 1963; K. Suzuki 1964). Aber auch der Gehalt an Ceramidoligosacchariden, die sich von den Gangliosiden der Reihen G_{Lact}, G_{GNTrII} und G_{GNT} ableiten, ist hier erhöht. Die Verteilung der Ganglio-side in verschiedenen Fällen amaurotischer Lipidose nach Analysen von K. Suzuki (1964) gibt Tabelle 3.

---

[1] Vgl. Tabelle der Abkürzungen, S. 194.

Tabelle 3. *Gangliosidverteilung* im Hirn*

	Alter (Jahre)	$G_{Lact}1$	$G_{GNTrII}1$	$G_{GNT}1$	$G_{GNT}2a$	$G_{GNT}2b$	$G_{GNT}3$	$G_{GNT}4$
normal	2	1	2,0	18,6	43,9	16,9	15,1	3,0
Tay-Sachs	2	1,3	83,4	2,1	8,1	3,2	1,6	1
spätinfantile Lipidose	2	2,6	3,8	80,4	8,7	2,4	1,0	1,1
normal	adult	1	1,5	13,0	25,4	33,2	23,8	4,2
adulte Lipidose	20	2,6	4,4	20,3	26,2	25,7	16,6	4,2

* Prozentverteilung der lipidgebundenen Sialinsäure.

Als charakteristisch für Morbus Gaucher wurde das Fehlen einer Gluco-cerebrosid-glucohydrolase beschrieben (R. O. BRADY, J. N. KANFER und D. SHAPIRO 1965). Bei dieser Krankheit wurde in der Milz eine Erhöhung des Gehaltes von $G_{Lact}1$ gefunden (J. MENKES 1964).

Über Gangliosidvermehrung wurde auch bei Niemann-Pickscher Krankheit (H. JATZKEWITZ, H. PILZ und K. SANDHOFF 1965) und beim Gargoylismus berichtet. Die Verteilung der Einzelkomponenten der Gesamtganglioside zeigt beim Gargoylismus eine bevorzugte Bildung von Gangliosiden mit kleinem Kohlenhydratteil (A. TAGHAVY, K. SALSMAN und R. LEDEEN 1964).

D. Stoffwechsel der Ganglioside

Die Bildung von Sphingosin aus Palmitinaldehyd bzw. Palmitinsäure und Serin konnte in vitro (R. O. BRADY und G. J. KOVAL 1957, J. ZABIN 1957, R. O. BRADY, J. V. FORMICA und G. J. KOVAL 1958) und in vivo (A. N. DAVISON, R. S. MORGAN, M. WAJDA und G. PAYLING WRIGHT 1959, J. CLÉMENT und G. DI COSTANZO 1964) durch Einsatz von markierter Palmitinsäure bzw. von Serin (3-C[14]) gezeigt werden. Die Biosynthese der Ganglioside ist in der jugendlichen Phase besonders rege (P. MANDELSTAM und R. M. BURTON 1959). Während dieser Zeit ist die relative Zunahme an Gangliosid drei- bis fünffach größer als die des Gesamthirngewichtes (J. F. GREANEY 1961). Viele neuere Arbeiten beschäftigten sich mit dem Einbau (in vitro und in vivo) von radio-markierter Glucose, Galaktose, Glucosamin und Galaktosamin in Ganglioside (E. G. TRAMS und L. J. LAUTER 1962, P. MANDELSTAM und R. M. BURTON 1959, R. M. BURTON 1963, R. M. BURTON, J. M. BALFOUR und J. M. GIBBONS 1964). Die Kohlenhydrate werden nahezu vollständig ohne vorherigen Abbau inkorporiert. Allerdings wird auch ein Teil ihrer Radioaktivität im Ceramidteil des Moleküls sowie in der Sialinsäure wiedergefunden. C[14]-markierte N-Acetyl-neuraminsäure ließ sich in vitro mit einem Präparat aus Rattennieren in Gangliosid einbauen (J. N. KANFER, R. S. BLACKLOW, L. WARREN und R. O. BRADY 1964). Dabei erwies sich Des-NANS-G_{GNTrII} als beste Vorstufe. Es wurde die Bildung von $G_{GNT}1$ gezeigt. Das bedeutet, daß außer der N-Acetyl-neuraminsäure auch noch ein Galaktoserest eingebaut wurde.

K. SUZUKI und S. R. KOREY (1963, 1964) untersuchten den Einbau von Glucose-C^{14} in verschiedene Ganglioside in einem zellfreien System einer Hirnmikrosomenfraktion und in vivo. Sie fanden, daß hierbei der absolute Betrag neueingebauter Sialinsäure mit der relativen Menge jedes einzelnen Gangliosids variiert. Das wird von den Autoren so gedeutet, daß alle Haupt-ganglioside, die sie untersuchten ($G_{GNT}1$—3), gleichzeitig entstanden, also keines die Vorstufe für die Bildung eines anderen darstellte. Die in durch Neuraminidase nicht abspaltbare Sialinsäure eingebaute Aktivität war größer als diejenige der durch das Enzym angreifbaren Sialinsäure. H^3-markiertes Psychosin wurde in vivo leicht in Ganglioside eingebaut (E. G. TRAMS und L. J. LAUTER 1962); dagegen konnte nach Injektion von synthetischem, C^{14}-markiertem Cerebrosid keine Aktivität in höher glykosierten Sphingo-lipiden beobachtet werden, obwohl z. B. die Existenz von G_{Gal} eine auch biosynthetisch nahe Verwandtschaft der Ganglioside zu den Cerebrosiden ver-muten ließe (J. KANFER 1965).

Cerebrosid abbauende Glucosidasen wurden in Hirn (R. M. BURTON, J. M. GIBBONS und Y. M. BALFOUR 1964) und Milz (R. O. BRADY, J. KANFER und D. SHAPIRO 1965) nachgewiesen.

$$\text{Cerebrosid} \rightarrow \text{Ceramid} + \text{Hexose}$$

Ferner wurde auch das Vorkommen einer N-Acylase (Ceramidase) in Hirn, Leber und Nieren gezeigt (SH. GATT 1963).

$$\text{Ceramid} \rightarrow \text{Sphingosin} + \text{Fettsäure}$$

Mit n-Butanol extrahierte Enzyme aus Hirn hatten eine Neuramidase-und Hexosaminidasewirksamkeit (Gangliosidase). Es wurde allerdings nur weniger als 1 μM Sialinsäure pro Gramm Frischgewebe pro Stunde abgespalten (S. R. KOREY und A. STEIN 1962, 1963 a und b).

Als Halbwertszeiten für die Stoffwechselrate der Ganglioside im Hirn wurden nach Injektion von Galaktose-1-C^{14} bzw. von Glucosamin-1-C^{14} bei Ratten durch Bestimmung der Gangliosid-Galaktose 24 Tage bzw. Gangliosid-hexosamin 10 Tage ermittelt (R. M. BURTON, Y. M. BALFOUR und J. M. GIBBONS 1964). Für Cerebrosid ergab sich unter gleichen Bedingungen eine Halbwertszeit von 45 Tagen.

E. Biologische Eigenschaften der Ganglioside

Die zellphysiologische Bedeutung der Ganglioside ist noch unbekannt. Folgende Wirkungen der Ganglioside legen die schon durch ihre Lokalisation gegebene Vermutung nahe, daß Gangliosie bei der Erregungsleitung in Nerven beteiligt sind.

H. McIlwain (1960) beobachtete, daß die Extraatmung von Hirngewebs-
schnitten bei elektrischer Reizung nach der Aufbewahrung in Krebs-Ringer-
Glucose-Lösung bei 0° stark herabgesetzt wird. Sie wird in den ersten 5 Std
etwa von 100 auf 15 %, später langsamer, erniedrigt. Die reguläre Atmung
bleibt dabei unbeeinflußt. Einen gleichen Effekt konnte er durch Zugabe von
basischen Proteinen und Polypeptiden (Histone, Protamin, Poly-L-Lysin)
erzielen. Ganglioside, in Mengen wie sie etwa dem Eigengehalt des Gewebes
entsprechen, vermögen die Extraatmung des Hirngewebes bei elektrischer
Reizung weitgehend wiederherzustellen. (Durch Zusatz von 120 μg Gangliosid
pro ml bis zu 80 % der ursprünglichen Extraatmung.) Es wurde gezeigt, daß
basische Proteine die Wiederaufnahme von Kalium nach Erregung blockieren,
sie tritt auf Zugabe von Gangliosid wieder ein. Clupein verursacht eine durch
Ganglioside verhinderbare Hypopolarisation. H. McIlwain erklärt diese Er-
scheinungen dadurch, daß bei 0° Histone den Zellkern verlassen und funk-
tionswichtige Gangliosidzentren durch Komplexbildung blockieren. Ganglioside haben keinen Einfluß auf die Kaliumaufnahme von Hirnschnitten bei
elektrischer Reizung. Merkwürdigerweise wird die ATP-ase einer frisch
präparierten Hirnmikrosomenfraktion durch Ganglioside inhibiert, während
hier basische Proteine die Enzymwirksamkeit steigern (A. Schwartz, H. S.
Bachelard und H. McIlwain 1962). H. McIlwain schließt aus seinen Be-
obachtungen, daß Ganglioside am Ionentransport durch die Nervenmembranen
beteiligt sind.

D. W. Wooley und B. W. Gommi (1964) nehmen an, daß Ganglioside
spezifische Receptorlipide für Serotonin darstellen. Nach ihnen kann die
Serotoninempfindlichkeit von Gewebe (z. B. von Magenstreifen) durch Einwirkung von Neuraminidase aus Clostridium perfringens[1] oder Vibrio cholerae[1]
in Gegenwart von Äthylendiamintetraessigsäure zerstört werden. Weder das
Enzym allein noch die Säure allein haben eine solche Wirkung. Zugabe von
Gangliosid bewirkt hier — ähnlich wie bei galaktosämischen Ratten — eine
Wiederherstellung der Serotoninempfindlichkeit. Die weitaus stärkste Wirkung wurde hierbei mit einem Gangliosid aus Hirn erzielt, das mit $G_{Lact}2$[2]
identisch zu sein scheint. Manche anderen Ganglioside waren praktisch un-
wirksam (D. W. Wooley und B. W. Gommi 1965). $G_{Lact}2$ trägt als Anion von
den bisher bekannten Gangliosiden die größte „lipophile elektrische Ladung"
(„Lipoph. el. Lad." $= \dfrac{\text{lipophiler Molekülteil}}{\text{hydrophiler Molekülteil}} \times$ el. Lad.). Serotonin kommt im
Hirn in den Nervenendigungen bzw. wahrscheinlich in den in ihnen ent-
haltenen synaptischen Vesikeln vor (V. P. Whittaker 1959). Die Bindung

---

[1] Die aus Clostridium perfringens (E. A. Popenoe und R. M. Drew 1957) gewonnene
Neuraminat-Glykohydrolase wird im Gegensatz zum Enzym von Vibrio cholerae (E. Mohr
und G. Schramm 1960) nicht durch Äthylendiamintetraessigsäure inhibiert (T. A. C.
Boschman und J. Jacobs 1965).

[2] Vgl. Tabelle der Abkürzungen, S. 194.

des Serotonins an die Synaptosomen wird durch Lysergsäurediäthylamid und 5-Bromlysergsäurediäthylamid in Konzentrationen von 10^{-7} m verhindert. Durch Vorbehandlung mit Neuraminidase geht diese Inhibitionsmöglichkeit durch die Lysergsäurederivate verloren. Auch hier bewirkt Gangliosidzusatz eine Wiederherstellung.

P. V. Johnston und B. J. Roots (1965) erreichten bei der Präparation für die Elektronenmikroskopie durch Zusatz von Gangliosid eine weitaus verbesserte Erhaltung der Membranstrukturen von Neuronen, die nicht von Glia umgeben waren. Eine durch die Ganglioside hierbei hervorgerufene Simulierung der sonst zwischen Glia und Neuron befindlichen Zwischenschicht soll nach Ansicht der Autoren diese Membranstabilisierung bewirken.

Im Hirn ist die regionale Verteilung für Ganglioside und sog. „gebundenes Acetylcholin"[1] unterschiedlich (R. E. Howard und R. M. Burton 1964). In den elektrischen Organen von Fischen, die Acetylcholin und Acetylcholinesterase in hoher Konzentration enthalten, werden keine Ganglioside gefunden (E. G. Trams und C. J. Lauter 1962).

J. A. Lowden und L. S. Wolfe (1964) weisen darauf hin, daß die regionale Verteilung der Ganglioside im Zentralnervensystem derjenigen an γ-Aminobuttersäure parallel geht. γ-Aminobuttersäure ist als Überträgersubstanz der Nervenleitung an Inhibitionssynapsen wirksam. Infolgedessen liegt der Schluß nahe, daß sowohl γ-A-Butyrat wie auch Ganglioside sich bevorzugt an Inhibitionssynapsen finden. Das ist deshalb interessant, weil gerade diese Synapsen durch Tetanustoxin blockiert werden[2] und weil andererseits Tetanustoxin von Gangliosiden spezifisch gebunden wird.

1898 fanden A. Wassermann und T. Takaki, daß Hirnhomogenat das Gift des Wundstarrkrampferregers inaktiviert. K. Landsteiner und A. Botteri (1906) versuchten die das Tetanustoxin bindende Substanz zu gewinnen. Aber erst in jüngster Zeit konnte W. E. van Heyningen (1959a—c; J. Mellanby und W. E. van Heyningen 1965) die Frage nach dem Receptor des Tetanustoxins beantworten. Er zeigte, daß das Toxin durch die im Hirn vorkommenden Ganglioside spezifisch gebunden wird. Ganglioside und Tetanustoxin sedimentieren einzeln in der Ultrazentrifuge verschieden schnell. Im Gemisch jedoch werden sie zusammen abgeschleudert. Das Bindungsvermögen hängt von der Struktur der Ganglioside ab. Es ist nicht proportional dem Gehalt an Sialinsäure und scheint besonders groß zu sein, wenn mehrere Sialinsäurereste miteinander verknüpft vorkommen. W. E. van Heyningen fand, daß je 1 mg verschiedener Ganglioside folgende Toxinmengen binden

---

[1] Zum Begriff des „gebundenen Acetylcholins" vgl. u.a. E. A. Hosein und L. Proulx (1965), Arch. Biochem. *109*, 129, E. A. Hosein, P. Rambaut, J. G. Chabrol und A. Orzeck (1965), Arch. Biochem. *111*, 540.

[2] Anmerkung. Diese Blockade wird als Hemmung der Freisetzung von Neurotransmittor-Substanzen an den präsynaptischen Strukturen der Inhibitionssynapsen angesehen (V. B. Brooks, D. R. Curtis und J. C. Eccles 1957).

kann (Toxinkonzentration 5 mg/ml): $G_{GNT}1$: 2,59 mg; $G_{GNT}2a$: 3,64 mg; $G_{GNT}2b$: 19,3 mg; $G_{GNT}3$: 20 mg. 1 Mol $G_{GNT}3$ bindet demnach bei hoher Toxinkonzentration unter Annahme eines Molgewichtes von ca. 70000 für das Tetanusgift etwa 0,6 Mol Toxin. Die sphingosinfreien sialinsäurehaltigen Oligosaccharide der Ganglioside vermögen Tetanustoxin nicht zu binden. Cerebroside, die allein kein Toxin zu binden vermögen, können in Komplexgemischen mit Gangliosiden von großem Einfluß sein. Bei 10 D_L an Toxin pro ml fixiert 1 mg Komplex (0,75 mg Cerebrosid + 0,25 mg Gangliosid) 2000 D_L von Tetanustoxin, während 1 mg eines anderen Komplexes (0,5 mg Cerebrosid + 0,5 mg Gangliosid) nur 30 D_L zu binden vermag, obwohl dieses Gemisch doppelt soviel Gangliosid enthält wie das erste (W. E. VAN HEYNINGEN, private Mitteilung).

Auch Staphylokokken- und Diphtherietoxin wird durch Ganglioside inaktiviert (W. E. VAN HEYNINGEN 1961a und b). Ganglioside bewirken weiterhin bei einigen Infektionen eine Resistenzsteigerung. Mäuse überlebten nach Gabe von Gangliosiden eine 100—200fache tödliche Dosis von Coli-Bakterien (O. WESTPHAL 1960). Schon 0,15—15 µg Gangliosid intracerebral jungen Mäusen 2 Std vor oder gleichzeitig mit PR8 bzw. NWS-Viren injiziert, verhinderten die sonst zu beobachtenden Krämpfe (O. BOGOCH 1959).

Ganglioside sind wie alle Glykosphingolipide in der Lage, die Bildung von spezifischen Antikörpern hervorzurufen (J. E. SOMERS, J. N. KANFER und R. O. BRADY 1964, A. L. SHERWIN, J. A. LOWDEN und L. S. WOLFE 1964, R. O. BRADY und E. G. TRAMS 1964). Die gegen Gangliosid gebildeten Antikörper wurden durch passive Hämagglutination von Erythrocyten nachgewiesen, die mit Gangliosiden überzogen waren (M. YOKOYAMA, E. G. TRAMS und R. O. BRADY 1963). R. O. BRADY u. Mitarb. benutzten diese Hapteninhibitionstechnik zur Aufklärung der Molekülstruktur des Zuckerteiles der Ganglioside. Ihre Befunde (J. E. SOMERS et al. 1964) bestätigen den auf chemischem Wege ermittelten Aufbau. Auch die Sialinsäurereste tragen zu der immunologischen Spezifität bei.

In den letzten Jahren fand man, daß neben den „klassischen" kohlenhydrathaltigen Proteinen auch Glykosphingolipide, vor allem aber solche des Erythrocytenstromas, Blutgruppeneigenschaften besitzen. Die meisten der in dieser Hinsicht untersuchten Verbindungen sind sialinsäurefrei.

Ganglioside sind wirksame Hemmstoffe in M- und N-Antigen-Antikörpersystemen (M. YOKOYAMA et al. 1963) wie auch bei der Reaktion von $Rh_0(D)$-Antikörpern mit $Rh_0(D)$-blutgruppentragenden Erythrocyten (M. C. DODD, N. J. BIGLEY und V. B. GEYER 1960, M. C. DODD, N. J. BIGLEY, G. A. JOHNSON, R. H. McCLUER 1964). Hier sind $G_{GNT}1$ und $G_{GNT}2b$ bessere Inhibitoren als $G_{GNTrII}1$ und $G_{GNT}3$. Für die $Rh_0(D)$-Spezifität wichtig ist demnach die Endgalaktose der Ganglio-N-tetraose. Die Wirksamkeit wird erhöht durch die

Tabelle 4. *Ganglioside, deren Derivate*

Formel
NANS (2→3) Gal 1→0^1—Sph— Fs
Gal 1→0^1—Sph—Fs
Glc 1→0^1—Sph —Fs
NANS (2→3) Gal <
NANS (2→3) Gal (β,1→4) Glc 1→0^1—Sph-—Fs
NANS (2→3) Gal (β,1→4) Glc <
NGNS (2→3) Gal (β,1→4) Glc 1→0^1—Sph—Fs
NGNS (2→3) Gal (β,1→4) Glc <
NANS (2→8) NANS (2→3) Gal (β,1→4) Glc 1→0^1—Sph— Fs
Gal (β,1→4) Glc 1→0^L--Sph-—Fs
Gal → Gal →0^1—Sph— Fs

und verwandte Glykolipide.

Herkunft	Fettsäure	Kurzbezeichnung	Bemerkungen
Hirn Mensch, normal		G_{Gal}	Vorkommen in geringer Menge [1]
G_{Gal} *Nieren* Mensch		Des-NANS-G_{Gal}	[1]* Cerebrosid [44]
Hirn *Rückenmark* Mensch			[45] [46]
Serum, Milz, *Leber* Mensch, normal *Milz*			[47, 18]
Mensch, „Gaucher"	Behensäure		[2, 46, 48, 49, 50]
G_{GNT}			[3, 4, 5]
G_{Gal}		Des-Sph-G_{Gal}	[1], aus NANS (2→3) Gal (β,1→4) Glc [6], aus G_{GNT}1 [6], aus Kuhcolostrum [7]
Hirn Mensch, normal „Gargoylismus" Rind Meningiom, Mensch *Milz, E. Stroma*		G_{Lact}	[1], Gangliosid $B_2{}^+$ [8]; $G_{M3}{}^+$ [9], G_6 [10]** [29] [56]
Mensch normal „Gaucher" Rind	Lignocerinsäure (Nervonsäure)		[11] [12] keine genauen Angaben [13] (Fettsäureanalyse) Gangliosid A
G_{Lact}1		Des-Sph-G_{Lact}1	[1] identisch mit 3'-Lactaminyl- lactose [14]
Milz, E. Stroma Rind	Lignocerinsäure (Behensäure)	G_{Lact}[NGNS]1	[11] und [15]$^+$
G_{Lact}[NGNS]1		(Des-Sph-G_{Lact}1 [NGNS])	[16] aus Kuhcolostrum [7]
Hirn Mensch, normal		G_{Lact}2	G'_{Lact} [1], G_{Lact}2 $\xrightarrow{-NANS}$ G_{Lact}1 $\xrightarrow{-NANS}$ Des-NANS-G_{Lact}*; [43]
G_{Lact}	Stearinsäure	Des-NANS-G_{Lact}	Ceramid-Lactosid aus Ganglio- sid [11]
G_{GNT}	Stearinsäure		[3, 5]
E. Stroma, Serum Mensch, normal	Lignocerinsäure		[17, 47] Component B [18, 19]
Milz/Leber Mensch, normal			[19] 20% -OH-Fs [18], 80% n-Fs
Rind Pferd Epidermal- Carcinom (H.Ep.3) *Niere*, Mensch		Cytolipin H	[21, 47] [47] [20] $C_{24, 22, 16}$ viel; $C_{20, 18, 14}$ wenig [44, 51]
Hirn, Mensch „Tay Sachs" *Niere*	Stearinsäure		[22]**
Mensch, normal „Gaucher"	C_{22}, C_{16}, C_{20}, C_{23}, $C_{24:1}$, C_{18}, C_{20}		[51] [25] keine Hydroxy-Fettsäuren
„Fabry"		GL 2	[24]**

Tabelle 4

Formel

$$\left.\begin{array}{l} \text{NANS} \quad (2\longrightarrow 3) \\ \qquad\text{und} \\ \text{Gal} \quad (1\longrightarrow 4) \end{array}\right\} \text{Gal } (1\longrightarrow 4) \text{ Glc } 1\longrightarrow 0^1 \text{ --Sph- } \cdot \text{ Fs}$$

$$\text{Gal} \quad (1\longrightarrow 4) \text{ Gal} \quad (1\longrightarrow 4) \text{ Glc } 1\longrightarrow 0^1\text{---Sph--- Fs}$$

$$\left.\begin{array}{l} \text{NANS} \\ \text{NANS} \end{array}\right\} \text{Gal } (\beta,1\longrightarrow 3)\,\text{Gal}(\beta^+,1\longrightarrow 3) \text{ Gal } 1\longrightarrow 0^1 \text{ ---Sph ---Fs}$$

$$\text{GalNAc } (\beta,1\longrightarrow 4) \text{ Gal } (\beta,1\longrightarrow 4) \text{ Glc } 1\longrightarrow 0^1 \text{ ---Sph---Fs}$$
$$\text{NANS} \begin{pmatrix} 3 \\ \uparrow \\ 2 \end{pmatrix}$$

$$\text{NANS} \quad (2\longrightarrow 3) \text{ GalNAc} \quad (1\longrightarrow 3) \text{ Gal} \quad (1\longrightarrow 4) \text{ Glc } 1\longrightarrow 0^1\text{---Sph---Fs}$$

$$\text{NANS}\left\{ \begin{array}{l} \text{GalNAc } (\beta,1\longrightarrow 4) \text{ Gal } (\beta,1\longrightarrow 4) \text{ Glc } 1\longrightarrow 0^1\text{---Sph--- Fs} \\ \qquad\qquad\qquad\qquad \begin{pmatrix} 3 \\ \uparrow \\ 2 \end{pmatrix} \\ \qquad\quad \text{NANS} \end{array}\right.$$

$$\text{GalNAc } (\beta,1\longrightarrow 4) \text{ Gal } (\beta,1\longrightarrow 4) \text{ Glc } 1\longrightarrow 0^1 \text{ ---Sph---Fs}$$

$$\text{GalNAc} \longrightarrow \text{Gal} \longrightarrow \text{Glc} \longrightarrow 0^1 \quad \cdot\text{Sph} \quad \text{Fs}$$

$$\text{GalNAc } (\beta,1\longrightarrow 4) \text{ Gal } (\beta,1\longrightarrow 4) \text{ Glc} <$$
$$\text{NANS} \begin{pmatrix} 3 \\ \uparrow \\ 2 \end{pmatrix}$$
$$\text{GalNAc } (\beta,1\longrightarrow 4) \text{ Gal } (\beta,1\longrightarrow 4) \text{ Glc} <$$

$$\text{Gal } (\beta,1\longrightarrow 3) \text{ GalNAc } (\beta,1\longrightarrow 4) \text{ Gal } (\beta,1\longrightarrow 4) \text{ Glc } 1\longrightarrow 0^1\text{---Sph } -\text{Fs}$$
$$\text{NANS} \begin{pmatrix} 3 \\ \uparrow \\ 2 \end{pmatrix}$$

$$\text{Gal } (\beta,1\longrightarrow 3) \text{ GalNAc } (\beta,1\longrightarrow 4) \text{ Gal } (\beta,1\longrightarrow 4) \text{ Glc } 1\longrightarrow 0^1\text{--- Sph- ---Fs}$$

$$\text{GalNAc } (1\longrightarrow 3) \quad \text{Gal} \quad (1\longrightarrow 4) \text{ Gal} \quad (1\longrightarrow 4) \text{ Glc } 1\longrightarrow 0^1\text{---Sph---Fs}$$

$$\text{Gal } (\beta,1\longrightarrow 3) \text{ GalNAc } (\beta,1\longrightarrow 4) \text{ Gal } (\beta,1\longrightarrow 4) \text{ Glc} <$$
$$\text{NANS} \begin{pmatrix} 3 \\ \uparrow \\ 2 \end{pmatrix}$$
$$\text{Gal } (\beta,1\longrightarrow 3) \text{ GalNAc } (\beta,1\longrightarrow 4) \text{ Gal } (\beta,1\longrightarrow 4) \text{ Glc} <$$

$$\text{NGNS } (2\longrightarrow) \{ \text{ Gal } (\beta,1\longrightarrow 4) \text{ GlcNAc } (\beta,1\longrightarrow 3) \text{ Gal } (\beta,1\longrightarrow 4) \text{ Glc } 1\longrightarrow 0^1\text{--- Sph- ---Fs}$$

(Fortsetzung)

Herkunft	Fettsäure	Kurzbezeichnung	Bemerkungen
Hirn Mensch, normal	Stearinsäure		[23]***
Serum, Leber, Milz Mensch, normal			[19]** [52]
Niere, Mensch, „Fabry"	81% $C_{22:0}$ + $C_{24:0}$ + $C_{24:1}$	GL 3	[24]
Hirn Mensch, normal	Stearinsäure	Gangliosid D	[33]
Hirn Mensch, normal „Tay Sachs"	Stearinsäure	$G_{GNTrII}1$	G_0 [6]. Identisch mit „Tay-Sachs"-Gangliosid [3, 26]**; FM-Gangliosid [27]**; G_{M2}^+ [9]; [28]; G_5 [10]**; [30]**
Hirn, Mensch	Stearinsäure	Gangliosid A	[23, 32] Konstitution fraglich
Hirn Mensch, normal	Stearinsäure	$G_{GNTrII}2$	G'_{GNTrII} [1], $G_{GNTrII}2 \xrightarrow{-NANS} G_{GNTrII}1$
G_{GNTrII}	Stearinsäure	Des-NANS-G_{GNTrII}	[11]*
G_{GNT}	Stearinsäure		[3, 5]
Hirn, Mensch, „Tay Sachs"			[3]**, [22]**, [53] wohl identisch mit Des-NANS-G_{GNTrII}
G_{GNTrII}		Des-Sph-$G_{GNTrII}1$	[6, 16]
G_{GNT} G_{GNTrII}		GNTr II	[6]
Hirn Mensch, normal	Stearinsäure	$G_{GNT}1$	G_I [5, 6], wohl identisch mit: Gangliosid A_2 (Mono-des-NANS-Gangliosid B_1) [35, 8]; G 4, [34]**; Major ganglioside, [3], 1-G, [27, 36]; G_{M1}^+, [9]; Gangliosid I a, [37]**; Derivat: Permethyl [5]
Hirn, Mensch, „Tay Sachs", Rind, G_{GNT}	Stearinsäure	Des-NANS-G_{GNT}	[6] [42]**
Niere, Mensch		Cytolipin K	[39]**; [54] wohl identisch mit „Main human globoside"
E. Stroma Mensch	$C_{23}H_{45}$. CO- (Nervonsäure) C_{23} (13%) C_{24} (79%)	„Main human globoside"	[38] [55]
$G_{GNT}1$		Des-Sph-$G_{GNT}1$	[6, 16]
G_{GNT} Rindermilz-gangliosid		GNT (Ganglio-N-tetraose)	[6] [11]
E. Stroma, Milz Rind	Lignocerinsäure oder Nervon-säure	$G_{LNnT}[NGNS]$	[40]

Tabelle 4

Formel

NGNS (2—→) { Gal (β,1—→4) GlcNAc (β,1—→3) Gal (β,1—→4) Glc <

        Gal (β,1—→4) GlcNAc (β,1—→3) Gal (β,1—→4) Glc <

        Ceramid-disialo-tetrahexosid, . . .

        Gal (β,1—→3) GalNAc (β,1—→4) Gal (β,1—→4) Glc 1—→0^1 —Sph— Fs
$$\text{NANS}\begin{pmatrix}3\\\uparrow\\2\end{pmatrix}\qquad\qquad \text{NANS}\begin{pmatrix}3\\\uparrow\\2\end{pmatrix}$$

        Gal (β,1—→3) GalNAc (β,1—→4) Gal (β,1—→4) Glc <
$$\text{NANS}\begin{pmatrix}3\\\uparrow\\2\end{pmatrix}\qquad\qquad \text{NANS}\begin{pmatrix}3\\\uparrow\\2\end{pmatrix}$$

        Gal (β,1—→3) GalNAc (β,1—→4) Gal (β,1—→4) Glc 1—→0^1—Sph— Fs
$$\text{NANS (2—→8) NANS}\begin{pmatrix}3\\\uparrow\\2\end{pmatrix}$$

        Gal (β,1—3) GalNAc (β,1—4) Gal (β,1—4) Glc <
$$\text{NANS (2—→8) NANS}\begin{pmatrix}3\\\uparrow\\2\end{pmatrix}$$

        Gal (β,1—→3) GalNAc (β,1—→4) Gal (β,1—→4) Glc 1—→0^1—Sph -- Fs
$$\text{NANS}\begin{pmatrix}3\\\uparrow\\2\end{pmatrix}\qquad \text{NANS (2—→8) NANS}\begin{pmatrix}3\\\uparrow\\2\end{pmatrix}$$

        Gal (β,1—→3) GalNAc (β,1—→4) Gal (β,1—→4) Glc <
$$\text{NANS}\begin{pmatrix}3\\\uparrow\\2\end{pmatrix}\qquad \text{NANS (2—→8) NANS}\begin{pmatrix}3\\\uparrow\\2\end{pmatrix}$$

        Gal (β,1—→3) GalNAc (β,1—→4) Gal (β,1—→4) Glc 1—→0^1—Sph—Fs
$$\begin{pmatrix}6\\\uparrow+\\2\end{pmatrix}\text{NANS}\qquad\begin{pmatrix}3\\\uparrow\\2\end{pmatrix}\text{NANS (8←—2) NANS}$$

$$\text{NANS}\left\{\begin{array}{l}\text{Gal (β,1—→3) GalNAc (β,1—→4) Gal (β,1—→4) Glc 1—→}0^1\text{—Sph—Fs}\\\text{NANS}\begin{pmatrix}3\\\uparrow\\2\end{pmatrix}\quad \text{NANS (2—→8) NANS}\begin{pmatrix}3\\\uparrow\\2\end{pmatrix}\end{array}\right.$$

$$\text{NANS}\left\{\begin{array}{l}\text{Gal (β,1—→3) GalNAc (β,1—→4) Gal (β,1—→4) Glc <}\\\text{NANS}\begin{pmatrix}3\\\uparrow\\2\end{pmatrix}\quad \text{NANS (2—→8) NANS}\begin{pmatrix}3\\\uparrow\\2\end{pmatrix}\end{array}\right.$$

   * chromatographisch identifiziert.
 ** keine vollständige Strukturangabe.
  + kein vollständiger Strukturbeweis.

Literatur zu Tabelle 4.

[1] KUHN und WIEGANDT (1964b).
[2] ROSENBERG und CHARGAFF (1958).
[3] SVENNERHOLM (1962).
[4] BOGOCH (1957).
[5] KUHN und EGGE (1963).
[6] KUHN und WIEGANDT (1963).
[7] KUHN und GAUHE (unveröffentlicht).
[8] KLENK und GIELEN (1963a).
[9] SVENNERHOLM (1963b).
[10] SUZUKI (1964).
[11] WIEGANDT (unveröffentlicht).
[12] SVENNERHOLM (1963).
[13] MENKES (1964).
[14] KUHN und BROSSMER (1959).

(Fortsetzung)

Herkunft	Fettsäure	Kurzbezeichnung	Bemerkungen
G_{LNnT}[NGNS]		Des-Sph-G_{LNnT} [NGNS]	[16]
G_{LNnT} Frauenmilch		LNnT (Lacto-N-neotetraose)	[40] [41]
Schwein, Hirn			[37]
Hirn Mensch, normal Rind		$G_{GNT}2a$	G_{II} [6,31,5] identisch mit Gangliosid B$_1$ [8, 35]; G$_3$ [34]** und 2-G [36]**; G_{DIa} [9]; Gangliosid I b [37]**; Derivat: Permethyl [5]
$G_{GNT}2a$		Des-Sph-$G_{GNT}2a$	[6, 43, 16]
Hirn Mensch, normal Rind	Stearinsäure	$G_{GNT}2b$	G_{III} [6, 31] identisch mit G$_2$ [34] und 3-G [27, 36]; G_{DIb}** [9]; Gangliosid III b [37]**
$G_{GNT}2b$		Des-Sph-$G_{GNT}2b$	[6, 31]
Hirn Mensch, normal Rind	Stearinsäure	$G_{GNT}3$	G_{IV} [6, 31] wohl identisch mit G_{T1}** [9]; Gangliosid III c [37]; G$_1$ [10]**
$G_{GNT}3$		Des-Sph-$G_{GNT}3$	[6, 31]
Hirn Mensch		4-G	[36] anscheinend nicht identisch mit $G_{GNT}3$ [27]**
Hirn Mensch, normal Rind	Stearinsäure	$G_{GNT}4$	G_V [6]; Gangliosid IV b [37]**; G$_0$ [10]**; [36] Strukturvorschlag
$G_{GNT}4$		Des-Sph-$G_{GNT}4$	[16]

[15] KLENK und PADBERG (1962).
[16] WIEGANDT und BASCHANG (1965).
[17] YAMAKAWA, KISO, HANDA, MAKITA und YOKOYAMA (1962).
[18] SVENNERHOLM und SVENNERHOLM (1963a).
[19] SVENNERHOLM und SVENNERHOLM (1963b).
[20] RAPPORT, GRAF und YARIF (1961).
[21] RAPPORT, GRAF, SKIPSKI und ALONZO (1959); RAPPORT, GRAF, ALONZO (1960).
[22] GATT und BERMAN (1963).
[23] KLENK und GIELEN (1960).
[24] SWEELEY (1963).
[25] PHILLPPART und MENKES (1964).
[26] KANFER, BLACKLOW, WARREN und BRADY (1964).
[27] JOHNSON und McCLUER (1963).
[28] KLENK, LIEDTKE und GIELEN (1963).
[29] TAGHAVY, SALSMAN und LEDEEN (1964).
[30] BOOTH (1963).
[31] KUHN und WIEGANDT (1963b).

[32] Klenk und Gielen (1961).
[33] Klenk und Gielen (1963 b).
[34] Korey und Gonatas (1963).
[35] Klenk und Kunau (1964).
[36] Johnson und McCluer (1964).
[37] Tettamanti, Bertona und Zambotti (1964).
[38] Yamakawa, Yokoyama und Kiso (1962).
[39] Rapport, Graf und Schneider (1964).
[40] Kuhn und Wiegandt (1964 a).
[41] Kuhn und Gauhe (1962).
[42] Gatt und Berman (1961).
[43] Wooley und Gommi (1965).
[44] Mårtensson (1963).
[45] Thierfelder und Klenk (1930).
[46] Suomi und Agranoff (1964).
[47] Makita und Yamakawa (1962).
[48] Brady, Kanfer und Shapiro (1965).
[49] Halliday, Deuel, Tragermann und Ward (1940).
[50] Marinetti, Ford und Stotz (1960).
[51] Makita und Yamakawa (1964).
[52] Wagner (1964).
[53] Makita und Yamakawa (1963).
[54] Makita (1964).
[55] Yamakawa, Yokoyama und Handa (1963).
[56] Seifert und Uhlenbruck (1965).

NANS der „vorderen" Galaktose, die in 8-Stellung ein weiteres Molekül NANS trägt:

$$\text{Gal } (\beta\ 1\longrightarrow3)\text{ GalNAc } (\beta\ 1\longrightarrow4)\text{ Gal } (\beta\ 1\longrightarrow4)\text{ Glc}$$

$$\text{NANS } (2\longrightarrow8)\text{ NANS} \left(\begin{matrix}3\\ \uparrow\\ 2\end{matrix}\right)$$

Die als Kohlenhydrat eines Gangliosids aus Rindererythrocyten bzw. Rindermilz aufgefundene Lacto-N-neotetraose besitzt im serologischen Test die Wirkungsspezifität von Pneumokokken Typ XIV-Polysacchariden.

O. Westphal (1960) studierte die Pyrogenität von Hirngangliosidpräparaten an Kaninchen. Gangliosidgaben von 150—200 µg/kg bewirkten eine Steigerung der Körpertemperatur um 0,6[1].

F. Isolierung und Reindarstellung der Ganglioside

Die Extraktion der Ganglioside aus Frischmaterial oder aus Acetontrockenpulver erfolgt meist mit Gemischen organischer Lösungsmittel, die in der Lage sind, die in situ vorliegenden Komplexe zu zerlegen. Man benutzt z. B. Mischungen von Chloroform/Methanol oder Tetrahydrofuran/Wasser. Bei einer anschließenden Verteilung gehen die Ganglioside je nach Verhältnis ihres hydrophoben zum hydrophilen Molekülteil mehr in die organische oder wäßrige Phase[2]. Eine weitere Abtrennung von Verunreinigungen gelingt mittels Ionenaustauscherharzen oder Ausfällung der Ganglioside z. B. in Form ihrer Bariumsalze.

Die Auftrennung der Gangliosidgemische in die einzelnen Komponenten geschieht durch Chromatographie z. B. an Cellulose, Sephadex oder Kieselgel

---

[1] Anmerkung. Dieser Befund sollte mit den heute zur Verfügung stehenden reinen Gangliosiden nachgeprüft werden.

[2] Vgl. Abschnitt A. 3. für den Einfluß von Calcium auf die Löslichkeitseigenschaften der Ganglioside.

mit Laufmittelgemischen wie n-Butanol:Pyridin:Wasser = 6:2:2, Chloroform:Methanol:Wasser = 60:30:7 oder n-Propanol:Wasser = 8:2 ect. Einzelheiten der Reingewinnung der Ganglioside und ihre quantitative Bestimmung wurden kürzlich zusammenfassend dargestellt (H. WIEGANDT 1965).

Herrn Prof. Dr. R. KUHN und Fräulein Dr. A. GAUHE danke ich sehr für die Korrektur dieser Arbeit.

Literatur

ABRAHAMSON, M. B., R. KATZMAN, and H. P. GREGOR (1964): J. biol. Chem. 239, 70, 1369.

BARKER, S. A., and G. J. PARDOE (1964): 147th meeting Amer. Chem. Soc., Abstr. of papers, p. 5C, Nr 12.

BLIX, G. (1938): Skand. Arch. Physiol. 80, 46.

— (1959): In: M. L. WOLFROM (Hrsg.), Proc. 4th Internat. Congr. Biochem., vol. I, p. 94. London: Pergamon Press.

— E. LINDBERG, L. ODIN u. I. WERNER (1956): Acta Soc. Med. upsalien. 61, 1.

BOGOCH, S. (1957): J. Amer. chem. Soc. 79, 3287.

— (1959): Nature (Lond.) 185, 392.

—, and E. S. BOGOCH (1959): Nature (Lond.) 183, 53.

BOOTH, D. A. (1963): Biochim. biophys. Acta (Amst.) 70, 486.

BOSCHMAN, T. A. C., u. J. JACOBS (1965): Biochem. Z. 342, 532.

BRADY, R. O., J. V. FORMICA, and G. J. KOVAL (1958): J. biol. Chem. 233, 1072.

— J. N. KANFER, and D. SHAPIRO (1965): Biochim. biophys. Res. Commun. 18, 221.

—, and G. J. KOVAL (1957): J. Amer. chem. Soc. 79, 2648.

—, and E. G. TRAMS (1964): Ann. Rev. Biochem. 33, 75.

BREYER, U. (1965): J. Neurochem. 12, 131.

BROOKS, V. B., D. R. CURTIS, and J. C. ECCLES (1957): J. Physiol. (Lond.) 135, 655.

BURTON, R. M. (1963): Biochemistry 2, 580.

— Y. M. BALFOUR, and J. M. GIBBONS (1964): Fed. Proc. 23, Nr 774, 230.

—, u. J. M. GIBBONS (1964): Biochim. biophys. Acta (Amst.) 84, 220.

—, u. R. E. HOWARD, S. BAER u. Y. M. BALFOUR (1964): Biochim. biophys. Acta (Amst.) 84, 441.

CLÉMENT, J., and G. DI COSTANZO (1964): Biochem. biophys. Res. Commun. 15, 163.

DAVISON, A. N., R. S. MORGAN, M. WAJDA, and G. PAYLING WRIGHT (1959): J. Neurochem. 4, 360.

DAWSON, R. M. (1965): Biochem. J. 97, 134.

DE ROBERTIS, E., A. PELLEGRINO DE IRALDI, G. RODRIGUEZ DE LORES ARNAIZ, and L. SALGANICOFF (1962): J. Neurochem. 9, 23.

— G. RODRIGUEZ DE LORES ARNAIZ, L. SALGANICOFF, A. PELLEGRINO DE IRALDI, and L. M. ZIEHER (1963): J. Neurochem. 10, 225.

DEUL, D. H., and H. McILWAIN (1961): J. Neurochem. 8, 246.

DIETZEL, P. B. (1957): In L. VAN BOGAERT, J. N. CUMINGS, and A. LÖWENTHAL (Hrsg.): Cerebral lipidoses. Oxford: Blackwell.

DODD, M. C., N. J. BIGLEY, and V. B. GEYER (1960): Science 132, 1398.

— — G. A. JOHNSON, and R. H. McCLUER (1964). Nature (Lond.) 204, 549.

DOMBRO, R. S., L. S. BRADHAM, N. K. CAMPBELL u. D. W. WOOLEY (1961): Biochim. biophys. Acta (Amst.) 54, 516.

EICHBERG, J., V. P. WHITTAKER, and R. M. C. DAWSON (1964): Biochem. J. 92, 91.

ESCOLÁ PICÓ, J. (1961): Arch. Psychiat. Nervenkr. 202, 95.

— (1964a): Acta neuropathol. 3, 269.

— (1964b): Acta neuropathol. 3, 309.

— (1964c): Nervenarzt 35, 461.

GALAMOS, D. S., u. V. M. KAPOULAS (1965): Biochim. biophys. Acta (Amst.) **98**, 278.
GAMMACK, D. B. (1963): Biochem. J. **88**, 373.
GATT, SH. (1963): J. biol. Chem. **238**, PC 3131.
— (1965): J. Neurochem. **12**, 311.
—, and E. R. BERMAN (1961): Biochem. biophys. Res. Commun. **4**, 9.
— — (1963): J. Neurochem. **10**, 43.
GOTTSCHALK, A. (1953): Nature (Lond.) **172**, 808.
— (1955): Biochem. J. **61**, 298.
— (1960): The chemistry and biology of sialic acids and related substances. Cambridge: Cambridge Univ. Press.
GREANEY, J. F. (1961): Fed. Proc. **20**, 343.
HALLIDAY, N., H. J. DEUEL, L. J. TRAGERMANN, and W. E. WARD (1940): J. biol. Chem. **132**, 171.
HANDA, SH. (1963): Jap. J. exp. Med. **33**, 347.
HENDRICKSON, H. ST., and C. E. BALLOU (1964): J. biol. Chem. **239**, 1369.
—, and J. G. FULLINGTON (1965): Biochemistry **4**, 1599.
HESS, H. H., and E. ROLDE (1964): J. biol. Chem. **239**, 3215.
VAN HEYNINGEN, W. E. (1959): J. gen. Microbiol. **20**, 310.
— (1963): J. gen. Microbiol. **31** 375.
— (1959a): J. gen. Microbiol. **20**, 291.
— (1959b): J. gen. Microbiol. **20**, 301.
— (1961a): Brit. J. exp. Path. **42**, 397.
— (1961b): J. gen. Microbiol. **24**, 121.
HONEGGER, G. G., u. T. A. FREYVOGEL (1963): Helv. chim. Acta **46**, 2265.
HOWARD, R. E., u. R. M. BURTON (1964a): Biochem. biophys. Acta (Amst.) **84**, 435.
— — (1964b): Biochem. Pharmacol. **13**, 1677.
IRVIN, R. L., and E. G. TRAMS (1961): Fed. Proc. **20**, 174.
JATZKEWITZ, H., H. PILZ, and K. SANDHOFF (1965): J. Neurochem. **12**, 135.
—, u. K. SANDHOFF (1963): Biochim. biophys. Acta (Amst.) **70**, 354.
JOHNSON, G. A., u. R. H. MCCLUER (1963): Biochim. biophys. Acta (Amst.) **70**, 487.
— — (1964): Biochim. biophys. Acta (Amst.) **84**, 756.
JOHNSTON, V. P., and B. J. ROOTS (1965): Nature (Lond.) **205**, 778.
KANFER, J. N. (1965): J. biol. Chem. **240**, 609.
— R. S. BLACKLOW, L. WARREN, and R. O. BRADY (1964): 146th meeting of the Amer. Chem. Soc. 1964, Abstr. of papers, p. 36 A, Nr 76, Biochem. biophys. Res. Commun. **14**, 287.
KARLSSON, K. A. (1964): Acta chem. scand. **18**, 565.
KLENK, E. (1935): Hoppe-Seylers Z. physiol. Chem. **235**, 24.
— (1942): Hoppe-Seylers Z. physiol. Chem. **273**, 76.
—, u. W. GIELEN (1960): Hoppe-Seylers Z. physiol. Chem. **319**, 283.
— — (1961): Hoppe-Seylers Z. physiol. Chem. **326**, 144.
— — (1963a): Hoppe Seylers Z. physiol. Chem. **330**, 218.
— — (1963b): Hoppe-Seylers Z. physiol. Chem. **333**, 162.
—, u. W. KUNA (1964): Hoppe Seylers Z. physiol. Chem. **335**, 275.
— U. LIEDTKE u. W. GIELEN (1963): Hoppe-Seylers Z. physiol. Chem. **334**, 186.
—, u. G. PADBERG (1962): Hoppe-Seylers Z. physiol. Chem. **327**, 249.
— W. VATER, and G. BARTSCH (1957): J. Neurochem. **1**, 203.
KÖNIG, H. (1962): Nature (Lond.) **195**, 782.
— D. GAINES, TH. MCDONALD, R. GRAY, and J. SCOTT (1964): J. Neurochem. **11**, 729.
KOREY, S. R., and J. GONATAS (1963): Life Sci. **2**, 296.
—, and A. STEIN (1962): Fed. Proc. **21**, 283.
— — (1963a): J. Neuropath. exp. Neurol. **22**, 67.
— — (1963b): Life Sci. **3**, 296.
—, and R. D. TERRY (1963): J. Neuropath. exp. Neurol. **22**, 2—104.

KUHN, R., u. R. BROSSMER (1959): Chem. Ber. 92, 1667.
—, u. H. EGGE (1963): Chem. Ber. 96, 3338.
—, u. A. GAUHE (1962): Chem. Ber. 95, 518.
— — (1965): Chem. Ber. 98, 395.
—, u. H. WIEGANDT (1963a): Chem. Ber. 96, 866.
— — (1963b): Z. Naturforsch. 18b, 541.
— -· (1964a): Z. Naturforsch. 19b, 80.
— — (1964b): Z. Naturforsch. 19b, 256.
LANDSTEINER, K., u. A. BOTTERI (1906): Zbl. Bakt., I. Abt. Orig. 42, 562.
—, and P. A. LEVENE (1925): J. Immunol. 10, 731.
LEDEEN, R., and K. SALSMAN (1965): Biochemistry 4, 2225.
LOWDEN, J. A., and L. S. WOLFE (1964): Canad. J. Biochem. 42, 1587.
MAKITA, A. (1964): J. Biochem. (Tokyo) 55, 269.
—, and T. YAMAKAWA (1962): J. Biochem. (Tokyo) 51, 124.
— — (1963): Jap. exp. Med. 33, 361.
— — (1964): J. Biochem. (Tokyo) 55, 365.
MANDELSTAM, P., and R. M. BURTON (1959): Fed. Proc. 18, 280.
MARINETTI, G. V., T. FORD, and E. STOTZ (1960): J. Lipid Res. 1, 203.
MARSHALL, W. E., and J. PORATH (1965): J. biol. Chem. 240, 209.
MÅRTENSSON, E. (1963): Acta chem. scand. 17, 2356.
McGUIRE, E. J., and S. B. BINKLEY (1964): Biochemistry 3, 247.
McILWAIN, H. (1960): Biochem. J. 76, 16P.
— (1963): The chemical exploration of the brain. Elsevier.
McKIBBIN, J. M., and W. R. VANCE (1964): Fed. Proc. 23, Nr 1642, 372.
MEINDL, P., u. H. TUBBY (1965): Mh. Chem. 96, 802, 816.
MELLANBY, J., and W. E. VAN HEYNINGEN (1965): J. Neurochem. 12, 77.
MELTZER, H. C. (1964): Fed. Proc. 23, Nr 768, 229.
MENKES, J. (1964): Biochem. biophys. Res. Commun. 15, 551.
MOHR, E., u. G. SCHRAMM (1960): Z. Naturforsch. 15b, 409.
PHILLPPART, M. P., and J. H. MENKES (1964): 146th meeting of the Amer. Chem. Soc.
    1964, Abstr. of papers, p. 9A, Nr 20.
POPENOE, E. A., and R. M. DREW (1957): J. biol. Chem. 228, 673.
QUARLES, R., and J. FOLCH-PI (1965): J. Neurochem. 12, 543.
RAPPORT, M. M., L. GRAF, and H. SCHNEIDER (1964): Arch. Biochem. 105, 431.
— —· V. P. SKIPSKI, and N. F. ALONZO (1959): Cancer (Philad.) 12, 438.
— — and J. YARIF (1961): Arch. Biochem. 92, 438.
RODRIGUES DE LORES ARNAIZ, G., and E. DE ROBERTIS (1964): J. Neurochem. 11, 213.
ROJAS, E., u. J. M. TOBIAS (1965): Biochim. biophys. Acta (Amst.) 94, 394.
ROSENBERG, A., and E. CHARGAFF (1958): J. biol. Chem. 233, 1323.
SAMBASIVARAO, K., and R. H. McCLUER (1963): Fed. Proc. 22, 300.
SCHMIDT, G., u. A. KIETA-FYDA (1964): VI. Internat. Congr. Biochem. New York,
    Abstr. VII −133, p. 594.
SCHWARTZ, A., H. S. BACHELARD, and H. McILWAIN (1962): Biochem. J. 84, 626.
SEIFERT, H., u. G. UHLENBRUCK (1965): Naturwissenschaften 52, 190.
SEMINARIO, L. M., N. HREN, and C. J. GÓMEZ (1964): J. Neurochem. 11, 197.
SHAH, D. O., and J. H. SCHULMAN (1965): J. Lipid Res. 6, 341.
SHERWIN, A. L., J. A. LOWDEN, and L. S. WOLFE (1964): Canad. J. Biochem. 42, 1640.
SMITS, G. (1961): Psychiat. Neurol. Neurochir. (Amst.) 64, 9.
SOMERS, J. E., J. N. KANFER, and R. O. BRADY (1964): Biochemistry 3, 251.
SPHINGOLIPIDOSES (1962): Symposium Cerebral Sphingolipidoses. A symposium on Tay-
    Sach's disease and allied Disorders. New York: Acad. Press.
STANACEV, N. Z., u. E. CHARGAFF (1962): Biochim. biophys. Acta (Amst.) 59, 733.
— — (1965): Biochim. biophys. Acta (Amst.) 98, 168.
SUOMI, W. D., and B. W. AGRANOFF (1964): Fed. Proc. 23, 375.
— — (1965): J. Lipid Res. 6, 211.

222 H. Wiegandt: Ganglioside

Suzuki, K. (1964): Life Sci. **3**, 1227.
— (1965): J. Neurochem. **12**, 629.
—, u. S. R. Korey (1963): Biochim. biophys. Acta (Amst.) **78**, 388.
— — (1964): J. Neurochem. **11**, 647.
Svennerholm, E., u. L. Svennerholm (1963a): Biochim. biophys. Acta (Amst.) **70**, 432.
— — (1963b): Nature (Lond.) **198**, 688.
Svennerholm, L. (1962): Biochem. biophys. Res. Commun. **9**, 436.
— (1963a): Acta chem. scand. **17**, 860.
— (1963b): J. Neurochem. **10**, 613.
Sweeley, C. C., and B. Klionsky (1963): J. biol. Chem. **238**, PC 3148.
—, and B. Walker (1964): Analyt. Chem. **36**, 1461.
Taghavy, A., K. Salsman, and R. Ledeen (1964): Fed. Proc. **23**, Nr 163, 128.
Tettamanti, G., L. Bertona u. V. Zambotti (1964): Biochim. biophys. Acta (Amst.) **84**, 756.
Thierfelder, H., u. E. Klenk (1930): Die Chemie der Cerebroside und Phosphatide. Berlin: Springer.
—, u. E. Walz (1927): Hoppe-Seylers Z. physiol. Chem. **166**, 217.
Trams, E. G., L. E. Giuffrida, and A. Karmen (1962): Nature (Lond.) **193**, 680.
—, u. L. J. Lauter (1962): Biochim. biophys. Acta (Amst.) **60**, 350.
Wagner, A. (1964): Clin. chim. Acta **10**, 175.
Wallenfels, K., G. Bechtler, R. Kuhn, H. Trischmann u. H. Egge (1963): Angew. Chem. **75**, 1014.
Wassermann, A., u. T. Takaki (1898): Berl. klin. Wschr. **35**, 5.
Westphal, O. (1960): In: R. Kuhn u. Mitarb., Angew. Chem. **72**, 805.
Wherret, J. R., and H. McIlwain (1962): Biochem. J. **84**, 232.
Whittaker, V. P. (1959): Biochem. J. **72**, 694.
— J. A. Michaelson, and R. J. A. Kirkland (1964): Biochem. J. **90**, 293.
Wiegandt, H. (1965): In: N. Zöllner u. D. Eberhagen (Hrsg.), Untersuchung der Lipide im Blut. Berlin-Heidelberg-New York: Springer.
—, u. G. Baschang (1965): Z. Naturforsch. **20b**, 164.
Wolfe, L. S. (1961): Biochem. J. **79**, 348.
Wolman, M., u. H. Wiener (1965): Biochim. biophys. Acta (Amst.) **102**, 269.
Wooley, D. W., u. N. K. Campbell (1962): Biochim. biophys. Acta (Amst.) **57**, 384.
—, and B. W. Gommi (1964): Nature (Lond.) **202**, 1074.
— — (1965): Proc. nat. Acad. Sci. (Wash.) **53**, 959.
Yamakawa, T., N. Kiso, S. Handa, A. Makita, and S. Yokoyama (1962): J. Biochem. (Tokyo) **52**, 226.
—, and N. Ueta (1964): Jap. J. exp. Med. **34**, 37.
— S. Yokoyama, and N. Handa (1963): J. Biochem. (Tokyo) **53**, 28.
— — and N. Kiso (1962): J. Biochem. (Tokyo) **52**, 228.
Yokoyama, M., E. G. Trams, and R. O. Brady (1963): J. Immunol. **90**, 372.
Zabin, J. (1957): J. Amer. chem. Soc. **79**, 5334.
Zilliken, F., u. M. W. Whitehouse (1958): In: M. L. Wolfrom and R. S. Tipson (Hrsg.), Advanc. Carbohyd. Chem. **13**, 237.
Zöllner, N., u. D. Eberhagen (Hrsg.) (1965): Untersuchung und Bestimmung der Lipide im Blut. Berlin-Heidelberg-New York: Springer.

Namenverzeichnis

Die gewöhnlich gesetzten Ziffern weisen auf die entsprechende Stelle im Text und die *kursiven* Seitenzahlen auf das Literaturverzeichnis hin.

Svennerholm, E., u. L.
Svennerholm 217,
222
Svennerholm, L. 192, 206,
216, 222
— s. Svennerholm, E. 217,
222
Sweeley, C. C. 217
— u. B. Klionsky 222
— u. B. Walker 199, 222
Szende, J. s. Issekutz, B. v.
152, 180

Taghavy, A., K. Salsman u.
R. Ledeen 207, 217, 222
Takaki, T. s. Wassermann,
A. 210, 222
Tanaka, R., u. K. P. Strick-
land 204
Tarding, F., u. P. Schambye
146, 187
Tarnowski, W. s. Hilz, H.W.
98, 180
Tarrant, M. E., R. Mahler u.
J. Ashmore 104, 187
Tasaki, I., u. J. J. Chang 11,
89
— s. Hild, W. 11, 18, 22, 84
Tasaki, T. s. Motokawa, K.
48, 87
Tata, J. R. 128, 187
Taxi, J. 67, 89
Teng, C. s. Hastings, A. B.
161, 179
Tennyson, V. M. s. Pappas,
G. D. 51, 87
Tepperman, H. M., u. J.
Tepperman 98, 103, 187
— s. Tepperman, J. 187
Tepperman, J., u. H. M.
Tepperman 187
— s. Tepperman, H. M. 98,
103, 187
Terner, C. s. Krebs, H. A.
161, 181
Terry, R. D. s. Korey, S. R.
205, 206, 220
— s. Torack, R. M. 60, 89
Tettamanti, G., L. Bertona
u. V. Zambotti 218, 222
Thierfelder, H., u. E. Klenk
218, 222
— u. E. Walz 190, 222

Thorn, G. W., A. E. Renold
u. G. F. Cahill jr. 167,
187
— s. Ingle, D. J. 120, 180
— s. Koepf, G. F. 120, 181
— s. Lewis, R. A. 119, 182
Tietze, F. s. Mortimore,
G. E. 151, 183
Tjaden, R. J. s. Perry, W. F.
107, 184
Tobias, J. M. s. Rojas, E.
196, 221
Tobin, J. s. Altszuler, N.
153, 174
Tomita, T. 47, 89
Torack, R. M., M. L. Duffy
u. J. M. Haynes 60, 89
— R. D. Terry u. H. M.
Zimmermann 60, 89
Toye, K. E., u. J. F. Manery
158, 187
Tragermann, L. J. s. Halliday, N. 218, 220
Trakatellis, A. C., A. E.
Axelrod u. M. Montjar
99, 187
Trams, E. G., L. E. Giuffrida
u. A. Karmen 191, 222
— u. L. J. Lauter 191, 195,
207, 208, 210, 222
— s. Brady, R. O. 211, 219
— s. Irvin, R. L. 201, 220
— s. Yokoyama, M. 211,
222
Treherne, J. E. 10, 27, 32,
62, 90
— s. Smith, D. S. 63, 89
Trischmann, H. s. Wallen-
fels, K. 199, 222
Tschirgi, R. D. 32, 50, 90
— s. Wolff, P. H. 49, 90
Tubby, H. s. Meindl, P. 199,
221

Ueta, N. s. Yamakawa, T.
199, 222
Uhlenbruck, G. s. Seifert, H.
204, 218, 221
Unger, R. H., A. M. Eisentraut, M. S. McCall u.
L. L. Madison 153, 187
— s. Madison, L. L. 146,
151, 152, 182

Ussing, H. H. 158, 187
Utter, M. F., u. D. B. Keech
113, 187
— — u. M. C. Scrutton
113, 187

Vallecalle, E. s. Svaetichin,
G. 47, 89
Vance, W. R. s. McKibbin,
J. M. 204, 221
Vater, W. s. Klenk, E. 220
Vaughan, M. s. Haugaard,
N. 151, 179
Vergara, F. E. s. Niemeyer,
H. 95, 96, 183
Vester, J. W. 155, 187
— u. M. L. Reino 155, 187
Villar-Palasi, Rosell-Peréz,
Richman u. Friedman
154
Villee, C. A., u. B. Hastings
167, 187
Villegas, G. M., u. R. Ville-
gas 28, 90
— s. Villegas, R. 19, 26, 90
Villegas, J. s. Svaetichin, G.
47, 89
Villegas, L. s. Villegas, R.
19, 26, 90
Villegas, R., L. Villegas, M.
Gimenez u. G. M. Ville-
gas 19, 26, 90
— s. Villegas, G. M. 28, 90
Viñuela, E., M. Salas u.
A. Sols 96, 187
— s. Salas, M. 96, 185
Virchow, R. 3, 5, 90
Volk, B. W. s. Harris, A. F.
201

Wagle, S. R., u. J. Ashmore
122, 187
Wagner, A. 218, 222
Wajchenberg, B. L. s. Hen-
nes, A. R. 167, 179
Wajda, M. s. Davison, A. N.
207, 219
Wakil, S. J. 100, 187
Waksman, B. H., u. R. D.
Adams 75, 90
Walberg, F. s. Mugnaini, E.
3, 5, 75, 87

Sachverzeichnis

ERGEBNISSE DER PHYSIOLOGIE
BIOLOGISCHEN CHEMIE UND
EXPERIMENTELLEN PHARMAKOLOGIE

REVIEWS OF PHYSIOLOGY
BIOCHEMISTRY AND
EXPERIMENTAL PHARMACOLOGY

HERAUSGEGEBEN VON

K. KRAMER   O. KRAYER   E. LEHNARTZ
MÜNCHEN   BOSTON   MÜNSTER/WESTF.

F. LYNEN   A. v. MURALT
MÜNCHEN   BERN

U. G. TRENDELENBURG   H. H. WEBER   O. WESTPHAL
BOSTON   HEIDELBERG   FREIBURG/BR.

SONDERDRUCK AUS BAND 57

ST. W. KUFFLER AND J. G. NICHOLLS

THE PHYSIOLOGY OF NEUROGLIAL CELLS

WITH 33 FIGURES

NICHT IM HANDEL

SPRINGER-VERLAG
BERLIN · HEIDELBERG · NEW YORK 1966

Inhaltsverzeichnis

57. Band

ERGEBNISSE DER PHYSIOLOGIE
BIOLOGISCHEN CHEMIE UND
EXPERIMENTELLEN PHARMAKOLOGIE

REVIEWS OF PHYSIOLOGY
BIOCHEMISTRY AND
EXPERIMENTAL PHARMACOLOGY

HERAUSGEGEBEN VON

K. KRAMER   O. KRAYER   E. LEHNARTZ
MÜNCHEN   BOSTON   MÜNSTER/WESTF.

F. LYNEN   A. v. MURALT
MÜNCHEN   BERN

U. G. TRENDELENBURG   H. H. WEBER   O. WESTPHAL
BOSTON   HEIDELBERG   FREIBURG/BR.

SONDERDRUCK AUS BAND 57

R. STEELE

THE INFLUENCES OF INSULIN ON THE HEPATIC
METABOLISM OF GLUCOSE

WITH 3 FIGURES

SPRINGER-VERLAG
BERLIN · HEIDELBERG · NEW YORK 1966

Inhaltsverzeichnis

57. Band

ERGEBNISSE DER PHYSIOLOGIE
BIOLOGISCHEN CHEMIE UND
EXPERIMENTELLEN PHARMAKOLOGIE

REVIEWS OF PHYSIOLOGY
BIOCHEMISTRY AND
EXPERIMENTAL PHARMACOLOGY

HERAUSGEGEBEN VON

K. KRAMER   O. KRAYER   E. LEHNARTZ
MÜNCHEN   BOSTON   MÜNSTER/WESTF.

F. LYNEN   A. v. MURALT
MÜNCHEN   BERN

U. G. TRENDELENBURG   H. H. WEBER   O. WESTPHAL
BOSTON   HEIDELBERG   FREIBURG/BR.

SONDERDRUCK AUS BAND 57

H. WIEGANDT

GANGLIOSIDE

MIT 5 ABBILDUNGEN

SPRINGER-VERLAG

BERLIN · HEIDELBERG · NEW YORK 1966

Inhaltsverzeichnis

57. Band

SPRINGER-VERLAG
BERLIN·HEIDELBERG·NEW YORK

Studies in Physiology

Presented to John C. Eccles

Edited by D. R. Curtis and A. K. McIntyre

with the collaboration of numerous experts

With 80 figures
VIII, 276 pages 8vo. 1965
Cloth DM 28,—

■ Prospectus on request!

This book is a collection of short reviews on physiological topics of current interest by former associates and postgraduate pupils of Sir John Eccles. The volume commemorates the award of the 1963 Nobel Prize in Medicine or Physiology, which Sir John shared with A. L. Hodgkin and A. F. Huxley. The authors, all of whom are well known for their original contributions to scientific literature, are from laboratories throughout the world, and the wide range of topics reflects Sir John's influence on, and interest in, the study of physiology.

As would be expected, the majority of the essays deal with problems of neurophysiology and neuropharmacology such as mechanisms of sensation, organization of reflexes and central pathways, the operation of nerve cells as revealed by intracellular recording, the release of transmitter substances, the electrical properties of membranes and the pharmacology of central synaptic transmission. In addition other topics discussed include the problems of diffusion, foetal physiology, water metabolism, autonomic ganglia, neurohypophyseal secretion, sweat and salivary glands.

Published earlier:

In the USA and Canada this book is distributed by Academic Press Inc., Publishers, New York

The Physiology of Synapses

By **John Carew Eccles**, Professor of Physiology, The Australian National University Canberra

In English. With 101 figures. XII, 316 pages 8vo. 1964. Cloth DM 36,—

Druck der Universitätsdruckerei H. Stürtz AG., Würzburg

The Physiology of Synapses

By John Carew Eccles
Professor of Physiology
The Australian National University
Canberra

With 101 figures. XII, 316 pages 8vo.
1964. Cloth DM 36,—

(Die Verbreitungsrechte dieses Werkes
für USA und Canada
liegen bei Academic Press Inc.,
Publishers, New York)

Contents: The development of ideas on the synapse. Structural features of chemically transmitting synapses Physiological properties of chemical transmitting synapses in the resting state. Excitatory postsynaptic responses to presynaptic impulses. Excitatory transmitter substances. The release of transmitter by presynaptic impulses. The generation of impulses by the excitatory potential and the endplate potential. The presynaptic terminals of chemically transmitting synapses. Excitatory synapses operating by electrical transmission. The postsynaptic electrical events produced by chemically transmitting inhibitory synapses. The ionic mechanism generating the inhibitory postsynaptic potential. Inhibitory transmitter substances. Pathways responsible for postsynaptic inhibitory action. Inhibitory synapses operating by electrical transmission. Presynaptic inhibition. The trophic and plastic properties of synapses. Epilogue. References. Subject Index.

Studies in Physiology

Presented to John C. Eccles
Edited by D. R. Curtis and A. K. McIntyre

With 80 figures. VIII, 276 pages 8vo.
1965. Cloth DM 28,—

Zentralnervensystem

Bearbeitet von Christof Stumpf
und Hellmuth Petsche
Redaktion Oskar Eichler

Mit 56 Abbildungen. XII, 316 Seiten Gr.-8⁰. 1962. (Handbuch der experimentellen Pharmakologie, 16. Bd.: Erzeugung von Krankheitszuständen durch das Experiment. Teil 7) Ganzleinen DM 112,—

SPRINGER-VERLAG
BERLIN · HEIDELBERG · NEW YORK